Ceramic Fabrication Technology

MATERIALS ENGINEERING

1. Modern Ceramic Engineering: Properties, Processing, and Use in Design: Second Edition, Revised and Expanded, *David W. Richer- son*
2. Introduction to Engineering Materials: Behavior, Properties, and Selection, *G. T. Murray*
3. Rapidly Solidified Alloys: Processes • Structures • Applications, *edited by Howard H. Liebermann*
4. Fiber and Whisker Reinforced Ceramics for Structural Applications, *David Belitskus*
5. Thermal Analysis of Materials, *Robert F. Speyer*
6. Friction and Wear of Ceramics, *edited by Said Jahanmir*
7. Mechanical Properties of Metallic Composites, *edited by Shojiro Ochiai*
8. Chemical Processing of Ceramics, *edited by Burtrand I. Lee and Edward J. A. Pope*
9. Handbook of Advanced Materials Testing, *edited by Nicholas P. Cheremisinoff and Paul N. Cheremisinoff*
10. Ceramic Processing and Sintering, *M. N. Rahaman*
11. Composites Engineering Handbook, *edited by P. K. Mallick*
12. Porosity of Ceramics, *Roy W. Rice*
13. Intermetallic and Ceramic Coatings, *edited by Narendra B. Dahotre and T. S. Sudarshan*
14. Adhesion Promotion Techniques: Technological Applications, *edited by K. L. Mittal and A. Pizzi*
15. Impurities in Engineering Materials: Impact, Reliability, and Control, *edited by Clyde L. Briant*
16. Ferroelectric Devices, *Kenji Uchino*
17. Mechanical Properties of Ceramics and Composites: Grain and Particle Effects, *Roy W. Rice*
18. Solid Lubrication Fundamentals and Applications, *Kazuhisa Miyoshi*
19. Modeling for Casting and Solidification Processing, *edited by Kuang-O (Oscar) Yu*
20. Ceramic Fabrication Technology, *Roy W. Rice*

Additional Volumes in Preparation

Coatings for Polymers and Plastics, *edited by Rose Ann Ryntz and Philip V. Yaneff*

Micromechatronics, *Kenji Uchino and Jayne Giniewicz*

Ceramic Fabrication Technology

Roy W. Rice
Alexandria, Virginia

MARCEL DEKKER, INC. NEW YORK • BASEL

Library of Congress Cataloging-in-Publication Data
A catalog record for this book is available from the Library of Congress.

ISBN: 0-8247-0853-9

This book is printed on acid-free paper.

Headquarters
Marcel Dekker, Inc.
270 Madison Avenue, New York, NY 10016
tel: 212-696-9000; fax: 212-685-4540

Eastern Hemisphere Distribution
Marcel Dekker AG
Hutgasse 4, Postfach 812, CH-4001 Basel, Switzerland
tel: 41-61-260-6300; fax: 41-61-260-6333

World Wide Web
http://www.dekker.com

The publisher offers discounts on this book when ordered in bulk quantities. For more information, write to Special Sales/Professional Marketing at the headquarters address above.

Current printing (last digit):
10 9 8 7 6 5 4 3 2 1

PRINTED IN THE UNITED STATES OF AMERICA

Preface

There is a spectrum of needs for reference, overview, and instructional material concerning the fabrication of ceramic and ceramic composite specimens and especially components. These needs range from, at one extreme, addressing basic scientific principles and parameters of different processing and fabrication methods to, at the other extreme, basic engineering aspects, including costs and related operational factors. Scientific principles are most extensively treated in various books that focus on individual, or a limited range of, established processing methods, mostly those based exclusively on pressureless sintering. Such books may address in a perfunctory manner, or not at all, important topics such as pressure sintering processes, melt processing and fabrication, and chemical reaction processes, especially the important subject of chemical vapor deposition (CVD). Some engineering aspects of some processes are treated in some books, but mostly in a limited way and often in older books. The counterpart of basic scientific and basic engineering aspects are detailed operational factors that address both cost and component performance trade-offs that are needed for a successful manufacturing process. However, these are addressed very little or not at all in the literature since they would be very extensive and generally proscribed in their treatment by proprietary concerns.

The concept and goal of this book is to provide a link between basic science and the ultimate, but nonexistent, detailed engineering/operational treatment of the subject. It is intended to complement several very useful books emphasizing scientific aspects by providing a more pragmatic engineering-oriented approach and a broader, more comprehensive perspective. The book includes industrially and technologically important topics such as pressure sintering, reaction processing and fabrication, and various fusion processes, as well as speciality processing/fabrication, e.g., for porous or composite bodies. This is not at the expense of the more extensively used powder consolidation and pressureless sintering, but some less used methods, such as electrophoretic deposition, and emerging ones, such as rapid prototyping/solid free-form fabrication, are also treated. Instead, a balance has been sought by focusing on overall and key engineering aspects, with more limited detailed discussion of processes that are extensively treated in other books. Important engineering factors are often addressed via summary descriptions of successful solutions to engineering challenges, e.g., at the extreme of processing parameters such as handling great shrinkages in sintering large parts.

The practical engineering aspect of the book is provided in three fashions. The first is the selection and balance of topics, as mentioned above, including substantial discussion of costs and trade-offs. Such discussion is extended to promising processes not yet used in production, to aid in their development and evaluation for niche, and possibly more extensive, opportunities for production. Examples of this broader, more pragmatic approach include substantial emphasis on processing and fabrication by methods other than pressureless sintering, as well as a chapter on densification with additives and one on use of additives in powder preparation and other processing and fabrication methods. Another important example of the broader approach taken in this book is attention to the capabilities and limitations of various processing and fabrication methods in terms of materials and microstructures, hence the effect on component performance, as well as component character, e.g., size, shape, and costs.

The first of three additional factors to note about this book is the referencing. There is a huge and still rapidly growing literature on topics included in the book, making a comprehensive presentation impossible. Literature searches of data bases can help provide information on specific topics, and were used some, but such searches cannot be effective as a means of assembling the bulk of the information for preparation of a book. This author has instead followed nearly all of the topics of this book, and in two companion books (*Porosity of Ceramics* and *Mechanical Properties of Ceramics and Composites: Grain and Particle Effects*, both titles, Marcel Dekker, Inc.) continuously for over 30 years. Much of this included obtaining and filing, on an ongoing basis, copies of the first, multiple, or complete page(s) of papers or reports of interest. This organized collection, which fills over 10 full-sized file cabinets, was the primary basis for references for this book (and the two companion ones), but the bulk of this information was still too voluminous to include. Thus, pertinent files were reviewed to select material to be used and referenced, with the primary selection criteria being the pertinence and importance of the results. The bulk of the references came from the author's files, but still generally constitute a few to several percent of his files. Other reviews and summaries along with earlier, especially landmark, as well as more recent, work indicating newer directions, giving other pertinent references, or both, have been included to the extent possible. Overall the author's perspective from continuous interest, contacts, and activity in improved fabrication and processing of advanced ceramics and ceramic composites has been the basis of selecting the topics covered and the literature referenced.

The second additional feature of this book to note is its relation to the two other books referenced above. The three books together summarize the linkage between fabrication/processing and most important properties of ceramics. In particular, this book notes the impact of fabrication and processing on microstructure and, to some extent, on properties, as a guide, while more detailed property effects via impacts of microstructure can be found in the two books noted above.

The third additional aspect to mention of this book is the evaluation of specific industrial practices, especially uses of specific processes. Such information is generally limited, especially more recent changes in usage, due to proprietary interests. Where such usage is not clearly documented or widely known, but is known to the author with a reasonable degree of certainty, it is indicated with qualifications such as probable, appears, or believed.

Many people have contributed in a variety of ways to the development of this book, especially colleagues at my three places of employment: The Boeing Co. (Seattle WA), the U.S. Naval Research Lab (Washington, DC), and W. R. Grace (Columbia, MD), particularly the following from Grace: Ken Anderson, Jerry Block, Rasto Brezny, Craig Cameron, Jyoti Chakraverti, Jack Enloe, Av Kerkar, and Tariq Quidir at W. R. Grace. Several people have aided by reading drafts of chapters or sections of them (numbers shown in parenthesis), providing comments, and sometimes additional references, as follows: Dave Lewis (U.S. Naval Res. Lab.) and Bob Ruh (Air Force Materials Lab.) (1–8); Jack Sibold (TDA Res. Inc.) (2); Ken Anderson (now with Zircoa), and Jyoti Chakraverti (now with Ferro Corp.) (4); Jack Rubin (consultant) (5); John Locher (Saphikon), Rich Palicka (Cercom Inc.), Ken Sandhage (Ohio State Univ.), and Fred Schmid (Crystal Systems) (6), as well as Curt Scott (now deceased) for several discussion and inputs. Finally, Drs. Steve Freiman and Sheldon Wiedrehorn and Mr. George Quinn of NIST are thanked for making me a visiting scientist there and hence giving me library access.

Roy W. Rice

Contents

Abbreviations

CVD	chemical vapor deposition
CVI	chemical vapor infiltration
EFG	edge film-fed growth (of single crystals or eutectic systems)
HEM	heat exchanger method (of single crystal growth and possibly of eutectic or polycrystalline bodies)
PVD	physical vapor deposition (e.g., evaporation or sputtering processes)
RBSN or RSSN	reaction bonded or sintered silicon nitride
RSSC	reaction sintered SiC
SFF	solid freeform fabrication, closely related to, and often synonymous with, rapid prototyping
v/o	volume percent
w/o	weight percent

1

Background and Overview

1.1 INTRODUCTION

Most books on making ceramic bodies focus on the dominant technology of consolidating and densification of (primarily chemically derived) powders, mainly via sintering [1-3]. These books provide valuable insight into the underlying scientific principles that control such processing, as well as provide useful information on many of the process parameters, but their perspective on choice of fabrication method(s) is a basic one rather than an engineering one. Thus, such books generally have limited or no information on many of the important engineering or cost aspects of producing ceramic components. Further, even within their more basic scope, they are generally focused on the most common methods, e.g., of liquid chemical preparations of powders and their die pressing and sintering. Generally, they provide limited or no information on other methods of producing ceramic components, e.g., of chemical vapor deposition (CVD) or various melt processing routes, and typically no information on the property and engineering trade-offs between different basic production methods or within variations of a given approach, such as sintering of bodies from different forming methods. Thus, while existing books address the use of additives in densification, they do so only in broad terms of liquid-phase sintering, not by discussing specific additive uses for sintering, and they do not address a number of other additive uses. Further, there is limited discussion of the shape, especially of component size, capabilities of the processing and fabrication technologies addressed, nor their cost aspects.

At the other extreme there are books that focus more on specific engineering aspects, for instance, specific formulations, including uses of both additives and of binders, but mainly for more traditional ceramics [4], for which such information is generally known, but is often proprietary for many newer ceramic materials. There are also some books that focus on specific powder fabrication/forming techniques [5-8], as well as on some other fabrication techniques, mainly CVD [9,10].

This book is intended to complement and supplement previous books by providing a much broader perspective on ceramic fabrication, which is defined as the combination of various process technologies to produce monolithic or composite ceramic pieces/components within given shape, size, and microstructure property bounds for a given composition. The focus is on higher performance monolithic ceramics, but with considerable attention to ceramic composites, especially particulate composites, as well as attention to some specialized bodies, e.g., those of designed porosity. This book is not intended to be an engineering fabrication "cookbook" since many of the technologies are not in production, and many that are may have various proprietary aspects. Instead, it is meant as a guide to the technological alternatives for practical application for those concerned with development of practical fabrication technologies beyond laboratory preparation of specimens for research purposes. Thus, while a broad range of topics is addressed for completeness, emphasis is given to technologies that are addressed less or not at all in previous books, but have known or potential practicality. Hence, while, both conventional and alternative powder-based fabrication are addressed, considerable attention is given to both CVD and melt processes, as well as to reaction processing. Further, the use of additives in all of these processes is reviewed, and specific attention is given to the issue of size and shape capabilities of different fabrication methods. Also, to the extent feasible, cost aspects are addressed, and examples of specific engineering extension of limits of given fabrication technologies are given. Finally, some overall trends and opportunities are discussed.

Before proceeding to the discussion of the various processing/fabrication technologies of subsequent chapters, four basic topics are addressed in the following three sections, the first being rational—why ceramics and opportunities and challenges to selecting candidate ceramics. Then, broad issues impacting ceramic development and application are discussed, followed by discussion and illustration of costs and trade-offs. Finally, some overall engineering factors are discussed, particularly sizes and shapes achievable, as well as possibilities of joining, and their associated costs and ramifications. These topics are treated in this chapter from a wide perspective, while some of these factors are discussed in more detail where specific fabrication technologies are addressed. These are large subjects that can only be illustrated and summarized here (especially production costs) to provide guidance and awareness of their parameters, variations, and importance.

1.2 WHY CERAMICS AND WHICH ONES

The first decision to be made in selecting material candidates for an application is to determine which types of materials to consider. This commonly entails both fabrication and cost issues discussed below, especially where ready availability is desired or required, and significant development is not realistic. However, a basic question for many needs, especially longer term ones, is, What material candidates have the best intrinsic property potential to meet the requirements of the application, especially if they are demanding? This is especially true for ceramics and ceramic composites, since there is such a diversity of materials and properties, with much of their potential partially or substantially demonstrated, but often untapped. This potential arises from both the extremes and the unique combinations of properties that are obtainable from the diversity of ceramic materials.

Perspective on diversity can be obtained by remembering that solid materials can be divided into nominally single-phase materials that are polymeric (mainly plastics or rubbers), metallic, or ceramic, or into two- or multiphase composites of constituents from any one of the three basic single-phase materials, or combinations of two or three of the single-phase materials. Ceramics, or more specifically *monolithic ceramics*, are thus defined as nominally single-phase bodies that are not composites nor metals or polymers. While this includes a few elemental materials such as sulfur, or much more importantly, the various forms of carbon, the great bulk and diversity of ceramics are chemical compounds of atoms of one or more metallic elements with one or more metalloid or nonmetallic elements.

The more developed ceramics are mostly compounds of two types of atoms, that is, binary compounds, which are typically classified by the nonmetallic or metalloid anion element they contain—for example, compounds of metals with nonmetals, such as oxides and nonoxides, the latter including borides, carbides, halides, nitrides, silicides, and sulfides. Key examples are listed in Table 1.1. However, there are a variety of known ternary ceramic compounds formed with a third atom constituent. Those that contain either two metallic and one metalloid or nonmetallic types of atom continue to be classified as carbides, oxides, etc., as for binary ceramics. However those containing one type of metallic atom with two types of atoms of either metalloid or nonmetallic designation or a combination of one of each, are named by their latter atoms, e.g., as carbonitrides and oxysilicides, for compounds containing carbon and nitrogen or oxygen and silicon atoms, respectively. There are also higher-order ceramic

TABLE 1.1 Some Properties of More Common Refractory Metals and Binary Ceramics[a]

Material		Density (g/cc)	MP (°C)	CTE (ppm/°C)	E (GPa)	Other
A) Refractory metals	Nb	8.4	2470	9	100	Ductile
	Ta	16.6	3000	8	190	Ductile
	Mo	10.2	2620	8	320	
	W	19.3	3400	7	420	
	Re	22	3180	7	480	Expensive
B) Borides	HfB_2	11.2	3250	6–7		
	NbB_2	7.2	2900	9		Decomposes
	TaB_2	12.6	3000	6–7	260	
	TiB_2	4.5	2900	7	500	
	WB_2		2900			
	ZrB_2	6.1	3000	8	450	
C) Carbides	HfC	12.7	3880	7	430	
	SiC	3.2	2600	6	450	Sublimes
	NbC	7.8	3700	7	450	
	TaC	14.5	3700	9	450	
	TiC	4.9	3140	9	450	
	ZrC	6.7	3450	8	420	
D) Nitrides	BN	2.2	3000	High crystalline anisotropy		Sublimes
	HfN	13.9	3300	7		
	TaN	14.1	3200	5		
	ThN	11.6	2800			α–emitter
	TiN	5.4	2950	10	260	
	ZrN	7.4	2980	8		
E) Oxides	BeO	3	2500	8	400	Toxic
	HfO_2	9.7	2750	11		
	MgO	3.6	2800	16	350	Hydrates
	ThO_2	9.8	3200	11	240	α–emitter
	ZrO_2	5.7	2715	12	230	

[a]MP = melting point, CTE = coefficient of thermal expansion, and E = Young's modulus.

compounds, that is, ceramic compounds consisting of four or more atomic constituents that are generally much less known. Such higher-order compounds offer opportunity for extending ceramic technology via more diverse properties.

The diversity of ceramics and their properties is significantly extended by the fact that the properties of a given ceramic compound can be varied, often substantially, by changing microstructure via differences in fabrication/processing, which is extensively discussed elsewhere [11,12]. The diversity is also significantly extended by addition of one or more other ceramic compounds that form a solid solution with the base ceramic compound. The limitations of such solid solution extension of properties are the limits of solubility due either to precipitation or reaction, or both. However, these limits on solubility also provide more specialized ways of extending the range of ceramic properties via ceramic composites, i.e., ceramic bodies consisting of two or more ceramic phases that have limited or no mutual solubility and a considerable range of chemical compatibility. More extensively, ceramic composites are made by consolidating mixtures of composite phases, which are classified by the character of the additional, usually second, phase, that is, particulate, whisker, platelet, or fiber composites, which are addressed in this book, generally in decreasing extent in the order listed.

The resultant diversity of ceramic properties from all of the ceramic compounds, their solid solutions, and composites is illustrated in part by a very abbreviated listing of some properties of the more refractory members of the more common and more extensively developed binary ceramic materials in Table 1.1. Note that other binary systems have refractory compounds, for example, sulfides and phosphides with melting points of 2000 to 2500–2700°C, and many systems with compounds having melting points of 1500–2200°C or above. Also note that, while ternary and higher-order compounds typically have lower melting temperatures than the more refractory binary compounds, this is not always true.

The property diversity of ceramics is further shown by the following observations addressing the six categories of functional properties: (1) thermal-chemical, (2) mechanical, (3) thermal conduction, (4) electrical, (5) magnetic, and (6) electromagnetic. Thus, there are a number of ceramics that have among the highest potential operating temperatures, approaching their melting points at and above those of their only other competitors, the refractory metals (Table 1.1), and have the highest energies for ablation, especially in the absence of melting, as is the case for important ceramics that commonly sublime without melting. Further, the diversity of ceramic compositions provides candidates for a diversity of environments, for example, halides for halide environments and sulfides for sulfur environments, as well as the formation or application of at least partially protective coatings that are chemically compatible with the ceramic substrate.

Considering mechanical performance, many ceramics have high stiffness and high melting points, reflecting the strong atomic bonding. While stiffness generally decreases with increasing temperature, as for other materials, it is typically an important attribute of many ceramics across the temperature spectrum. High bond strengths of many refractory ceramics also correlates with their high hardnesses, which tends to correlate some with armor performance and especially with much wear and erosion resistance, as well as with compressive strengths that can also be of importance at high temperatures, but are typically more important at modest temperatures [2,11,12]. Tensile strengths, though being particularly sensitive to microstructural and thus to fabrication process parameters, also correlate in part with elastic moduli, and can be quite substantial over a broad range of temperatures. Also note that some ternary compounds (such as mullite and perhaps higher-order compounds) can have much higher creep resistance than their more refractory binary constituents. At the extreme of mechanical precision, many ceramics offer the highest degrees of precision elastic stability, i.e., dimensional stability under mechanical and thermal loading, which is typically most pronounced and important at modest temperatures [12].

Different ceramics have among the lowest intrinsic thermal conductivities and others the highest ranges of thermal conductivities, with even more extremes shown for electrical conductivity or resistivity. This includes both highest-temperature superconductors (TiN and TiC) prior to the discovery of much higher-temperature ternary and higher-oxide superconductors of extensive interest for about the past 10 years. Some ceramics also have the highest resistance to dielectric breakdown, hence the ability to be good insulators even under very high electrical fields, as well as other important electrical properties [2,11,12]. These include high-temperature semiconductors for a variety of applications and ionic conductors for diverse applications, such as advanced fuel cells and batteries, as well as sensors. Of particular importance for many technological applications are ferroelectric and related electrical properties, especially in some ternary ceramics, which like many other properties are most often of particular importance at or near room and moderate temperatures. This is commonly also the case for their important magnetic and electromagnetic properties, but elevated temperature performance of such functions can also be important.

While a single property may drive applications, unique combinations of properties are commonly an important factor. Thus, for example, good magnetic properties in nonconductive, i.e., dielectric, ceramics is an important factor in their magnetic applications, while application of the transparency of dielectric ceramics to ultraviolet (UV), visible, infrared (IR), microwave, and other electromagnetic waves is often made due in part to the temperature capabilities of many ceramics. These and other applications are also often partly driven by the substantial hardnesses of many ceramics as reflected in their resistance to wear, erosion, and ballistic impact, for example, for transparent armor windows.

The challenge of fabricating ceramic components for various applications requires that the properties and performance sought be obtained in a reliable and cost-effective fashion. However, both these goals of suitable reliability and cost are dependent on the impacts of component composition as well as size, shape, and dimensional-surface finish requirements on fabrication routes, and the parameters of the processing techniques within these routes.

The properties sought are usually determined by the composition of the body, mainly by the compound selected. However, there may be uniform, heterogeneous, or both variations in either the composition of the ceramic compound sought, other constituents or impurities in solid solution or as second phases, or combinations of these. Thus, some compounds are very stable in composition during use, but others are less so, while fabrication processing parameters may often present greater problems of composition stability. For example, some oxides such as Al_2O_2, BeO, and SiO_2 are quite stable, while others such as oxides of Ce, Ti, and Zr are less so, such that they may be reduced from their normal oxygen stoichiometry in varying degrees, depending on the extent of reducing conditions, temperatures, and times of exposure. Component sizes and shapes are key factors in resultant composition gradients and their effects. Such reduction can be very detrimental to some uses, especially electrical and electromagnetic and sometimes mechanical properties, as well as some possible use in special cases. Similar, though often less extreme effects of stoichiometry deviations may occur with useful nonoxide ceramics. Both the presence of impurities and use of additives can be important issues, since increased purification typically means increased costs and may have other ramifications on fabrication, as additives may be important in fabrication but present limitations in use. Chapter 3 addresses the use of additives in preparation of ceramic raw materials that, while having some desirable effect, may retain some impurities, variations in composition, or both. Chapter 5 extensively addresses use of additives in fabrication.

Another basic impact of fabrication on properties is its effects on microstructure, which arise for both intrinsic and extrinsic reasons, the latter often reflecting effects of chemical or physical heterogeneities in the body. The reason for this is that microstructure plays an important role in many properties, with the most critical microstructural factor being porosity. While porosity is critical to some important applications such as catalysis or filtration, and can also aid some other properties and applications, it commonly significantly reduces many important properties, such as mechanical and optical ones [11]. Thus, a fraction of a percent of porosity that scatters visible light may render a potentially transparent ceramic window ineffective for its purpose, while the ~ 5% porosity left from much sintering can reduce many mechanical properties by 10–25%. Next most significant is grain size (G), with many mechanical properties increasing as G decreases, by ~ 50% or more as G goes from ~ 100 to ~1 μm, but other properties may be unaffected by G or increase with increasing G [12]. In composites the dispersed particle size plays a similar role as G in monolithic ceramics, but the matrix grain size also still has similar effects in composites as in monolithic ceramics, though it may be more restrained in grain growth by the dispersed particles. Heterogeneities of grain and particle structure and porosity, as well as impurities or additives, can also be important in limiting levels and reliability of properties. All of these microstructural effects are impacted by the ceramic, the fabrication route, and processing parameters selected for a given application.

The above diversity of ceramic performance based on both the diversity of ceramic materials and on impacts of fabrication via effects on microstructure is a double-edged sword. On the one hand, performance diversity provides wider opportunity for application. On the other hand, it can dilute resources between possible competing candidates, which may jeopardize success of any candidate. It can also mean that the candidate that may be implemented is not the best one overall, but the one that required less development; however, once established, it is harder to replace with a potentially superior candidate. Such trade-offs are impacted by specific economic factors (see Sec. 1.4), as well as by larger political and economic factors. (see Sec. 1.3).

1.3 POLITICAL AND ECONOMIC FACTORS IMPACTING DEVELOPMENT AND APPLICATION OF ADVANCED CERAMICS

A major change reducing opportunities for development and application of ceramics (and other advanced materials) was the end of the cold war's reduction of military-aerospace funding. It probably also reduced opportunities for advanced processing and materials development, such as of ceramics with potential for extremes of performance, by shifting the balance from more impetus on performance to more on availability/affordability. This made such funding decisions driven even less by technology push, which is rare in general industry, where market pull dominates as the driving force for new technology. Thus, for example, implementation of some military systems, such as phased-array radar, was paced by the commercial development of cost-effective applicable technology developed in volume for home microwave ovens. Also, for perspective, it should be noted that the primary justification for one of the earlier major ceramic turbine engine programs was driven primarily by geopolitical concerns of the cold war for the availability of elements such as Cr, Co, and Mn critical for super-alloys in hot sections of metal turbine engines, to potentially be replaced by Si from sand and N from the air. Improved fuel efficiency was also cited as a benefit, but was not a major driving force, especially when oil costs were not high. Such efficiency provides much less driving force in the consumer market as shown by poor sales of fuel-efficient cars in the past, except when gas was scarce. The high sales of higher-fuel-consuming sport utility vehicles in recent years in part resulted from low fuel costs. More recent justifications for ceramics in turbines have focused on reduced erosion, hence less maintenance and longer life for ground-based turbine auxiliary power units, (APUs) i.e., turbine-driven electrical auxiliary power units, and also possibly lighter weight for airborne APUs.

Two other related changes have been the revolutionary changes in medical and biological technology and electronics, especially in telecommunications and personal computers. While both offer opportunities for ceramic applications, for example, of bio- and electronic ceramics respectively, these are generally modest and also have some negative effects. The ceramic opportunities in these areas are limited on the one hand by distribution and liability issues for bioceramics (especially in the United States which has less use of bioceramics than Europe), and on the other hand by the short, rapid product development cycles of many electronic systems, which make it very difficult to implement newer technologies such as use of ceramics often represent. Further, these areas are absorbing large amounts of government and industrial funding, which leaves less money for other technologies. Additionally, the broad and growing availability of computer design technology has made design changes using existing materials, such as metals, much more rapid and cost-effective, thus allowing significant product improvements via new or improved design rather than new materials implementation, the latter typically being a much longer and more costly process. Thus, the addition of an additional set of valves in each cylinder of many automobile engines and the aerodynamic designs of cars, especially truck tractors, reduced driving forces for more fuel-efficient ceramic engine technology.

Another important factor is regulation, along with other public policy factors. Government engine emission controls generated the market of at least $300 million per year for ceramic exhaust catalyst supports. Other existing and pending emission controls also provide further opportunity for ceramics (e.g., for burners), as well as for some competition from metallic burners and catalyst supports. Taxes can also be a factor, for example, an earlier tax on larger auto engines in Japan provided a financial impetus for development and sales of Si_3N_4 turbocharger rotors, and elimination of the tax greatly reduced the ceramic turbocharger market—which, if it does grow in the future could experience competition from other materials, such as metals, allowing variable pitch or carbon-carbon for lower mass. The high petroleum fuel taxes in most other developed countries were a major factor in the development of better fuel-efficient cars, which allowed other countries to expand their market share in the United States. Subsequent U.S. auto fleet fuel-efficiency standards helped provide a more uniform incentive for improvement of U.S. automobile efficiency in the face of fluctuating fuel costs.

Two other important and related factors that are commonly not adequately recognized are that there is always competition for any material application, and that the competition is not static. Thus, many ceramic applications must compete with application of other materials, such as plastics or metals, as well as lower cost ceramics in bulk form by themselves or via coating technology, alternative designs, or both. Thus, various ceramic, intermetallic, or metallic coatings can compete with bulk ceramic components for a variety of wear and other (e.g., biomedical) applications. Low costs for many traditional ceramics due to use of low-cost mineral constituents and especially of processed materials, in particular, Al_2O_3 due to the economies of scale that are a result of the aluminum industry and generally lower processing costs for oxide ceramics, provide competition for higher performance ceramics, especially nonoxides. Metals are clearly the major competition for ceramics in engines, where air-cooling designs allow metals to be used beyond their normal temperature limits, but will most likely be replaced by ceramics (possibly also air-cooled) in some applications. In other cases it is strict cost competition—ceramic cam followers were considered for a number of years by Chrysler to replace metallic needle bearings in some of their auto engines. Ceramic cam followers have not been implemented in Chrysler auto engines due to reduced costs of the metal bearings stimulated by the potential of ceramic competition. (*Note:* The ceramic followers had to be lower cost than the metal bearings to be considered for automotive implementation; ceramic cam followers have been implemented in some diesel engines where cost-performance trade-offs are different.)

The above example of ceramic cam followers in auto engines also illustrates the fact that it is difficult to displace an established technology that can still be improved. This is also shown by other examples, such as solar cell panels for power in space, which have repeatedly been extended to larger sizes and higher powers beyond previously projected limits. However, changing of the balances between competing technologies over time is important, but can be complex. Thus, radomes used on missiles and aircraft used to be polymeric based composites below Mach 1, and ceramic above it, but the former have been improved over time for use to Mach 2–3, thus potentially reducing the market for ceramic radomes. However, the upper missile velocities for which ceramic radomes are used have also been substantially extended, thus maintaining considerable use of ceramic radomes. Similarly, plastic electronic packages have increased in their temperature-environmental and other capabilities allowing them to replace some ceramic packages, but application of ceramic packages in more demanding environments has also increased, leaving ceramic packages still a large and growing business. However, commercial AlN electronic substrates and packages for higher heat dissipation, though present as items of commerce, are well short of earlier expectations due to AlN's higher cost, competition from other heat dissipating materials and methods, as well as reductions in power to operate some semiconductor devices, and hence reduced needs for high heat dissipation. The higher cost of AlN, especially versus that of Al_2O_3, can be reduced with increased volume, as for most material. However, important questions for all materials are whether costs can be reduced enough to attract high-volume use, and are there intermediate cost/volume applications to bridge the gap(s) between lower and higher volume applications. Note that this issue of progressively expanding markets to provide an opportunity to progressively increase production volume and thus reduce costs can be particularly problematical for new technologies as discussed by Christenson [13]. Thus, major technological developments often start as niche markets, which may not be of interest to the business discovering them but may ultimately replace them. For example, buggy makers were generally not interested in automobiles, which were considerably developed by others before they started to replace the horse-drawn buggy.

It is useful to briefly take a long-term and broad perspective on ceramic engine programs. These actually began around the end of World War II with Allied intelligence indicating possible work in Germany to use ceramics to enhance performance of their jet fighter aircraft introduced late in the war [14], leading to a U.S. program following the war. The U.S. turbine blade candidates were a sillimanite (—Al_2O_3—SiO_2) and a BeO "porcelain" (~ 85% BeO), then later $MoSi_2$, and especially, an ~ 80% TiC-20% Co cermet. All were unsuccessful (the latter giving cermets their reputation for often giving the brittleness of ceramics at lower and the poor deformation resistance of metals at high temperatures, rather than the hoped for combination of higher toughness of metals at lower and ceramic creep resistance at higher temperatures). This earlier turbine program apparently lead to industrial investigation of ceramics for piston engines, for example, cylinder liners, apparently focused on alumina-based ceramics. Thus, one could say that these types of ceramic programs have been investigated for ~ 50 and ~ 40 years, respectively, without success, and one could add that ceramic ball or roller bearings have been investigated for nearly 40 years and have only begun recently to achieve moderate commercial success. These earlier programs showed the need for better ceramic materials, development of which, especially silicon nitrides and related materials, stimulated subsequent programs and were further improved by them.

Thus, viewed in a broader perspective, the above programs are interrelated and moderate successes, and both bearing and piston engine cam applications of silicon nitride are, in part, derivatives of turbine engine programs, as was the temporary success of silicon nitride turbocharger rotors. This is also at least partially true of the increasing use of other ceramics (e.g., ZrO_2), in other diesel engine fuel-wear applications, which took fewer years from investigation to application and are growing with progression to newer engine models. Vehicles using hybrid combustion-battery or fuel-cell propulsion offer important opportunities for ceramics that benefit from past ceramic engine efforts. (A hybrid turbine-electric drive was proposed by this author as a follow-up to ceramic engine programs, but was withdrawn based on advice that it was too expensive to have two power sources, an assumption that will be tested by the hybrid vehicles in production and development.) Overall, some possible stalls of uses, such as silicon nitride turbocharger rotors or thermal shock-resistant oxides as exhaust port liners, may prove permanent or temporary. Expansion of the ongoing applications and other developing opportunities, however, indicate growing engine application of ceramics—for example, silicon nitride or carbon-carbon valves or high-performance graphite pistons.

It is useful to note that electronic applications of ceramics are greater and have experienced much better growth than initially projected, especially in their earlier inception, (application forecasts were often short of actual results, while forecasts for engine applications of ceramics have typically been far above actual results). Thus, technology push can be successful, especially for new developments (but probably requires substantial government support), while market pull developments are typically faster and more likely to be successful, especially in modest time frames.

1.4 COST AND PROFIT FACTORS

While there are factors that impact the markets for ceramics on a broader, often long-term basis as outlined above, more fundamental to the success of any specific product is first its specific potential market and its producibility at acceptable costs. The market outlines the character of the product, such as its technical requirements, scale of potential production, product pricing, and possible interrelation of these, but much may be speculative and uncertain, especially for new applications. The producibility is directly related to fabrication (i.e., determining whether the perceived or known component performance, size, shape, dimensional, and other requirements can be met), as well as what its costs are likely to be and how they compare with potential prices and thus what the potential profitability may be. Again it must be emphasized that all of these factors may be quite uncertain for a new product but better defined for a manufacturer entering an existing market for an existing component. Many of the issues, especially those of marketing, are well beyond the scope of this book and are thus addressed little or not at all. The focus is on those issues most directly related to fabrication—focusing on costs—with some limited comments on prices and profitability.

Cost of technology development and especially of actual production implementation are critical to a product's successful introduction and success. Actual production costs for a given part are typically proprietary, so much of the information available comes from general production knowledge and especially from the increasing use of computer modeling of fabrication costs. Cost evaluations via modeling are a critical factor in successful development and implementation of ceramic component fabrication, both in the early stages of consideration as well as during development and implementation, since this indicates both possible fabrication routes and the more costly steps that need particular attention. However, such evaluations are quite dependent on key operational parameters (e.g., the specific fabrication route, processing parameters, and volumes of production), as well as factors such as excess material used, sharing of production resources, and especially yields achieved (the percent of components manufactured that are suitable for sale versus those that have to be scrapped). These factors are typically highly proprietary, and thus not available publically, as is true for much data to verify cost models. However, there are some generally known cost aspects as well as specific cost modeling studies that provide useful guidelines, recognizing that there are exceptions to all trends that require specific evaluations of specific cases. Again, the purpose here is not a tutorial on modeling methods, which is a large subject in itself, but to give some sources of such information, and especially familiarize readers with factors, variations, and some basic guidelines.

First, consider overall trends, a major one being what technical market into which the specific component will fall, with a major differentiation being whether the component is an electronic one, especially a multilayer electronic package, or for some other use, such as for thermal-structural functions. Electronic packages (and to a lesser extent other electronic and some electrical components as well as ceramic cutting tools) sell at prices much higher per unit weight, reflecting both higher production costs and their small size, hence limited per part cost, and probable more value added. An overall assessment of the advanced ceramics industry in the United States outlined by Agarwal [15], that is apparently more focused on structural ceramics, noted that ceramic processing is typically batch and labor intensive. He attributed 40–50% of manufacturing costs of high-performance ceramics to inspection and rejection (basically to yield), versus 5–10% for high-performance metals, thus again emphasizing this as a major cost issue for ceramic production. Agarwal cited typical 15–20% of total costs for ceramic finishing, mainly for machining, and only 5–10% for metals. However, for precision parts (e.g., those in engines), machining costs can be substantially higher (e.g., machining costs were a major factor in projected high costs of toughened zirconia components for possible use in more efficient diesel engines (see Table 1.2). The potentially high costs of much machining of ceramic components is a major reason for emphasizing near net-shape fabrication, though an important exception has been ceramic ball manufacturing for bearings as discussed later.

While machining costs can be very high, and often the dominant single process cost if extensive machining is required, it is important to again emphasize another factor that often determines the viability of a product, namely its yield—the percentage of output that ends up in useable product. Resulting yields

TABLE 1.2 Estimated Manufacturing Costs for PSZ Diesel Engine Components[a]

Component	Headface plate	Piston cap	Cylinder liner
Body costs ($)	1.55 (1%)	1.36 (1%)	7.84 (3%)
Forming costs ($)	2.88 (1%)	8.08 (3%)	38.09 (15%)
Firing costs ($)	17.45 (9%)	12.97 (5%)	47.48 (19%)
Grinding costs ($)	173.93 (89%)	259.03 (91%)	161.32 (63%)
Total costs ($)	195.81	281.44	154.74

[a]Costs of subsidiary operations [16] shown as a percent of the total component cost in parentheses. Note that in another related study modeling costs of some of the first two listed and other related components directed at design and manufacturing changes to reduce machining, showed overall costs substantially reduced to somewhat under $100 each, with grinding costs reduced to 37–68% of total costs (but increasing the other fixed costs as a percent of other their total reduced costs [16,17]).

at each fabrication step vary in amount and cost impact, with the limited amount of recovery often being greater, hence somewhat less costly, in earlier processing stages, such as powder, and much greater in later stages, such as firing and machining. However, the cumulative product yield, which can be well < 50% in earlier stages of product production and possibly < 80% in later stages of production of complex components, is most critical, and is often the determining factor in product success and of constraints on costs of individual fabrication steps, especially more costly ones [18-23].

Turning to other specific fabrication costs, raw materials are a factor; Agarwal cites them as 5–10% of total costs (for structural ceramics, metals, and polymers), which is a common range, but subject to a variety of conditions, they are sometimes lower or higher (see Table 1.2). Thus, for example, Rothman and coworkers [18,19] report that higher cost SiC powder ($22/kg, giving material costs at 22%) could be used for making small disk seals if a high overall yield (86%) was assumed, versus ~ $3/kg powder (giving material costs at ~5%) with 40% overall yield. The impact of raw materials costs depends greatly on the amount used—expensive materials such as silver, gold, and platinum are used in electronic ceramics, but in small quantities, such that many packages can be sold for a few dollars each. More generally, consider a range of small, structural components, such as a Si_3N_4 balls for bearing applications: A $^1/_4$-in. diameter ball requires a modest amount of Si_3N_4 powder, while a larger $^1/_2$-in. diameter one requires eight times as much powder, so the sensitivity of such balls to raw material costs increases substantially as ball size increases. Thus, raw materials used for small balls may not be economically viable for larger ones. This issue of raw materials costs is important because many desirable powders have been developed, but their high costs severely limit their economical viability. For example, Schoenung, and coworkers [20-21] conducted substantial modeling of ceramic costs for a variety of advanced engine uses, such as cam rollers, valves, and guides, showing that zirconia toughened components never or barely allowed the costs to get down to target component prices at substantial product volumes of 5–10 million/yr with \$13/kg powder, whereas \$4.50/kg powder cost allows component costs to reach target prices at production quantities of 2–4 million/yr. However, also note that many components, such as many in engine applications, have an approximately fixed physical volume, in which case zirconia suffers a disadvantage of requiring nearly twice the weight of powder per component relative to other candidate ceramics such as Si_3N_4. On the other hand, other powders such as those of Si_3N_4 are commonly much more expensive than \$4.50/kg, thus not changing the raw material cost limitations much if at all favorably. Schoenung and coworkers' evaluations of similar Si_3N_4 components also showed similar material cost limitations—\$44/kg powder costs not even coming close to target prices even at production volumes of 10 million/yr, and even \$11/kg powder costs barely reaching the upper target price range at volumes of ~ 7 million/yr.

Das and Curlee [24] have also shown the importance of reducing Si_3N_4 powder costs (from ~ \$44/kg) along with machining costs making cam roller followers and turbocharger rotors more cost competitive with costs of metal components. However, their assertion that such higher cost ceramic engine components will be implemented where the improved benefits are adequately communicated must be viewed with considerable uncertainty. Morgan [25] correctly cites potential cost savings from broader use of advanced chemical preparation of ceramic raw powders and their processing, but does not address the key issue of how various production steps can be successfully made to go from the typical modest starting markets and common less-efficient batch processes used at such earlier levels of production to achieve potential large-scale lower costs.

Quadir et al. [26] corroborated that lower raw materials costs, including additives, are important in developing a lower cost Si_3N_4 (e.g., for wear, more modest temperature applications, and thermal shock resistance) and that comminution is an important powder cost factor where it must be used.

Tooling costs can vary from very modest to quite substantial depending on several factors, but particularly on the fabrication process selected. Thus, tooling costs for colloidal processing such as electrophoretic deposition and tape or slip casting, as well as isopressing, can be quite limited (though times for thicker deposits, tape lamination, drying times, die storage, and loading isopresses can be important cost factors). Pressure casting can entail more expensive tooling (and again deposition time issues). Die pressing and extrusion tooling costs can be modest (e.g., a few ten thousand dollars), since shapes are often simple, but even limited complexity and multiple cavity dies (for faster production and better use of presses) can substantially increase costs. Injection molding tooling can be substantially more expensive since it can form complex parts, a key virtue of injection molding, with tooling costs often reaching \$50,000 or more. Such die cost are thus a factor in the choice of forming methods since modest levels of production, e.g., a few thousand components, often cannot cost effectively support more expensive tooling.

Some have assumed explicitly or implicitly that energy costs of sintering or other fabrication/densification processes are an important problem, avoiding some fabrication and seeking other approaches such as use of some highly exothermic reaction processes, such as self-propagating synthesis, to eliminate energy costs of densification. However, the energy costs (the "fuel bill" for most industrial ceramic processing), for example of sintering and hot pressing, are commonly < 5% of component costs (see Table 1.2). Thus, while savings on these costs are of value since these normally increase profits, they are not a major factor in determining process selection. (Further such energy savings can be greatly overshadowed by the high costs of the raw materials to yield the energy saving of such reaction processing [27,28]. Also, as discussed in later chapters, energy costs for other fabrication methods such as CVD and even melt processing are commonly similar.) Such energy savings would be of more impact if they also reduced the costs of heating facilities—their size, maintenance, and plant space used, (which are often factored into firing costs)—but these savings by such reaction processing also appear limited [27,28]. On the other hand, these or other reaction processes can give beneficial raw materials and (unexpected) comminution costs as discussed below and in subsequent chapters.

As noted above, there are other important factors in firing costs for ceramics, such as the furnace atmosphere, temperatures, volume, and more. Thus, oxide ceramics often have lower costs, since they can be fired in air, and at more moderate temperatures than nonoxides, the combination of which allows for both larger furnaces and especially continuous ones, such as tunnel kilns for oxides, both of which can increase cost effectiveness. While nonoxides are typically fired in batch kilns with significant lower through-capacity due to heat-up and cool-down times, some of the significant advantages of continuous firing of oxides can be realized by continuous firing of nonoxides. Thus, Wittmer and coworkers [29] showed substantial savings (e.g., 50–70% lower firing costs for silicon nitride fired in a (continuous) belt furnace than for firing in batch kilns). Though such belt furnaces typically do not have near the throughput of typical air fired tunnel kilns for firing oxides, they show substantial potential for continuous firing of nonoxides with sufficient production volume to justify higher belt furnace costs.

There has been a preliminary attempt at addressing cost aspects of hot forming of ceramics. Thus, Kellett and Wittenauer [30] discuss the possibilities of superplastic forming of nanograin ceramics such as Al_2O_3-ZrO_2 and Si_3N_4 as a means of lowering costs by producing components of near net shape, requiring less machining. Their focus was on effects of deformation rates on costs, reporting that strain rates of 10^{-3} to 10^{-5} sec^{-1} translate into forming times respectively of 4 min versus 6 hrs giving part costs of $4 and $400, respectively, for the case chosen. They also noted steps to improve these, in particular grain growth inhibition. (*Note:* Large deformation at higher strain rates, e.g., ~ 10^{-1} sec^{-1}, have recently been reported for a nanograin oxide composite; See Sec. 6.3.)

The impact of time required for individual process steps, as well as the overall cost factors and constraints in processing materials, can be seen by recognizing that there are just over 31.5 million seconds in a year, i.e., about 10.5 million for each 8-hr shift, with no days off, and about 7 million seconds for an 8-hr shift with typical days off. Thus, whatever the annual production expected, this time constraint is a fundamental factor in production costs since it defines how many parts must be produced per unit time and the impact of this on the productivity at each step to achieve the production goal. For example, if a million parts per year is the goal, then, depending on the shifts and days of production, a part must be produced every 7–31 sec (and even faster to allow for production losses, yields of < 100%). Thus if a process consists of 10 steps, on average a part must take no more than 0.7–3.1 sec in each step. Most steps in a process take longer, often much longer per part, than the allowed average times, which means that many parts must be made simultaneously to achieve the targeted average time per part. For example if parts are formed in a press with a single cavity die, then the number of such presses needed will be the actual time for forming each part divided by the average time allowed for the forming step, which means that process steps that take longer times and have low or no multiplicity of simultaneous part formation from each press requires large numbers of presses and their associated costs.

Turning briefly to costs of producing ceramic powders (and whiskers), there are similarities to producing ceramic bodies, for example, impacts of starting materials. Thus, Schoenung [31] has modeled the costs of making Si_3N_4 powder by direct nitridation of Si versus by laser-stimulated CVD gas-phase nucleation of powder. For the assumed parameters, she showed nitridation yielding powder costs of \$17–230/kg, mainly \$25–50/kg, depending on seed powder costs, time of nitriding, and especially comminution costs; while the costs of the laser-CVD powder were more, > \$100/kg, driven heavily by the high cost of silane gas (assumed to be \$160/kg).

Schoenung [32] has also modeled costs of producing SiC whiskers via vapor solid (VS) reaction of SiO_2 and carbon, showing that costs could not be reduced to < \$50/kg for the assumptions made, with raw materials cost and yields being important factors. Another indicator of the impact of precursor costs for ceramics is shown by the cost per kg of three common oxides in Table 1.3, which contrast with costs from other more conventional precursors of <\$1 to < \$3/kg. The costs of additives used in processing—commonly to aid densification and properties (see Chap. 5)—can also be significant. Thus, use of rare earth and other oxides for PSZ and TZP as well as Si_3N_4 often measurably increase raw materials costs, especially more expensive additives such as yttria versus others

TABLE 1.3 Costs of Common Oxide Ceramics from Commercial Sols[a]

Ceramic/sol precursor	Colloidal	Alkoxide
SiO_2	3.74	11.66
Al_2O_3	8.80	9.02
ZrO_2	27.50	27.50

[a]Costs in $/kg circa 1986 assuming 100% theoretical ceramic yield.

such as ceria, and especially magnesia or calcia. Another indicator of the frequent importance of comminution costs are commercial prices for abrasive grade SiC grits. The ratio of the costs of finer grits (400, 600, and 1000) to that of 240 grit material of two different purity grades from each of two major manufacturers were 1.6, 2.4., 4.2, and 1.1–1.2, 2, and 5.3–5.8. Since all were ground from the same SiC, the substantial increases in costs as grit size decreased is primarily due to comminution cost and secondarily to classification.

Finally, note that the typical model of decreasing costs per unit of production, e.g., per unit weight of powder or per component, progressively decreasing smoothly toward an approximately limiting cost at high volume (and hence also commonly with the passage of time), while a useful guideline, is often an oversimplification, as shown by costs of graphite fibers [23]. Thus, various perturbations of such smooth, continuous decreases occur, due to discrete changes in infrastructure, process technology, or both. Examples would be adding another or larger or more expensive piece of processing equipment such as a continuous furnace, the acquisition and installation of which require substantial levels of production before it can be justified. A specific example indicated in one ceramic study was that a relatively simple-shaped turbine vane could be hot pressed to net shape in modest quantities at lower cost than by injection molding and sintering, due to the much higher cost of the injection molding die versus the hot pressing tooling. Thus, the break-even level of production for both processes was several thousand per year, raising the dilemma of potentially starting with a more expensive process (injection molding) and hoping that volumes rose sufficiently to justify its cost, or of starting with hot pressing at low volume, then changing to injection molding if volume rose high enough, but with some cost penalty for switching processes (R. Palicka, Cercom, Vista, CA, personal communication, 2000). (Fig. 1.1).

Turning briefly to price and related profitability, it is essential to remember that price is determined by competition at one or more levels, which may change with different aspects of a given type of application. The most common and fundamental level is at the specific component material-fabrication level,

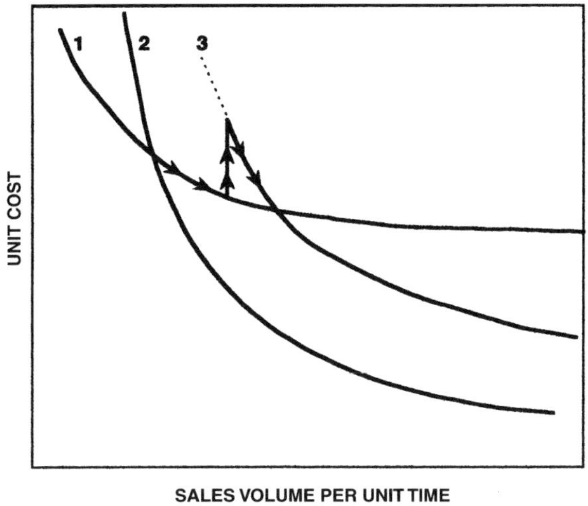

FIGURE 1.1 Schematic of cost impacts of changing manufacturing methods. Curves 1 and 2 reflect two different processes, e.g., hot pressing to net shape versus injection molding and sintering, respectively, with one reflecting a more efficient process at low volumes, but more inefficient at high volume. Thus, if process 2 is initially selected and production/sales volumes increase beyond the crossover point, this was a good selection, but not if the crossover point is not reached. On the other hand, if the process that is more efficient at lower volumes (curve 1) is chosen and volumes rise past the crossover volume, some added costs of changing the process would be required, shown schematically as a shift of curve 2 to curve 3 (probably with other changes in curve 3 not shown for simplicity).

but competition at the subsystem or complete system level can also be important, and may vary with the specific character of a given application. This is briefly illustrated for turbochargers, where at the fundamental level, ceramic (silicon nitride) turbocharger rotors compete with established use of metal rotors. This may be on a direct ceramic-metal rotor cost difference, or on the basis of overall turbocharger cost versus performance, but in either case the cost of metal components individually or collectively is a major factor in the competitive balance.

However, cost competition of different technologies may change as a function of component size, performance requirements. and market size, as well as other factors. As noted earlier, an important factor that drove use of ceramic turbocharger rotors for smaller auto engines was a Japanese tax on engine horsepower above a certain level, putting a premium on more performance from smaller engines. This shifted the balance in favor of ceramic rotors by putting a greater premium on faster response due to the lower ceramic versus metal density

(3.2 versus ~ 5.8 g/cc) until the tax was repealed. However, other circumstances could impact turbochargers—carbon-carbon rotors should be feasible with still faster response due to still lower density (e.g., 1.6-1.8 g/cc) and possibly less cost than silicon nitride rotors. On the other hand, variable pitch blades may be advantageous on some larger turbochargers, such as for diesel truck engines, again probably favoring metal rotors. Larger markets for turbochargers could also bring competition from other devices and related fabrication technologies, since a turbocharger is really only a way of temporarily increasing the volume of air delivered to the engine for faster acceleration; other devices to do this may be feasible, such as a small air compressor, storage tank, and valving to draw compressed air from the tank as needed, which could completely change the materials and fabrication technology picture.

Price constraints from existing technology can be seen in broad terms by considering resultant pricing of individual components or systems in which ceramics might be used. Thus, prices paid for major system purchases such as cars, tanks, and ships can provide some guidance. For example, the sales price of a 1200 kg car is ~ $20,000, which translates to ~ $17/kg, which, allowing for profit and assembly costs, means that the average purchase price for the individual components and basic materials (mainly steels) is generally < $10/kg. Thus, while small amounts of much more expensive materials, such as some ceramic components, can be tolerated, larger quantities of use rapidly become seriously price constrained by the existing technology. Such evaluation readily indicates cost constraints of substantial substitutions of ceramic for metal armor for tanks. While the competiting technology is commonly that of metals, it can also be other materials, including other ceramics. A particular case in point is that the modest use and growth of AlN for electronic packages with higher heat dissipation than Al_2O_3 is, in part, due to the lower costs of Al_2O_3 packages, providing incentives for designs that reduce the thermal dissipation in packages or alternative ways of accomplishing the dissipation.

Finally, once the commitment to a product has been made, whether or not the uncertainties have been adequately addressed, the issue is whether suitable profitability, in both amount and timing, can be reached. This is a function of the development costs, which include research and development, especially production and market development, as well as interest costs, and of profits, that is volume and price-cost differential (Fig. 1.2) [23,33]. Interest is a factor since to be adequately profitable, a new product needs to not only pay back the costs of its development, but do so with a return of interest that makes the development a worthwhile investment by the company versus other possible investments. For a substantial new product the interest costs can be a significant factor, which can be easily estimated by the rule of 72, that is, the time in years to double an interest cost or return multiplied by the annual interest rate is 72. Thus for an annual interest of 10%, the amount of money to be returned by product sales to the com-

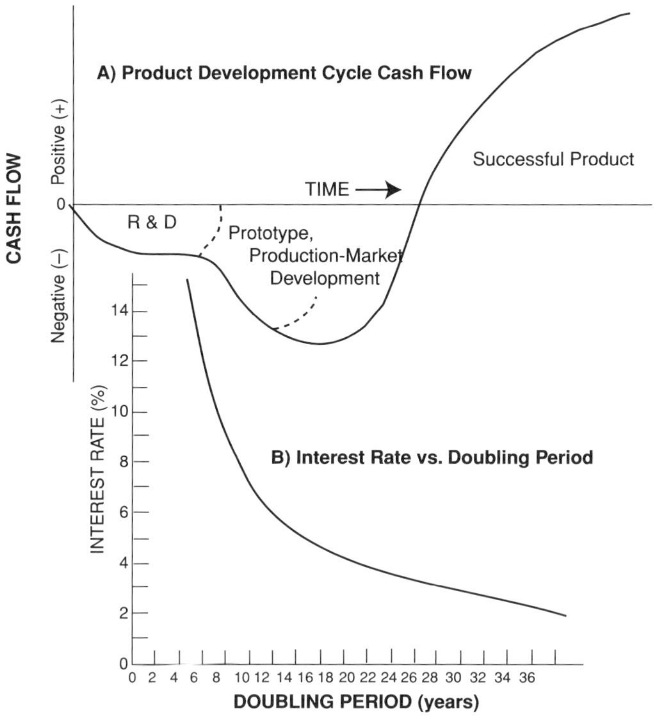

FIGURE 1.2 Cash flow factors for product development. (A) Schematic of the cash flow for a particular product development to be successful. Note that R & D costs are often much less than those for production and market development, and that positive cash flow only results after a substantial time with suitable profit from the product. (B) Interest rate versus the period (in yrs) to double the principle illustrating the rule of 72 and the significant impact that interest has on product payback. (From Ref. 23.)

pany would be doubled over a period of 7.2 years. Note that since such product investment costs are paid back only out of profits on sales of the product, payback times of a few to several years can be common.

1.5 OVERVIEW OF CERAMIC FABRICATION TECHNOLOGY

Ceramic fabrication technology consists of a diversity of processes that can be combined in various ways with varying material and microstructure (property), size and shape, and cost constraints and opportunities. The combinations of different processes to form a fabrication route to a component are determined primarily by the process of producing a solid component of suitable character. The dominant process for fabrication of higher technology ceramics is sintering, mostly without pressure, of preformed bodies made by various powder consolidation techniques using powders from various preparation processes (see Chaps. 2 and 3). Typical powder consolidation-forming processes are die or isostatic pressing, extrusion, injection molding, and various colloidal processes, which along with sintering and its variations, are discussed in Chapter 4, emphasizing practical issues. Such sintering-based fabrication is very diverse with many variations as outlined in Fig. 1.3. While all of the above forming processes have considerable shape capability, their ranking in terms of decreasing shape complexity capability is approximately injection molding, colloidal forming, extrusion, and die- or isopressing.

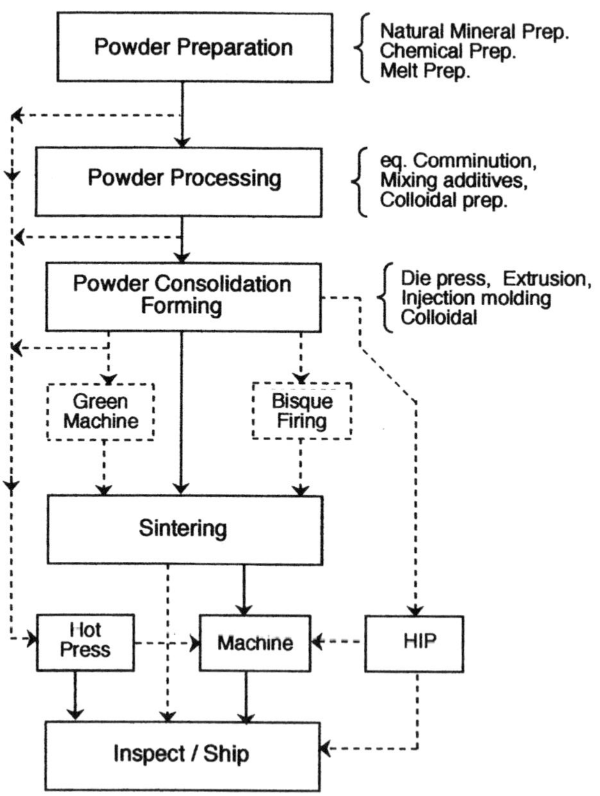

FIGURE 1.3 Schematic outline of sintering-based fabrication of ceramics and many of its variations, which are shown in boxes outlined by dashed lines and by dashed lines connecting various steps.

Variations in sintering-based fabrication include both newer methods of heating not shown in Fig. 1.3 that have been under investigation and can effect the sintering process, as well as many variations shown in Fig. 1.3. The latter include sintering under uniaxial pressure (i.e., hot pressing), which has seen considerable production use, and sintering under hydrostatic pressure (HIPing), done either following sintering or instead of sintering, which has seen some production use. Both of these pressure-sintering processes generally reflect higher costs, but have growing areas of application, and opportunities for further development as discussed in Chapter 6. There are also hot-forming processes, such as press forging, that have seen some investigation by starting with a sintered body or combining powder consolidation and hot forming of simple shapes. Such forming, though facing important time-cost issues (as discussed in Sec. 1.4), may have some specialized applications. Various reaction processes carried out in conjunction with either sintering without pressure or via various processes with pressure, especially for fabrication of ceramic composites are also addressed. Substantial discussion of this and other shifts in fabrication methods for monolithic versus composite ceramics is presented.

Other fabrication methods for producing ceramics without sintering, and may or may not entail use of powders (for example, polymer pyrolysis, deposition, and melting, as well as other reaction processes) are discussed in Chapter 6. There are also other fabrication methods for some speciality bodies, that is, fibers, balloons, beads, and bodies of designed porosity, including foams, which may entail various combinations of these methods (polymer pyrolysis), those entailing sintering, or both, which are discussed in Chapter 7. Deposition of coatings by various vapor processes is briefly discussed, while the use and significant potential of CVD for bulk monolithic and composite ceramics are more extensively discussed. Similarly, coating via various melt spray deposition techniques are noted, while making large bulk bodies by such methods are more extensively treated. Bulk melting and solidification are actually used to produce both some of the largest individual components as well as product volume of ceramics produced in view of its wide use for both the glasses and refractories, but has been more limited in its use for higher tech materials, mainly to single crystal growth. Earlier development of press forging of single crystals to produce shaped, polycrystalline optical (e.g., IR) windows, which has seen some use in production, is discussed along with other possible extensions of single crystal growth and melt casting.

Many issues discussed include effects of atmosphere on both calcining and sintering of powders as well as of adequate outgassing of anion impurities and adsorbed species, not only in various sintering processes but also other fabrication processes such as melting—for example, single crystal growth. Other methods of making powders and coarser particles, including melt forming, for fabrication of ceramic bodies for use by themselves or in nonceramic matrix composites are discussed. Important shifts in fabrication methods for various types of ceramic composites are addressed, as is the array of various rapid prototyping-free form fabrication methods emerging. As noted earlier, preparation of raw materials, densification of powders, and control of microstructures via use of additives is extensively discussed. The size and shape potential and limitations of various fabrication routes are also addressed, including joining ceramics to themselves, as well as other materials. This shows some important advantages of hot pressing, CVD, and especially of melt processing. Some aspects of surface machining and other methods of surface finishing as well as joining are also discussed. Again, in all of these subjects the focus is on practical aspects including costs.

1.6 SUMMARY AND CONCLUSIONS

In summary, the large and growing families of ceramics have many known individual, as well as combinations of, properties of great technological importance. While the ending of the cold war substantially reduced the driving force for application of ceramics, especially more performance-driven development, and tax and regulation driving forces may come, change, and go, there is still substantial need and opportunity for ceramic applications, and this is likely to increase in the future. However, much of the development must be more focused and conscious of costs and economic constraints. There is even more limited opportunity for technology push applications, and competitive materials costs, such as for metals, will commonly be a limitation, except for the advantage that ceramics often have lower densities than common metal competitors, so less mass of ceramic is used in the many cases where component volume is dictated by the design.

These changes provide added needs for evaluating and modeling costs of various ceramics and their fabrication/processing. While detailed evaluation of these costs as a function of production methods and factors such as volume is the ultimate arbiter, some issues or flags for more attention were noted. Thus, yields are often a major factor, especially in earlier stages of manufacturing, but can be a sporadic problem in long term production. Machining costs can often be very high, being an important motivation for near net-shape fabrication, but details of methods and volumes are important. Direct energy costs are often modest, especially in comparison to many impressions (e.g., $< 5\%$ of total production costs) but must be considered, often with attention to broader factors such as throughput and duplication of firing facilities. Tooling costs for forming methods such as die pressing, extrusion, and injection molding can be substantial, with high tooling costs requiring high levels of production over which to spread those costs. Raw materials, commonly $\sim 10\%$ or less of total costs still must be considered, especially if their percentage of total costs is substantially higher, with additive costs often being a factor needing attention.

The key for successful production of ceramic is profitably producing components that perform with suitable reliability at cost-effective prices. This means that much change typically has to occur in transitioning from laboratory preparation to production, which commonly means changes in raw materials and fabrication and processing. Thus, for example typical laboratory use of many fine but expensive powders that are ideal for processing is often not cost-effective; the focus needs to be not on what is ideal—that is, the finest, most uniform, purest powder with the optimum ceramic phase content—but on what are the true needs to achieve the product goal. A great deal of development of the actual manufacturing steps, including evaluation of alternative approaches. is commonly necessary.

REFERENCES

1. M.N. Rahaman. Ceramic Processing and Sintering. New York: Marcel Dekker, Inc., 1995.
2. D.W. Richerson. Modern Ceramic Engineering. New York: Marcel Dekker, Inc., 1992.
3. J.G.P. Binner, ed.Advanced Ceramic Processing and Technology. New Jersey: Noyes Publications, 1990.
4. F. Singer, F.F. Singer. Industrial Ceramics. London: Chapman and Hall, 1984.
5. J.S. Reed. Introduction to the Principles of Ceramic Processing. New York: John Wiley & Sons, 1988.
6. J.A. Mangles, G.L. Messing, eds. Forming of Ceramics, Advances in Ceramics, Vol. 9. Westerville, OH: American Ceramic Soc., 1984.
7. W.A. Knepper, ed. Agglomeration. New York: Interscience Publishers, 1962.
8. P.J. James. Isostatic Pressing Technology. London: Applied Science Publishers, 1983.
9. H.O. Pierson. Chemically Vapor Deposited Coatings. Westerville, OH: American Ceramic Soc., 1981.

10a. F.S. Glasso. Chemical Vapor Deposited Materials. CRC Press Inc., Boca Raton, FL, 1991.

10b. H.O. Pierson. Handbook of CVD, Noyes Pub. Inc., Park Ridge, NJ, 1992.

10c. R.F. Bunshah. Handbook of Deposition Technology for Films and Coatings. Noyes Pub. Inc., Park Ridge, NJ, 1994.

11. R.W. Rice. Porosity of Ceramics. New York: Marcel Dekker, 1998.
12. R.W. Rice. Mechanical Properties of Monolithic and Composite Ceramics, Grain and Particle Effects. New York: Marcel Dekker, 2000.
13. C.M. Christensen. The Innovator's Dilemma: When New Technologies Cause Great Firms to Fail. Boston, MA: Harvard Business School Press, 1997.
14. R.W. Rice. An Assessment of the Use of Ceramics in Heat Engines. NRL Memo, Report 4499, 1981.
15. J.C. Agarwal. Process economics and strategies in the advanced ceramics industry. Adv. Cer. Mat. 1(4):332–334, 1986.
16. J.C. Bentz. Ceramic Mmanufacturing development for adiabatic engine components. Cummins Engine Co. Report AMMRC TR 84-24, for AMMRC Contract DAA646-83-C-0002, 1984.

17. J.C. Bentz. Ceramic Manufacturing Methods and Technology Development for Adiabatic Engine Components. Cummins Engine Co. Final Report MTL TR 89-75, for AMMRC Contract DAA646-83-C-0002, 1989.
18. E.P. Rothman, H.K. Bowen. New and old ceramic processes: manufacturing costs. MIT Ceramics Processing Res. Lab. Report No. 63, 6, 1986.
19. E.P. Rothman, J.P. Clark, H.K. Bowen. Cost modeling of structural ceramics. Adv. Cer. Mats. 2(1):34–38, 1987.
20. J.M. Schoenung, E. Rothman, H. Bowen, J. Clark. Simulation of the potential market for ceramic engine components. In: W. Bunk, H. Hausner, eds. Ceramic Materials and Components for Engines, Proceedings of the Second International Symposium. Lübrck-Travemünde, FRG. Verlag Deutsche Keramische Gesellschaft, 1090–1098, 1986.
21. J.M. Schoenung. An Engineering and Economic Assessment of the Potential for Ceramics in Automotive Engines. MIT Ceramics Processing Res. Lab. Report No. 84, 1987.
22. K. Subramanian, P.D. Redington. Total cost approach for ceramic component development. Cer. Eng. Sci. Proc. 14(1–2):309–320, 1993.
23. R.W. Rice. Performance and applications of structural ceramics: status and needs. In: D.J. Viechnicki, ed. Thirty-Seventh Sagamore Army Research Conference, Structural Ceramics. Army Materials Technology Lab. Publication, 15–61, 1990.
24. S. Das, T.R. Curlee. The cost of silicon nitride powder and the economic viability of advanced ceramics. Am. Cer. Soc. Bul. 71(7):1103–1112, 1992.
25. P.E.D. Morgan. Structuring chemical technology to produce cost-effective ceramic products on a large scale. Am. Cer. Soc. Bul. 72(7):65–70, 1993.
26. T. Quadir, R.W. Rice, J.C. Chakraverty, J.A. Breindel, C. Cm. Wu,. Development of lower cost Si_3N_4. Cer. & Eng. Sci. Proc. 12(9–10):1952–1957, 1991.
27. R.W. Rice. Assessment of the application of SPS and related reaction processing to produce dense ceramics. Cer. & Eng. Sci. Proc. 11(9–10):1226–1250, 1991.
28. R.W. Rice. Summary assessment of the application of SPS and related reaction processing to produce dense ceramics. In: Z. A. Munir and J. B. Holt, eds. Combustion and Plasma Synthesis of High-Temperature Materials. New York: VCH Publishers, Inc., 1990, pp. 303–308.
29. D.E. Wittmer, J.J. Conover, V.A. Knapp, C.W. Miller, Jr. Continuous and batch sintering of silicon nitride. Am. Cer. Soc. Bul. 72(6):129–137, 1993.
30. B.J. Kellett and J. Wittenauer. Commercialization Issues in Superplastic Forming of Nanocrystalline Ceramics, Cer. & Eng. Sci. Proc. 17(3):101–108, 1996.
31. J.M. Schoenung. Analysis of the economics of silicon nitride powder production. Am. Cer. Soc. Bul. 70(1):112–116, 1991.
32. J.M. Schoenung. The economics of silicon carbide whisker fabrication. Cer. & Eng. Sci. Proc. 12(9–10):1943–1951, 1991.
33. E.E. Conabee. The business of technology: integrating marketing, R & D, manufacturing, and sales (marketing perspective). Cer. Eng. Sci. Proc. 10(7–8):685–692, 1989.

2

Preparation of Ceramic Powders

2.1 INTRODUCTION AND BACKGROUND

Ceramic powders are the basic starting materials for the majority of fabrication processes for producing components and samples of both monolithic and composite ceramics. Thus, sintering processes, whether with or without pressure, are the most used processes for fabrication of monolithic and composite ceramics. These processes not only start with powders, but generally depend critically on the nature of the powder for both component shape fabrication and subsequent sintering to be successful in their goals. Meeting these two goals of forming and densifying components generally imposes conflicting demands on powder character, for example, particle size, thus requiring adequate control of the powder preparation to provide powders of suitable compromise character. Very fine particle sizes are desirable for easier, more complete sintering, but can present problems of anion contamination, as well as green body fabrication limitations (see Sec. 2.2 and 8.2.1). Fabrication of many ceramic composites requires some similar and some different requirements on the particle character used as the dispersed phase in particulate composites. More severe densification challenges are found with composites with dispersed whiskers, platelets, or fibers. Fabrication of such composites shifts the emphasis in fabrication from pressureless to pressure sintering and some other processes as one goes progressively from monolithic ceramics to ceramic particulate composites. While some preparation of whiskers and platelets is briefly noted in this chapter, such preparation is typically dependent on additives (see Chap. 3), and preparation of fibers is via specially modified or designed processes (see Sec. 7.2).

It is desirable to have powders tailored to the specific ceramic fabrication process of interest, for example, pressureless sintering. However, this is often not practical, especially in development and limited production stages, since it is often more cost-effective to use available powders, possibly by modifying them (e.g., by comminution), the fabrication process, or both. There are a variety of other uses of ceramic powders that have different requirements, some often less demanding than in sintering. The latter includes the large field of raw materials for melt processing of ceramics, much of which is less demanding in terms of physical character of the powder (see Sec. 6.7). The broad field of abrasives for sawing, grinding, lapping, sanding, and polishing, though often using diamond, also often uses other ceramics, such as SiC and alumina, and other processes than typical preparation of sinterable powders. Another important application of ceramic powders is as feed material for plasma and other melt spraying processes, where there have been major shifts in processing technology for feed materials in the past 20 years. (see Sec. 7.5). There are also needs for dispersed ceramic particles for a variety of metal as well as some polymer matrix composites that differ in character from those for ceramic composites and the particles needed; for instance, in terms of size, shape, and single crystal or polycrystalline character.

It should be noted that the earlier powder preparation for traditional ceramics such as glasses and various porcelain bodies were primarily extraction, cleaning, and comminution of natural minerals such as clays, talc, silica sands, quartz, limestone, and feldspars. Such traditional ceramics are still important, but are not central to this book, the interested reader is referred to other sources [1,2]. The technology of interest in this book and chapter, that of fine or higher technology ceramics, entails more chemical processing in preparing of the ceramic powders, which is thus the focus of this chapter. However, physical aspects of powder processing such as comminution are still important, and are thus noted here, but are often accomplished in conjunction with other steps, such as milling for mixing of other ceramic, and organic (e.g., binder) constituents, and thus not extensively addressed here.

There is now a diverse and expanding array of various methods of preparing ceramic powders, many focused on conventional wet chemical processing, as well as use of other physical and especially chemical methods. The latter have resulted from increased chemical input to ceramic powder preparation, which has significantly broadened technical opportunities, but has also often left issues of practicality and cost. Only key aspects and examples of the processes can be addressed here, since most of these topics are large subjects in themselves, but at least some of the practicality issues will be addressed. There are other reviews, generally not as comprehensive, but with different perspectives, and with some different examples [1-9].

Processes based on conventional wet chemical processing of various salt precursors for oxide ceramics are addressed first along with their conversion (i.e., calcining) to ceramic powders. Then, extensions of conversion of oxide salt precursors to ceramics via processes such as freeze-drying and spray pyrolysis are discussed, followed by extension to other chemical processing of precursors, via sol-gel and preceramic polymers and their conversion to ceramics. This is followed by various melt, vapor, and other reaction processing of oxide powders, especially ternary oxides; then processing of nonoxide powders, especially by various reaction processes, including extensively used carbothermal reduction, and other processes (e.g., wet chemical, melt, or vapor-phase based) are addressed. While there has only been limited use of ternary nonoxide or mixed nonoxide-oxide compounds, their preparation is briefly addressed. Then, the important emerging technologies of coating ceramic (and metal) powder particles with a ceramic (or metal) coating are addressed, along with a summary of powder characterization.

2.2 PROCESSING ESTABLISHED BINARY OXIDE POWDERS VIA CONVENTIONAL CHEMICAL SALT PRECIPITATION AND CALCINATION

The most common method for commercial production of powders of binary oxide ceramic compounds with higher purity than typically occurs naturally is to obtain salts that can be thermally decomposed to the desired oxide compound. Common salts used are hydroxides, carbonates (and combinations of these, i.e., basic or bicarbonates), nitrates, sulfates, formates, acetates, and citrates. Digestion of a parent mineral in an acid or base is often a basic step in such processes, as is precipitation of the desired salt from a water-based solution. The final step is thermal decomposition of the salt to the desired oxide. The selection of the chemical process is impacted primarily by cost and the powder product character. Costs include those of the parent mineral and its processing, the acid or base, and the steps and facilities in the digestion, precipitation, and thermal decomposition (often referred to as *calcining*). Key aspects of required powder product character are chemical purity and particulate size and shape.

Application of the above factors in some respects is a straightforward selection of the chemical route guided by both its costs and resultant powder character, but can be more complicated. Thus, there are increasing limitations due to environmental factors, costs, or both; for example exhaust emissions from decomposition of salts, such as sulfates and especially nitrates. Key factors in the salt selection are: (1) the absence of salt melting (melting is frequent for some common salts such as nitrates, especially hydrated ones), since this is incompatible with powder production; and (2) salt decomposition temperature, as is readily obtained from reference sources [10,11], which is typically somewhat below the actual calcination temperature. These temperatures are important since in some special cases of differing crystal structures of the resultant oxide, too high a calcining temperature can preclude obtaining lower temperature crystal phases that may be desired. More generally, too high a calcining temperature can preclude obtaining a sufficiently fine (e.g., particles of a few microns to submicron in size) and unagglomerated powder, preferably of dense, single-crystal particles, due to varying, often excessive, particle growth and sintering.

Two complications that are not fully documented or understood are (1) What is "too high" a decomposition temperature (which varies for different ceramics)? and (2) How much greater should the calcination temperature be than the decomposition temperature? Thus, for example, the common calcination temperatures of the order of 1000°C or more for alum-derived Al_2O_3 would be excessive for basic-carbonate derived MgO, despite MgO being a more refractory oxide than Al_2O_3 (though $MgCO_3$ does lose CO_2 until ~ 900°C. More serious is that much of the important details of the results of calcining are determined by the local atmosphere in the powder being calcined [12–20]. This, in part, explains some of the above variations but also has important operational ramifications. Within the powder mass being calcined, gases such as H_2O, CO_2, and SO_2 are inherently produced, primarily based on the salt being calcined and secondarily on volatilization of species absorbed or adsorbed on the starting salt powder particles. These, especially the former gases, affect the calcination process and resultant powder character in three interrelated fashions that are often not adequately recognized:

1. Reduced the decomposition rate (in proportion to the partial pressure of gases given off in decomposition) and hence increases the time, temperature, or both of calcination.
2. Higher local partial pressures of active gases increase the rates of oxide powder particle growth, sintering, or both (and hardening agglomerates and entrapping porosity), thus reducing resultant powder sinterability.
3. Higher local partial pressures of active gases also increase the opportunity for their adsorption and possible reaction with the high surface area powder, to form or reform salts during its cooling from calcination temperatures [19], often with negative consequences as discussed in this section and Section 8.2.1.

Thus, for example, Anderson and coworkers [20] noted in their review of calcination of oxide powders that water vapor at pressures of 10^{-3} to ~ 5 mm Hg profoundly affects the course of many decomposition reactions and can increase the rate of crystallite growth by more than two orders of magnitude.

The above effects of local gaseous atmosphere within the powder mass being calcined has significant, widely neglected effects on calcining practice, since much of the atmosphere arises from the calcining process itself. Thus, the calcining furnace atmosphere only affects the actual atmosphere in the powder mass to the extent that the gas from calcining diffuses out of, and the furnace atmosphere diffuses into, the powder mass. This counterexchange of gaseous atmospheres is a function of the powder masses being calcined, their physical size and shape, particle packing (which is a function of particle size and shape, humidity, and powder mass), the amount and character of the gases released, the stage of calcining, and the gas flow and circulation in the furnace and around the powder masses. Many of these factors also further complicate the process due to the time for thermal diffusion to occur through the powder mass on both heating and cooling. The net result is considerable variation of resultant powder character, particularly between different calcining systems of different character, as well as scale. However, even within a given calcining furnace there are variations, for example, due to gradients within a given powder mass.

A key engineering decision to be made is the type of furnace to be used: a batch (e.g., a general box furnace) or a rotary calciner (or possibly a fluidized bed). Rotary calciners offer the most uniform calcining environment as well as continuous operation, but often need larger volumes to be cost-effective and may give cross-contamination if used with different materials without sufficient cleaning. Besides being inherently noncontinuous, batch furnaces generally require bulk masses of powder to be contained in crucibles or trays (the latter are preferred). The size and shape, as well as loading of these, are sources of variation in powder results, especially from laboratory scale trials, but also in production. That variation of powder character is a factor in subsequent processing is shown by past experiences. Rice [21,22] found in hot pressing transparent MgO from commercial powders that some powder lots would produce transparent MgO within normal hot pressing parameters, while other lots would not; so lots were sampled on a trial basis, and only material from lots successfully densified in trials were accepted, leading to substantially higher success rates than use of random lots. A likely immediate reason for this variation was use of batch calcining. However, a broader issue with regard to this problem is that many ceramic powders, such as MgO, are manufactured for a variety of uses, many of which may have little if any thing to do with requirements of various ceramic fabrication methods. Thus, the substantial and variable "loss on ignition" of the MgO and many oxide powders can be a source of problems (see Sec. 8.2.1), but are not necessarily a problem for many other uses of such powders. An example of this multiple use of powders is that Linde A (alum-sulfate derived) α-alumina powder was used in the production of high-pressure sodium vapor lamp envelope production, despite the fact that the major market for it was as powder for polishing abrasives. However, once the ceramic market for lamp envelopes developed sufficiently to be a sizeable and continuing market, modifications were made in the production process to yield a more consistent "ceramic grade" powder with larger particle size for lamp envelope production.

Consider now some general examples of binary ceramic oxide powder production, starting with Al_2O_3, which is produced on a large scale and derives important economies of scale from the parallel use of much of the preparation technology for the aluminum metal industry. The dominant and lower cost process is the Bayer process of digesting and solubilizing a major portion of the aluminum hydrates in bauxite ore via formation of $NaAlO_2$ in heated NaOH. The resultant liquor is separated by dilution and cooling, with the aluminate fraction being further cooled and seeded to produce a significantly purer Bayer aluminum hydrate that can be calcined to 98.5 to 99.7% pure Al_2O_3 (with 0.1–0.6% Na_2O) depending on process details [23,24]. While there are a variety of routes to higher purity alumina powders, a common one is forming aluminum sulfate by digestion of Bayer hydrate in sulfuric acid, then often converting this to an ammonium alum by reaction of the sulfate with anhydrous ammonia [25-28]. The highly hydrated alum is commercially calcined to produce various crystalline Al_2O_3 phases, mainly of γ or α structure, usually calcined, respectively, at ≤1100°C and ≥1150°C (for 2 hrs or more) [23-28]. Particle sizes depend on calcination and comminution, but are commonly approximately 0.05μm and 0.3μm, respectively (and ~ 1 μm for ceramic production grade).

BeO is commonly obtained by calcining either $BeSO_4$ or (nitrate derived) $Be(OH)_2$ (e.g., the latter at ≥900°C [29–31]. These yield powders commercially designated respectively as UOX and AOX, the latter being somewhat less pure and the former containing substantial needlelike particles that can limit sintering and yield preferred orientation in some fabrication (see Chap. 4), while AOX has a more equiaxed particle structure, producing isotropic bodies.

CaO is commonly calcined from $Ca(OH)_2$ or $CaCO_3$, the former at temperatures of ~600°C or above giving particle sizes of 0.1–0.4 μm, while the latter is calcined at higher temperatures (e.g., 1000°C) giving coarser particles [32,33]. Similarly, MgO is derived from $Mg(OH)_2$ or $MgCO_3$, but at somewhat lower temperatures with similar or somewhat finer particle sizes. However, calcination of hydrated basic carbonate—$MgCO_3 \bullet Mg(OH)_2 \bullet 4H_2O$—is probably more widely used industrially. Similar resultant MgO particle sizes of ~ 0.1 μm for both laboratory and commercial powders indicate commercial calcination temperatures of ~ 550°C [13,14,20,34,35]. Various additives and impurities can also effect calcination—by reducing or increasing by a few percent the aragonite-calcite transition temperature if starting from aragonite, and in this case decreasing the decomposition temperature by similar levels, or increasing it by ~ 7% (for ~ 1 atom % Sr) [33]. Both MgO and especially CaO powders present serious hydration problems that must be addressed in their use.

ZnO powders of uniform size distribution and uniform coating of the particles with dopants (i.e., Bi_2O_3, MnO, CoO, Sb_2O_3, and Cr_2O_3) needed for varistor behavior were aqueously prepared via precipitation of $Zn(OH)_2$ then decomposing this at > 55°C by Haile and coworkers [36].

Briefly, further consider some of the above factors, namely, powder particle crystal phase, physical, and impurity character. As noted above, oxides that can exist in more than one crystal structure will exhibit the crystal structures appropriate to the calcination temperature, that is, as long as the calcination temperature is below the transformation temperature for the phase of interest, within constraints of relations between starting salt crystal structure and resultant oxide crystal structure. Thus, calcining of aluminum alums yields primarily γ- and secondarily η-Al_2O_3 at 1000–1100°C, primarily δ- and secondarily θ-Al_2O_3 at 1100–1150°C, and α-Al_2O_3 at 1150°C and above [23–28]. This is generally true of the three trihydrates of Al_2O_3 and of one of two of the monohydrates, but is not true of the other monohydrate, diaspore, which is unique in being the only hydrate that transforms only and directly to α-Al_2O_3 on decomposition. Similar trends of calcining temperature versus oxide product crystal structure as for most aluminum hydrates also commonly hold for other oxides of differing structures (e.g., TiO_2). However, there are again variations (e.g., particle surface energy or other factors may affect results), such as for pure (unstabilized) submicron particles of ZrO_2 forming in the intermediate temperature tetragonal versus the lower temperature monoclinic structure. (Mixed crystal phases can aid or inhibit subsequent sintering.)

Turning to the physical character of the calcined product, a critical factor is particle size, which as seen above can be quite fine, that is, submicron and thus in or approaching nanometer scale. Particle size may in the limit be the crystallite size, but is often much larger, where particle size is frequently controlled by post calcining milling. It should be noted that while very fine particles are desirable for lower, easier, and possibly more complete sintering and finer resultant grain size, such particles also present practical limitations on manufacturability. Thus, for most ceramic production powder particle sizes ranging from a few tenths of a micron to < 10 μm are desired, which are obtained by higher temperature calcining.

Other important physical characteristics include particle shape (morphology) and related orientation, as well as the amount and character of porosity. Thus, in some cases, mainly those where fine, individual, separated crystallites of the precursor are converted to similar fine, individual separated crystallites of product (e.g., ~ 4 μm cuboidal particles of anhydrous $Al_2(SO_4)_3$ yielded somewhat smaller cuboidal particles of η-Al_2O_3 on calcining at 1000°C and still somewhat smaller cuboidal particles of α-Al_2O_3 on calcining at 1250°C [28]. Such calcined crystallite morphology was also noted above for sulfate versus hydroxide precursors of BeO. Some porosity may occur in such individual crystallites, but this is much more common in the frequent case where oxide crystallites from calcination are smaller than the salt particles being calcined. Gas trapped in porosity in the calcined oxide particles may present densification problems, in which case vacuum calcining may be sufficiently beneficial to warrant added costs and constraints, for example, the use of rotary calcining. Commonly resultant calcined particles are polycrystalline in character and may have various amounts and character of porosity related in part to the starting precursor and resulting oxide crystallite sizes, growth and orientation of the latter due to the topotactic (crystallographic) relations of precursor and product [37,38].

Chemical purity of the calcined product is also important—not only based on the impurity effects on the resultant ceramic product, but also on the effects on the nature of the calcined powder and on fabrication of the ceramic components from the powders produced. While much of the reduction of particularly undesirable impurities is addressed in the chemical preparation of purer precursors, some purification may occur during or after calcining. However, impurities may be introduced during and after the powder processing. Clearly, additives used in powder processing (see Chap. 3) must be selected based on their either being sufficiently benign or removable. However, an important impurity concern that is often not adequately addressed is contamination in preparation and storage of powders. Milling of precursor or calcined powders can introduce impurities, the extent and impact of which can be limited by choices of milling media and parameters. Another source of impurities can be debris particles from the calcining furnace, possibly more so in rotary or fluidized bed calciners versus normal furnace batch calcining, since the use of crucibles in the latter case can limit furnace contamination, especially with crucible stacking, lids, or both, but this can limit desired outgassing of the powder mass. However, chipping from crucibles, possibly more so with coarser, rougher exteriors of crucibles, e.g., as shown from silica crucibles for calcining production of MgF_2 of IR windows and domes [39,40], can be a problem.

An important source of impurities that is widely neglected, but can have significant effects, especially on bodies of thick cross sections and from pressure sintering are anion species left from adsorbed species on the precursor powder particles and especially those from decomposition during calcining. These include H_2O and CO_2 that are available from either the atmosphere, decomposition, or both, as well as S–O, N–O, and P–O species from sulfate, nitrate, and phosphate precursors, which are all part of the ubiquitous "loss on ignition". The latter arise from incomplete decomposition, and rereaction on cooling, and all such impurities can arise from entrapment in closed pores in the calcined particles, and adsorption on powder surfaces on cooling [6,41,42]. An example of the data from limited measurements of such anion species is shown in Table 2.1 for sulfate derived alumina (most evidence of such impurities comes from measurements and effects on ceramic bodies made from powders). Inadequate subse-

TABLE 2.1 Residual Sulfur Content of Al_2O_3 Powder from Calcined Ammonium Aluminum Sulfate[a]

Powder[b] source	Surface area (m^2/g) (~T, °C)[c]	Particle size (μm)[d]	Sulfur Content (w/o)[e]
Linde A	17 (> 1150)	0.3	0.02
Meller 1	17 (> 1150)	0.3	0.02
Linde B	80 (1100)	0.05	0.05-0.06
Meller 2	95 (≥ 1000)	0.05	0.55
BM	– (1000)	> 0.005	0.42

[a]Compiled by Rice [42].
[b]Commercial powders, except for BM [26].
[c]Estimated calcination temperatures for the four commercial powders and reported temperature for BM powder.
[d]Size data from commercial literature, except for BM powder, where the crystallite size was the only data available.
[e]Sulfur content in weight percent (w/o); form of the sulfur present, e.g., as elemental sulfur, silfides, sulfites or sulfates, was not determined.

quent removal of these anion species before or during densification often results in unsuitable components (see Sec. 8.2.1). Another important, and quite variable, source of contamination is from other ceramic or organic powders used in production, e.g., from dust that may have an opportunity to settle on the precursor, and especially the calcined powder. Dust may also include many sources of organic contamination, such as hair, dandruff, and smoke. Additionally, rodents and bugs can be sources of contaminants in industrial settings (contaminants such as hair, bug bodies or parts, as well as rodent and bug feces have been found in reagent grade ceramic powders) and are another ubiquitous component of "loss on ignition".

2.3 PRODUCTION OF OTHER SINGLE- AND MIXED-OXIDE POWDERS VIA SALT PRECURSOR DECOMPOSITION

Aqueous precipitation of precursor salts, followed by calcining them to produce oxide ceramic powders, has important applications as illustrated above for some important ceramics. It also has application to a substantial number of other single oxides, mixtures of oxides, and mixed-oxide compounds; however, it also has substantial limitations. Melting on some salts, especially hydrated ones, as for some nitrates, is one limitation, which was noted earlier. Another, much broader limitation is in the preparation of mixtures of oxides, whether for direct preparation of ternary or higher oxide compounds (e.g., $MgAl_2O_4$), doped oxides (e.g., ZrO_2 with stabilizer), or with additives for composites such as Al_2O_3-ZrO_2, due to limitations on coprecipitation of two or more constituent oxide species. Many of these limitations can be at least partially overcome by separate preparation of the constituent oxides, then mixing them and reacting them by solid-state reactions as discussed in Section 2.4, but this may be at some cost for added steps and some powder limitations of chemical homogeneity and homogeneous, fine particle size. Partly balancing these limitations are that several salts can be considered in precipitation processes giving opportunities to obtain salts that are compatible for coprecipitation and are free of melting or other limitations. Thus, for example, both barium [43] and strontium [44] hexaferrites powders have each been produced via precipitation from chloride solutions using mixtures of sodium hydroxide and carbonate. Similarly, $LaNiO_3$ powder was prepared via hydrous chlorides dissolved in water then coprecipitation as oxalates using a water-alcohol solution of oxalic acid [45]. Another example is the preparation of ZrO_2-Al_2O_3 composite powder by coprecipitated from nitrate solution using ammonium hydroxide [46].

There are, however, modifications of the precipitation process as well as more basic changes (considered below) that considerably diversify solution processing of ceramic powders. A modest modification referred to as *homogeneous precipitation*, entails the chemical source of precipitation being contained in a compound included in a solution of the salt constituents. The release of the precipitation agent (e.g., thermally via decomposition of urea) then causes more uniform precipitation and may allow a somewhat broader range of precursor chemistry. Thus, pure ZrO_2 [47] and oxides of Y, La, Ce, and Nd [48] have each been produced by this process, producing fairly uniform oxide particles ranging from < 1 μm to 1 μm or more, and spherical to polyhedral morphology depending on material and processing. Such processing has also been used for preparation of various ferrite and related ternary oxides [49], as well as mullite [50] and $MgAl_2O_4$ [51].

An alternative to the above precipitation processing is *emulsion processing*, which entails dispersing droplets of a liquid precursor in an immiscible fluid. This entails use of a salt solution, commonly a water solution, but other solvents and compatible soluble sources of the desired ceramics are also feasible. The solution is mixed with another liquid which is immiscible with the solvent of the solution (e.g., an oil for water-based solutions), but contains a suitable surfactant so the solution forms an emulsion with the immiscible liquid, i.e., the solution forms small spherical (e.g., 0.1–0.3 μm) droplets in the immiscible liquid. The number and size of droplets formed depends on the solute and immiscible liquid, the surfactant, and the mixing, especially high-shear mixing. In the solution manifestation, most of the immiscible liquid (e.g., oil) and all of the solution solvent (water) is removed by vacuum evaporation. Then, the resultant slurry is pyrolyzed in an atmosphere of limited or no oxygen so enough of the remaining immiscible liquid forms a char continuing to separate the individual dried spherical particles resulting from the droplets. The char-dried particle mixture is then calcined to both oxidize away the char and convert the precursor salt to the oxide sought, avoiding as much as possible interparticle sintering and other agglomeration, for example, via a belt or rotary furnace (calciner). This is generally followed by some milling to reduce agglomeration (e.g., giving particles similar in size to the original droplet sizes). Maher and coworkers [52] reported successful preparation of several oxide ceramics, including both single and multiple oxides of interest for electrical and electronic applications, including high-temperature superconductors. In another manifestation, use of either a binder with the precursor or a precursor that polymerizes via thermal treatment or use of a catalyst allows the droplets to be rigidized in the emulsion state, so they can simply be sieved out of the immiscible liquid. This eliminates the vacuum evaporation step, and may allow reuse of the immiscible liquid, but may often result in larger droplets or particles. While this is often more applicable to nonoxides derived from preceramic polymers, it has some applicability to oxides.

Another alternative to precipitation of ceramic precursor salts to prepare ceramic powders is via *sol-gel processing*. This basically entails conversion of a "sol solution" to a rigid solid or to solid particles in a liquid via gelling, the latter similar to precipitation. This has a variety of manifestations depending, in part, on whether the sol is based on alkoxides or on stabilization of colloidal particles of hydrated oxides or hydroxides in water. While each type of sol has its limitations in terms of oxides to which it is applicable, a substantial number of oxides can be made by one or the other approach. This and combinations within (and possibly between) sol approaches and specifics of gellation methods result in considerable diversity for making powders (as well as for making fibers, films, coatings, and bulk bodies). The basis of the sol determines what oxide sol compositions can be made, their oxide yield, compatibility for processing mixed oxide compositions, and limited additive levels and types (e.g., for colloidal stabilization) and costs (Table 1.2).

Both sol sources can often produce similar compositions at similar costs, with colloidal sols commonly gelled by water removal and alkoxide sols by reaction with water (of which only very small amounts are needed since the resulting polymerization for gelling produces more water for further polymerization). Thus, gelled beads can be produced by dripping colloidal sol droplets into a fluid medium that will extract enough water (or other solvent) fast enough from the sol droplets to gel them. An important earlier manifestation of this was the demonstration of the feasibility of producing uniform spherical particles from a few microns to a few millimeters in diameter that could be calcined to the desired oxides, then sintered to approximately theoretical density. Thus, by dripping colloidal sol droplets into a water-absorbing fluid, very uniform spherical beads of possible nuclear reactor oxide fuels were demonstrated, oxide catalyst support beads commercially produced [53] and making feed material for melt spraying demonstrated [54]. Spray-drying and related extensions of this are also feasible [55]. Alkoxide sols can also be gelled in bulk (similar to a casting operation), then ground into powder as discussed below (or directly further dried and fired into a bulk body; Sec. 6.6). However, narrow size distributions of uniform submicron spherical oxide powders can be produced in tailored reactors [56] or by in situ production of free water in solution [57]. In either case, whether a binary, mixed oxide, or higher-order oxide compound is gelled, the actual oxide crystallites obtained upon calcining dried gel particles are submicron in nature [58].

Despite the diversity of powders produced and the quality of those produced as fine, submicron, uniform crystallite particle sizes that allow sintering to high densities [59] at lower temperatures than many other powders, sol-derived powders are limited in their commercial applications, primarily because of costs. Crushing of bulk gelled and dried pieces, then calcining these pieces to produce oxide powders usually presents limitations on resultant sintering since this produces porous agglomerates, with the pores between the agglomerates being more resistant to sintering. This can be overcome by hot pressing, for example, as shown by Becher's use of this approach to prepare extremely uniform Al_2O_3-ZrO_2 composites [60] that were apparently the first to show both strength as well as toughness increases in such composites; but hot pressing is generally constrained because it is often more expensive than pressureless sintering.

The use of sols to produce alumina-based abrasive particles is a major commercial application of sol-gel processing that is illustrative of how and where processes that entail more expensive aspects, in this case, sol-gel processing, can be commercially viable in specialized applications. The process is to gel alumina-based sols with appropriate additives such as MgO or ZrO_2. Gelling of sols in trays gives bulk gelled bodies that are comminuted to yield the desired size abrasive particles after calcining and sintering of crystallites within the individual particles, sintering of the abrasive particles to one another being purposely avoided [61,62]. Despite some potential advantage of gel versus fused abrasive comminution and avoiding the problem of sintering comminuted gel-derived particles to one another, the process was seen as being about tenfold more costly than competing fused abrasives. Changes in composition, mainly use of much cheaper MgO for ZrO_2, and resultant processing improvements, some due to composition changes, allowed commercial introduction to replace some of the fused abrasive particles for some applications of coated abrasive products, particularly for finishing some steels. Note that this initial success focused on applications where limited amounts of the sol-derived abrasive particles were needed by mixing them with the normally used (cheaper) fused abrasive particles and in coated abrasives (i.e., sandpaper-type abrasive products where only a partial single layer of abrasive particles is needed, so abrasive materials costs are less significant to product costs). Three things were needed to get the abrasive used in some grinding wheels, where much greater abrasive volumes are used and abrasive costs are much more critical. The first was further lowering of gel-derived abrasive costs since just the raw materials costs were ~ $4/kg, while finished fused abrasive costs were $0.50–0.70/kg. Second was further improving the gel-derived abrasive performance versus its competition (fused Al_2O_3-ZrO_2); and third getting the customers to recognize that the resultant product was more cost-effective in some applications despite higher abrasive cost. Note that improving gel-derived abrasive performance was in part obtained by seeding to enhance formation of alpha alumina, which was done before such seeding became a popular research topic.

The above gelling of alkoxide-based sols to produce oxide powders, which occurs via polymerization, is but one chemical system in which polymerization plays a role. Lessing [63] has reviewed two others, namely polyesters of the popular Pechini type and cross-linked poly(acrylic acid) polymers, with the former having been used on over 100 different mixed-oxide compounds. This starts with various common precursor, water-soluble salts, such as chlorides, carbonates, hydroxides, isopropoxides, and nitrates chelated with a hydroxycarboxylic (usually citric) acid. The solution of these two ingredients is, in turn, dissolved with either ethylene or dietylene glycol by heating (80–110°C) and stirring to achieve a clear solution. Further heating then results in (reversible) polymerization of the solution and removes water freed in polymerization. Calcining, often in air, results first in charring of the polymer resin (e.g., at ~ 400°C), then oxide compound formation at 500–900°C. Lessing discusses a number of practical aspects of these polymer processes, such as advantages and disadvantages of foaming that may occur during pyrolysis, temperature control issues raised by the amount of organic material to be pyrolyzed, interaction with other processes such as freeze-drying (discussed below), and other practical issues, including costs. The latter are driven substantially by starting salt costs.

The first of four significant modifications of deriving oxide powders from salt precursors was freeze-drying, originally demonstrated by Schnettler and coworkers [64] for powders to sinter for ceramic applications. This entails rapid freezing of the salt precursor solution, commonly by spraying solution droplets into a cold liquid such as N_2 or hexane surrounded by an acetone-dry-ice bath, then subliming off the solute, usually water, often as water of hydration. This allows a substantially broader range of salts and their mixtures for doping or forming ternary or higher compounds to be processed. (Directional freezing can also be used to produce fibrous or cellular pieces.) There are still some limitations, primarily melting of some constituents during sublimation drying, though this can frequently be subdued with additives, e.g., ammonium hydroxide [65]. Application of this process to oxides of high surface area has been shown [66] and studied in some detail by Hibbert and coworkers [67–69]. Though apparently not yet applied on a significant industrial scale, evaluation of the process by Rigterink [70] shows potential for being a production process, while more theoretical considerations provide guidance for further development [71].

Removal of water from salt precursors, and frequently attendant melting problems, is also accomplished chemically by spraying droplets of the salt solution into a liquid that will extract the water. Thus, for example, O'Toole and Card [72] reported forming submicron spherical particles of Y_2O_3-ZrO_2 by spraying droplets of sulfate solutions into an absolute alcohol.

Another alternative is to thermally remove the water in a fashion that minimizes negative effects of melting, mainly the formation of large, hard agglomerates. Benign water removal has been demonstrated in a number of cases by spraying droplets of the precursor solution into a heated, stirred liquid that is not miscible with, nor decomposed by, the precursor or its products. The liquid, often an oil material such as kerosene, is heated sufficiently to evaporate the water, and encapsulate the residues of the dried droplets as particulates, which can subsequently be removed by filtering and then calcined, ground, and so forth. Richerson [4] has summarized some of this work, and Richardson and Akinic [73] describe preparation of small (1–3μm) spherical granules of yttria with crystallites of 0.1 μm size.

The fourth alternative and extension to solution-calcining processing is atomization of a solution and thermal treatment in the aerosol state. This can be just spray-drying, but is often extended to temperatures to also include calcination in the aerosol state. The process incorporating calcining in the aerosol state has a variety of names in the literature, with spray pyrolysis being the most widely used, as noted in the substantial reviews by Messing and coworkers [74] and Kladnig and Karner [75]. This process, which clearly limits the effects of locally produced calcining atmosphere on products, as discussed in the previous section, is quite versatile and has been applied to a variety of materials. It can be readily applied to almost any solution, as well as slurries or emulsions of single or mixed compositions. Depending on atomization capabilities, submicron to multimicron particles can be produced, possibly retaining some of the sphericity of the aerosol droplets, but serious shape variations and distortions as well as hard agglomeration can occur. Melting of intermediates can result in substantially larger (otherwise often nm) crystallite sizes, as well as hard, calcined agglomerates. Spray pyrolysis has considerable commercial use indicating its potential for cost-effectiveness as reviewed by Kladnig and Karner [75]. An important example of this is the spray-roasting of pickling liquors, which are a waste product of the steel industry. These are spray-roasted (after reduction of silica contents) to produce ferric oxide for ferrites as well as regeneration of HCl to be reused in pickling more steel. From a research and development standpoint, spray pyrolysis is a useful tool for making powders, often of a nanometer scale, via a liquid-based precursor by providing a route to processing powders from novel precursors, e.g., as demonstrated in work of Laine and coworkers [76,77].

Another, newer, less-developed oxide powder preparation process that incorporates calcining with dehydration and related steps is a combustion synthesis process that received considerable laboratory investigation, by Kingsley and coworkers [78] and others [79,80]. It entails use of metal salt solutions with an anion that is a good oxidizer, commonly nitrates, along with a fuel (e.g., amides or hydrazides, commonly urea) to be oxidized. After mixing the ingredients in an appropriate container, they are heated in a furnace to several hundred degrees for dehydration, decomposition, and then combustion (which can temporarily reach temperatures > 1500°C), all of which occurs in less than 5 min. During this process there is boiling, then foaming that results in a frothy or fluffy oxide, which is usually in its higher temperature phase (e.g., alpha alumina), but commonly of fine particle sizes, e.g., 0.1–2 μm with high surface area. Even finer powder particles, e.g., ~ 15 nm, are feasible, and mixed and higher-order oxides can be made [79]. Scaling to practical yields poses various issues, with safety being an important one, along with powder character and yield, as well as whether it is feasible to eliminate the furnace for calcining. However, the fact that reaction can also be brought about by microwave heating [80] suggests that the combustion reaction may be highly localized, which along with in-line mixing, may allow better control for safety purposes.

2.4 DIRECT PRODUCTION OF OXIDE POWDERS

There are several processes that yield oxide powders directly, without calcination and its costs and limitations. The first of these is hydrothermal preparation, which can yield a number of important single-, doped-, or mixed-oxide powders, many of which are important for electronic ceramics. (Though primarily investigated and applied to oxides such processing has potential application to at least some nonoxide ceramics.) Resultant oxide powders generally consist of single-crystal, morphological particles with limited agglomeration and limited particle size and shape variation for a given set of processing parameters (Fig. 2.1). Such processing commonly yielding average particle sizes of ~ 1–3 μm generally can be reasonably controlled, and size can often be lowered, e.g., to ~ 100 nm, usually by seeding. Common feed materials are oxides, hydroxides, chlorides and nitrates, sometimes with additives to aid the process (e.g., control pH), with earlier work focused on use of water, for example, at 100–350°C under pressures of < 15 MPa and residence times of 5–60 min in batch or continuous reactors, though more extreme parameters have been used experimentally. More recent work has included use of solvents other than water, such as glycols. Processing conditions can often be varied to yield different crystalline phases of the product oxide—for example, alpha alumina can be directly produced. These factors as

FIGURE 2.1 Micrographs of hydrothermally produced α-Al_2O_3, single-crystal (sapphire) powder particles. (A) Thin platelets and (B) mixed thick platelets and approximately equiaxial polyhedra. Multifaceted double-terminated pyrimidal particles are also produced depending on processing conditions, including the speed and extent of stirring the liquid. See also Fig. 3.3 for similar particle (crystal) morphologies obtained as a function of different liquid-phase processing parameters. (Photos courtesy of Prof. J. Adair, Penn. State U.)

well as practical issues are discussed in extensive literature, such as a review by Dawson [81], and papers by Adair and coworkers [82]. Considerable work on scaling and economics of the process has resulted in some commercial applications of hydrothermally prepared $BaTiO_3$ where the finer, more uniform, but more costly, powder can be cost-effective in view of better performance and limited quantities used. Note that the different particle morphologies readily produced should produce different extents and character of preferred orientation in components made from such powders as a function of fabrication methods and parameters. Characterizing such effects is important in optimizing uses of these morphological particles.

Another process of promise for a number of single and particularly mixed oxides for both structural and, especially, electrical-electronic applications is that which is carried out in molten salt baths [83-85]. Besides producing many of the same or similar oxides as hydrothermal processing, it also often yields morphological shaped, dense single-crystal particles (e.g., of 1 to several μm in size). However, it also yields much finer morphological or equiaxed particles, often as more agglomerated powders, which may often be friable agglomerates. It has broad applicability due in part to the diversity of salt systems that can be used. Two examples of such synthesis are

$$BaCO_3 + TiO_2 \Rightarrow BaTiO_3 + CO_2 \tag{2.1}$$

and

$$2PbO + ZrO_2 + TiO_2 \Rightarrow PbTiO_3 + PbZrO_3 \tag{2.2}$$

carried out, respectively, in molten $NaKCO_3$ at 800°C [83] and in molten NaCl-KCl mix at 1000°C [84]. The criteria for a suitable salt media, besides its chemical suitability for the reaction, is a low melting point, often aided by using mixed salts [e.g., Eq. (2.2) above], and good solubility of the salt, preferably in water, for easy recovery of the product powder. Besides scaling issues, costs of the operation, crucibles, product recovery, and salt recovery are likely to be important in determining use of this versatile process.

An important and broad method of preparing oxide powders is reaction processing, that is, processing that entails one or more chemical reactions as an important aspect of the powder preparation. Actually, reaction processing is an important aspect of most other powder making processes, including many or all aspects of the processes discussed above and those below. However, of specific interest are solid-state reactions to produce doped or alloyed oxide powders, and, especially, powders of ternary or higher oxide compounds from binary oxides or their precursors. The most commonly used method of production of doped or ternary or higher order oxide powders is via such solid-state reactions. Key issues are the uniformity and intimacy of the mixing of the reactant oxide constituents and the fineness of the reactant powder particles. Using one or more of the oxide constituents in precursor form, often preferably a soluble one, can often aid in both the uniformity and intimacy of mixing as well as the resultant oxide particle size; for example, due to mutual inhibition of particle growth of each oxide constituent by the other constituent(s). Other issues can be obtaining or maintaining desired stoichiometry, for example, due to vapor losses, which can often be countered by excess source of the vapor lost, calcining in crucibles with lids (which a balance between desired outgassing of the powders or precursors and vapor losses), or both. Common overall issues are agglomeration and grain growth, which are usually addressed by milling after the reaction and sometimes part way through the reaction. While either powder processes may produce better, e.g., more uniform, powders, at least on a laboratory scale, solid-state reaction continues to dominate industrial production and use of doped, mixed, ternary, or higher oxide compound powders because of costs.

Another type of reaction processing that has been investigated is self-propagating high-temperature synthesis (SHS) and related processing of the inorganic constituents. This entails reactions sufficiently exothermic that once ignited by local heating in one area of a compact, the reaction will generally propagate throughout the body with no external heating. While mostly used for nonoxide compounds or composites of oxides and nonoxides (see next section), it has some application to oxides. In particular, Xanthopoulou [89] has recently reviewed such SHS processing of inorganic pigments, which are often complex doped or mixed oxides, can advantageously be produced by SHS.

Consider vapor-phase processing, which almost always involves some reaction in the processing in each of its two main manifestations of chemical vapor deposition and plasma reaction. Other manifestations of vapor-phase processing, such as evaporization and condensation of oxides [90,91], are laboratory processes not discussed further. Earlier plasma reactions focused on arc plasmas, especially the tail flame of high intensity arcs [92-94], while producing very fine, e.g., nm-scale, particles of a number of oxides, mainly binary ones, these reactions do not appear to have had any industrial use. Key limitations are that most oxides are not conductive, so making electrodes of mixtures of oxides and carbon, while allowing arc vaporization, raises many issues. Arcs between metal electrodes have also been used (Fig. 2.2), but were extremely limited in length of operation due to melting [94]. More recently, much of the interest in plasma processing has focused on induction-generated plasmas, which can reactively form very fine, nm-scale, powder particles of a number of oxide powders, again mainly binary ones [93]. There has also been investigation of simply heating metals, of low to moderate vaporization temperatures, e.g., Al or Mg, to generate metal vapor that could then be burned with oxygen in a "torch" that has apparently been sold on a modest commercial scale [93].

The most extensive use of vapor-phase processing of oxide powders is via chemical vapor deposition (CVD) to produce vapor-phase nucleation and growth of oxide powders, rather than formation and growth of the oxide on a surface as used to form ceramic coatings or freestanding bulk ceramic bodies (see Sec. 6.6). Many precursors, especially for metal and vapor-phase reactions, are feasible. Organometallic compounds are commonly more expensive, often substantially so, and often pose some safety (toxicity) issues, but can result in oxide formation at more moderate temperatures, for example at 500–1000°C. Use of metal halides, especially chlorides, oxidized by water vapor are particularly common, often being used in the range of 1000–1500°C. Such reactions are used

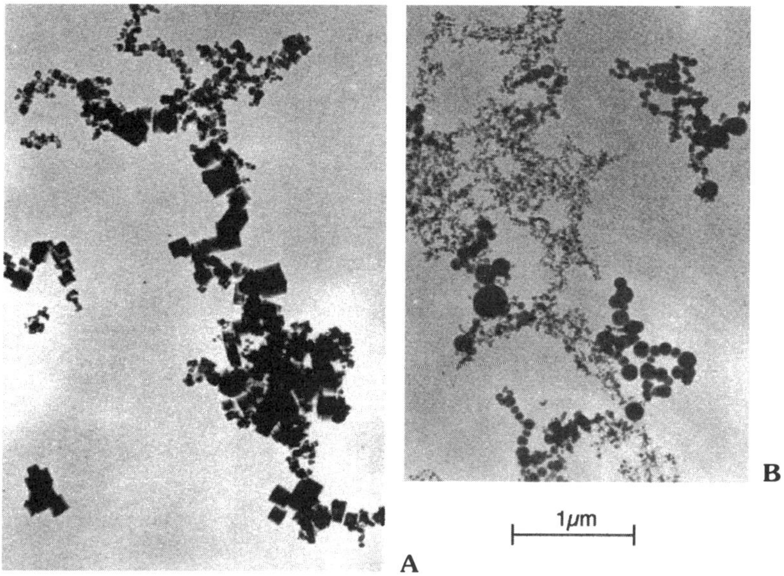

FIGURE 2.2 Powders produced by arc vaporization of Mg or Al metal electrodes in O_2 to produce (A) MgO whose cuboidal character indicated condensation below the melting point, (B) Al_2O_3 whose spherical character indicates condensation above the melting point.

to make high-tonnage quantities of fine particles of oxides such as Al_2O_3, SiO_2, and TiO_2 for a variety of uses, such as fillers, pigments, and for making ceramics. (However, note that halide, e.g., Cl, residues may remain [95] and have been cited as the cause of limitations on sintering [42], but such problems are probably economically solvable). Note that such CVD processing overlaps with other processes not only because of its use of reactions, but also because these may be stimulated by microwaves or other plasma generation methods. Also, while CVD processing of oxide powders is particularly applicable to binary oxides, it can have considerable applicability to doped, composite, and ternary or higher compound powders, as shown by Suyama and coworkers for powders in the TiO_2-ZrO_2 [96]. Al_2O_3-ZrO_2 composite powders [97] and ternary titanate [98] and mullite [99] powders are other examples of CVD versatility.

Another major, but not widely recognized, method of producing oxide powders is via melting, primarily by arc skull melting. This typically utilizes a water-cooled cylindrical steel shell container that is open on the top, closed on the bottom (from which the cylindrical shell is removable), and typically ~ 2 m in diameter with an aspect ratio of ~ 2 or more. This container is partly loaded with oxide powder into which horizontal graphite starting bars are placed such that they will contact the large vertical graphite electrodes (usually three for three-phase heating) that enter from the top of the container and terminate on the starting bars. Following further filling of much of the container with the powder to be melted, power is applied to the electrodes, which heats the starting bars to the point where enough of the surrounding ceramic powder is melted before the sacrificial starting bars are consumed by oxidation. Subsequent heating continues via arcing from the electrodes to the molten ceramic and electrical conduction in the molten ceramic. Besides having a ceramic composition that melts with sufficient congruency, and without excessive vaporization, adequate electrical conductivity in its molten state, and sufficient resistance to reduction under the harsh reducing conditions from the consumption of the starting bars and the presence and partial consumption of the electrodes are needed. An important operational factor is having powder that is coarse enough such that outgassing of adsorbed and entrapped gas is not explosive. It is also important that the character and packing of the powder in the container is such that upon melting, settling of the molten pool into the unmelted powder below is at a limited, reliable rate so the electrodes can be advanced to maintain electrical contact with the melt, and a suitable fraction of the powder load can be melted in a reasonable run-time (e.g., a few hours). After cooling for several hours, the cylindrical shell is removed, then the unmelted material around the solidified melt is removed, followed by breaking up the solidified molten mass, often initially manually with sledge hammers, then through varying degrees of comminution.

Despite the batch nature and related manual aspects of the process (which are major factors why much of the production has gone offshore), it is a widely used process which produces large tonnages of refractory grain, primarily for the refractories industry. Major oxide grains produced are Al_2O_3, MgO, SiO_2, and ZrO_2 (which undergoes some to substantial, but generally not destructive, reduction). Al_2O_3 is also used for abrasives, with two grades being made, brown and white, based or raw materials and resultant purity. As noted above in conjunction with sol processing of Al_2O_3 based abrasives, production of ~Al_2O_3 - ZrO_2 eutectic compositions via fusion and quench casting (e.g., in graphite book molds) is an important application of resultant fused grain for higher performance abrasives. (*Note:* The reduction of the ZrO_2 from the arc melting and casting in graphite appears to partially stabilize, hence toughen it, but oxidation may be a factor in the wear of the abrasive as indicated by loss of strength and toughness due to cracking on oxidation [100,101].) Higher quality fused MgO grain is also used as the electrical insulator, but thermal conductor, in electrical heating elements (e.g., as used in home electric stoves and some other home and industrial heater appliances). (Sized MgO grain is apparently vibratorialy filled between the central heating wire and the outer steel tube, with the latter then being swaged to reduce its diameter and increase both the thermal contact between the MgO grain and the metal tube as well as the MgO density (to ~ 90% of theoretical) and thermal conductivity.)

Two other more recent applications of fused-oxide powders have occurred. First, fused grain has replaced some use of spray-dried agglomerates of conventionally produced oxide powders to become the dominant source of powders for melt spraying of ceramics, especially for powders of doped or mixed oxides, e.g., such as Al_2O_3-TiO_2, and ZrO_2 with stabilizers, as well as ternary or higher-order oxides such as $MgAl_2O_4$. Eliminating binders needed for spray-drying, which are a possible source of problems in plasma spraying, is one of the advantages of fused powders for melt spraying. However, a major advantage is the compositional uniformity and stability of melt-derived powders versus often incomplete melting and mixing heterogeneities of mixed spray-dried powders. More recently there has been some commercial sale and use of melt-produced PSZ powders for production of PSZ bodies. Though such powders have apparently been somewhat more expensive than conventionally produced powders, e.g., due to costs of comminuting the solidified fused ingot, the fusion-derived powders not only offer more homogeneous composition, but also environmental stability. Thus, conventional powders of ZrO_2 mixed with low-cost CaO or MgO stabilizers are unstable in the presence of moisture, and hence in aqueous milling, air storage, or slip casting, while the fused powders of such compositions are stable, and replacement of the CaO or MgO with stable precursors such as carbonates in conventional powders still pose some issues. (*Note:* Substantial cost reductions should be feasible for PSZ powder production if ZrO_2 extraction from zircon and fusion of desired compositions can be combined, by boiling off much or all of the SiO_2, which is apparently already done to some extent. A key issue could be the partition of stabilizer between the ZrO_2 and the SiO_2. All fusion-derived powder costs should be substantially reduced if thin sheets are cast, e.g., as for fused Al_2O_3-ZrO_2 abrasives, and, especially, if streams of molten droplets can be splat cooled to reduce comminution costs, as well as calcination costs to reoxidize reduced materials such as ZrO_2.)

Two other processes for, and application of, melt-derived ceramic particles should be noted. The first is for finer (sand) milling media, e.g., used extensively in the paint industry. Approximately spherical, dense, wear-resistant ceramic particles, mainly ZrO_2 or Al_2O_3, of various sizes from < 1 mm to > 1 mm desired diameters are produced by various agglomeration techniques and sintering. However, a possibly superior product is also apparently produced by melt quenching such size droplets of zircon, which generally produces smoother, more spherical particles (Fig. 2.3A) which contributes to wear resistance. The quenching freezes in the decomposed ZrO_2-SiO_2 composition, which presumably provides some ZrO_2 toughening and limits microstructural scale (Fig. 2.3B), which also aids wear resistance.

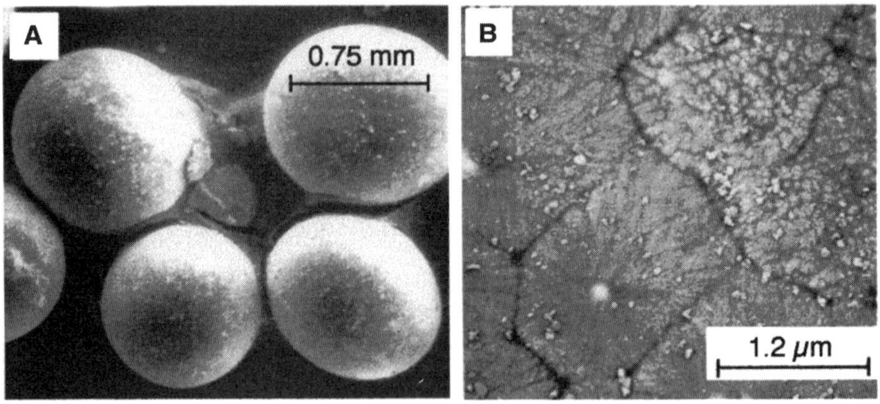

Figure 2.3 Sand milling media apparently made by quenching molten droplets of zircon. (A) Lower magnification showing generally good particle sphericity consistent with forming from molten droplets. (B) Higher magnification of typical surface showing the same structure found in quenched zircon [102], as well as some fine particle debris, apparently from use. Note that broken particles showed one to a few larger pores near the particle center, consistent with melt forming and quenching.

Finally, melt quenching has shown promise for producing desirable ceramic composite particles for processing tough ceramics, especially at or near eutectic composition. While other opportunities are discussed in Section 6.7.3, of greatest relevance here are Al_2O_3-ZrO_2 particles. Rice and coworkers [97,98] showed that hot pressing Al_2O_3-ZrO_2 abrasive, approximately eutectic, particles discussed above produced promising specimens, but both difficulty of getting fine enough particles, and serious loss of strength on oxidizing the partially reduced (hence partially stabilized) ZrO_2, resulting in destabilization and cracking, were problems. Homeny and coworkers [103] subsequently formed particles of similar compositions, but of finer size and microstructure, by passing them through a plasma torch, resulting in bodies of promising strengths and toughnesses. Melt quenching is used for commercial production of ZrO_2 by dissociating $ZrSiO_4$ particles by passing them through a plasma torch and leaching out the SiO_2.

2.5 PROCESSING OF NONOXIDE POWDERS

The preparation of nonoxide ceramic powders, though having some similarities to those for oxide ceramics, has both some different preparations or combinations of processing as well as different emphasis of methods, all of which reflect the diversity of their chemical character, relatively more limited development of their processing, or both. There is much less use of salt precursors and more of processes directly producing the nonoxide powder, as well as other processes.

An important example of an analog to a salt precursor process is the production of Si_3N_4 via forming of a silicon imide, $Si(NH)_2$, by reaction of $SiCl_2$ in solution in liquid NH_3 (i.e., under pressure to sustain the latter in the liquid state) followed by calcination to decompose the imide to Si_3N_4 [5,59]. UBE Industries, Ltd., Japan, has commercially produced a fine Si_3N_4 powder (Fig. 2.3) via this process, which has been used to produce good quality bodies, but is one of the more expensive Si_3N_4 powders. Crosbie and coworkers [104] have described modifications to the process to reduce costs and limit problems of residual chloride associated with the imide intermediate and resultant carbon contamination of the resultant Si_3N_4 and its negative effects on the oxidation resistance of resultant Si_3N_4 bodies.

Somewhat analogous preparations of precursors for AlN and TiN have been reviewed and reported by Ross and coworkers [105], some of which are based on electrochemical processing. Thus, AlN has been prepared in liquid NH_3 via:

$$AlBr_3 + 3KNH_2 \Rightarrow Al(NH_2)_3 + 3KBr \tag{2.3}$$

with the Al containing product above losing NH_3 to form oligomers at room temperature, and with calcination of the resultant oligomer product yielding

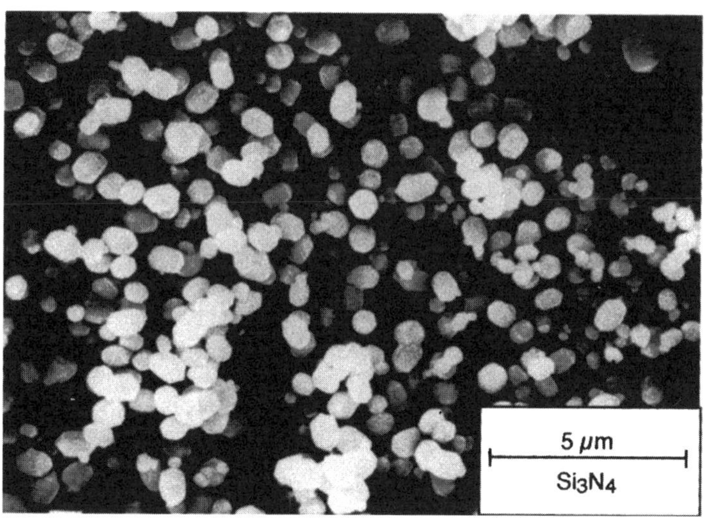

FIGURE 2.4 Micrograph of UBE Industries' commercially produced very uniform imide derived Si_3N_4 powder. Contrast with Fig. 2.5. (Photo courtesy of Dr. T. Yamamura of UBE Industries).

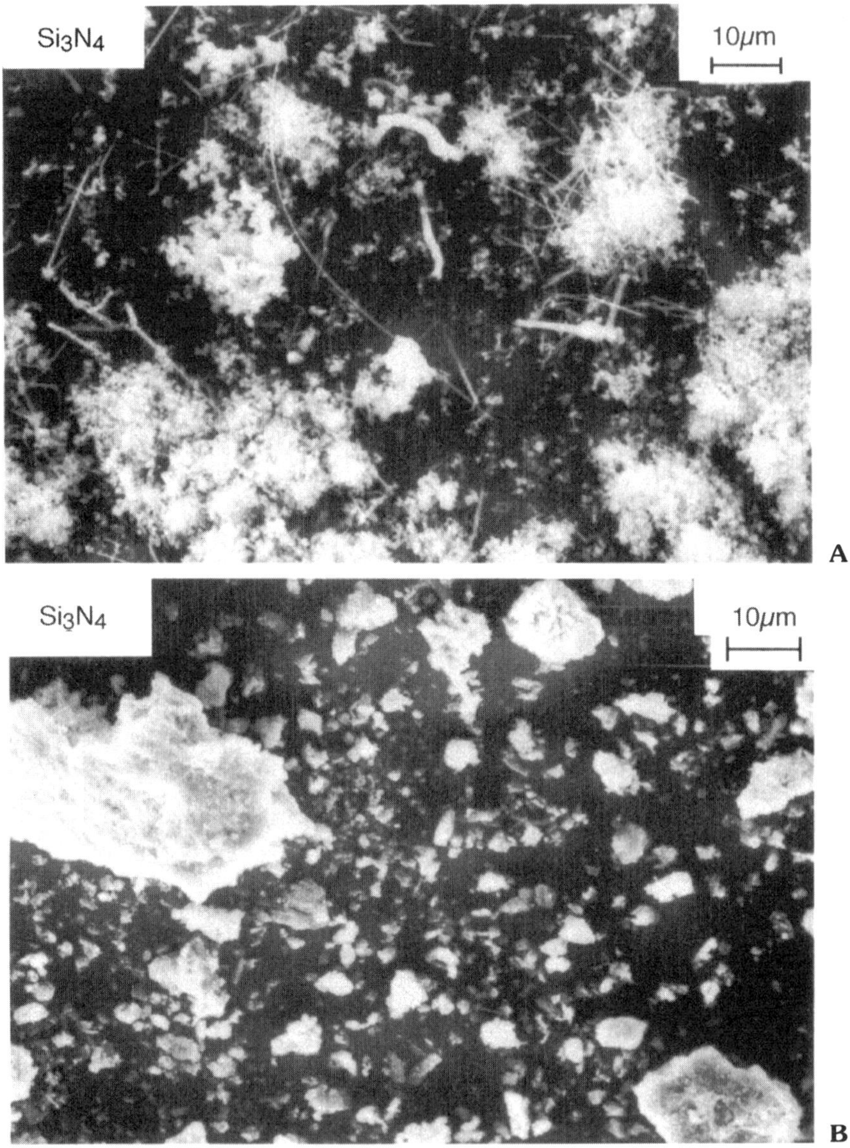

FIGURE 2.5 Comparison of Si_3N_4 powders from (A) CVD and (B) nitridation or carbothermal preparation. Note the inclusion of whisker material in A, which often occurs to various degrees in such CVD-derived material and larger agglomerates, which often occur in powder from conversion, such as nitriding Si or carbothermal preparation.

AlN. Ross and coworkers modified the process to use electrolysis to form the intermediate oligomer, which when calcined at 1100°C, yielded extremely fine (10–25 nm) AlN particles in the very small quantities made. The similarly prepared TiN had crystallite and particle seizes respectively of 60 and 480 nm. Note that other electrochemical preparations of nonoxide ceramics in molten salts are also discussed in Section 3.2 where the role of additives in the preparation processes are noted.

Examples of the extension of similar solution-based reactions to those used for oxides, but instead to directly yield a nonoxide ceramic powder rather than a precursor salt on precipitation, are those of Ritter and Frase [106]. They report reactions of Na and various chlorides in an organic solvent (possibly heated) to produce powders of compounds such as B_4C, SiC, and TiB_2, the latter via:

$$10\ Na + TiCl_4 + 2\ BCl_3 \Rightarrow 10\ NaCl + \text{"}TiB_2\text{"} \qquad (2.4)$$

They reported that the NaCl could be distilled off and the amorphous "TiB_2" precursor crystallized to TiB_2 at ~ 700°C, but no details on the powders, e.g., their purity, particle size, agglomeration, and possible costs, were given. These preparations have similarities and differences from that of homogeneous precipitation of fine (e.g. ~ 3 μm to submicron) ZnS particles by thermal decomposition of thioacetamide in acidic aqueous solutions [107], a key difference again being the direct precipitation of ZnS, not a precursor. Another similar process is the reaction of BF_3 and NH_3 at a low temperature in an aqueous solution that is then treated with NaOH to precipitate BN to be dried and heated to 800°C in N_2, but probably contains boria and borate products as reviewed by Ingles and Popper [108]. However, many details such as processing yields, rates, and costs as well as product quality and consistency are unknown.

Closer analogs are often found between sol-gel processing of oxide and nonoxide powders since a variety of organometallic compounds can form gels or other polymerizing nonoxide precursors. Many of these entail more conventional alkoxide-based sol processing with water-initiated polymerization where the organic part of the alkoxide is selected to pyrolyze in an inert atmosphere to very fine homogeneously distributed carbon to react with the metal oxide product, for example, SiO_2 to yield SiC [109–111]. This is a fairly common type of route, that is, using chemical processing to improve more conventional reaction, carbothermal, processing as discussed below. However, there have been a variety of laboratory demonstrations of polymerizing organometallic precursors that thermally decompose to, at least approximately, single-phase nonoxide compounds or mixtures of them (e.g., polysilanes reacting with NH_3 to produce Si_3N_4 [112] or directly produce it from polysilazanes, or directly produce SiC from polycarbosilanes). However, in many of these cases, particularly the latter ones, the intermediate product is a polymeric mass, as from many conventional sol-gel processes. These often produce particles that may have very fine crystallites, but are agglomerated in particles whose size is mainly a function of comminution of the intermediate polymer or the pyrolyzed product. Again there may be some specialized uses for such particles such as abrasives as for alumina-based gels, but otherwise such powders are normally not advantageous for making dense ceramics by pressureless sintering whether of oxide or nonoxide composition. On the other hand, processing of powders from liquid precursors or solutions to form droplets, for example, by spraying, that are then rigidized by polymerization may prove fruitful, as suggested by aerosol decomposition of a polymeric precursor of BN [113].

The first of two related reaction processes that collectively are the most significant sources of typical nonoxide ceramic powders, especially on a commercial scale, is direct elemental reaction, which in turn consists of two main approaches. The first, and primary commercial one, is the forming of important nitride ceramics, mainly AlN and Si_3N_4, via direct nitridation by heating Al and Si powders respectively in a mainly N_2 (often with H_2) or NH_3 atmosphere. Processing is commonly assisted by use of additives (see Sec. 3.2), such as Fe in Si_3N_4 (where it is often an impurity in the Si, from comminution) to aid the nitriding reaction and halide salts; for example, LiF or CaF_2, for AlN, apparently to aid in penetrating the surface oxide layer on the Al particle surfaces. In the case of Si, the reaction becomes exothermic as temperatures approach that for melting Si (just over 1400°C) so keeping the temperature below this level by controlling reactant gas flow and furnace temperature is important, since Si melting results in coalescence of much of the Si and incomplete nitridation. Even without such coalesence some grinding and renitriding of the comminuted material may be necessary, especially for higher quality Si_3N_4, which is widely used. However, by far the highest tonnages of Si_3N_4 powder made by this process are used to make Si_3N_4 refractories. AlN powder made by nitriding Al metal has been sold commercially for fabrication of high-quality AlN (e.g., for high thermal conductivity bodies), but the volume and history of this are substantially less than for Si_3N_4. Such direct reaction of the elements can in principle be used, and have been tried, for other nonoxide (especially binary) ceramics, such as borides, carbides, and silicides, besides other nitrides, but is often limited by elemental powder costs—for example, of Ti, Zr, and especially B—as well as frequently by processing details for these other materials being less forgiving than in making AlN or Si_3N_4 (e.g., in limiting melting problems).

The second, more recent processing of nonoxide powders of binary ceramics is by self-propagating high-temperature synthesis (SHS) that was popularized by substantial investigation in the Soviet Union [114]. This entails processing of ceramics from the elements whose compound formation is sufficiently exothermic that if the reaction is ignited by local heating in one area of a powder compact, it will propagate through the compact in much the same fashion as a fuse for firecrackers, dynamite, and other explosives burns by propagation along the tube of fuse material. Many of these reactions can be very vigorous (depending in part on the other factors such as the particle size of the reactants), and thus require safety precautions, which has been a factor in limiting their use. Originally, these reactions were seen as being desirable due to lower costs since the natural exotherm of the reactions eliminated furnace and heating costs, as well as their very transient nature being beneficial, for example, to produce finer particles and possibly different phases. However, cost benefits from self-heating appear to be marginal, but in at least some cases, the transient nature of the reactions may be advantageous, e.g., little or no longer range melting and resultant agglomeration. Thus, for example Golubjatnikov [115] showed that SHS processing (of Si_3N_4 powder) had some cost advantage, primarily due to lower comminution costs due to greater friability of the reacted powder mass. This is again a reminder that, while general trends can often be discerned from principles and experience, specific process evaluation may hold some surprises. While such SHS powder preparation is used mostly for preparation of binary compounds of nonoxides, it has also been used for some of the more limited work on ternary nonoxide ceramics, e.g., of Ti_3SiC_2 by Lis and coworkers [116].

Consider now the second and much more broadly applicable and used method of traditional reaction processing of mainly binary nonoxide ceramic powders, namely carbothermal reduction. This simply entails intimate mixing of oxide powders of the desired metals, metalloids, and carbon (or a source of it) to reduce the oxides, and if producing a carbide, to react with the reduced metal to form its desired carbide. Fine, uniformly, and intimately mixed reactive ingredients are important to react to the desired products with little or no residual oxide or excess carbon, at temperatures and times to limit excessive particle growth and sintering. Removal of residual undesired phases can sometimes be done with limited negative effects, but are an added cost and pose their own contamination problems. Fine carbon powders or liquid precursors such as sugar (dissolved in water) or furfuryl alcohol can be useful and are of modest cost, especially sugar [117], which has been used in a number of cases.

The first of a few examples are preparation of Si_3N_4 by carbothermal reduction of SiO_2 (which basically avoids the issue of Si melting) in a N_2 or NH_3 atmosphere, the latter being somewhat more reactive, generally producing mostly α Si_3N_4 (~ 2 μm) at ~ 1400°C [118]. Either fluidized-bed reactors [119] or rotary calciners [120] can be useful whether one of the reactants is a gas or all are solid (e.g., as for SiC) and may reduce agglomeration common in static bed reactors (see Fig. 2.5B). The phase of the oxides can aid in some cases; for example, γ Al_2O_3 is beneficial for making AlN at ~ 1500°C because of its finer character, but with effects of the starting skeletal structure of different Al_2O_3 phases [117,121]. On the other hand, anatase or rutile precursors for TiN have limited differences other than via some benefit of finer TiO_2 particle size and negative effects of purposely added particle TiO_2 coatings for pigment-grade material in making TiN at ~ 1150°C [122]. Reactions can be affected, often significantly, by various parameters, particularly temperature—e.g., SiC formation is via a solid-state carbon-SiO_2 reaction below 1400°C, while above this temperature gaseous reaction of SiO and C becomes dominant [123]. Vacuum processing or other control of CO pressure and continuous mixing (e.g., via a fluidized bed or rotary calciner) can also be important. While the above examples are binary compounds, more complex compositions can be made, such as sialons [124], sometimes using natural clays as lower cost raw materials [125]. Processing of ceramics such as TiB_2, SiC, and Si_3N_4 in pilot plant or production scale are reviewed by Shepard [93].

There are three extensions of carbothermal processing that should be noted. First, while such processing reduces or precludes melting of elemental precursors such as Al and especially Si, there are important cases where a low melting precursor is used, with the use of B_2O_3 for boron containing compounds being particularly important. Thus, for example, 4 of the 7 preparations of BN reviewed by Ingles and Popper [108] used B_2O_3 as the B source. B_2O_3 (or boric acid) is also the typical source of B in a variety of reactions involving carbothermal or other reductions, the latter being a second and larger extension of such reaction processing. Complications that may result from forming liquid phases during reaction are limited by actual or effective encapsulation of the initial solid particles that will melt so melted particles cannot coalesce. Such encapsulation may be via other solid constituents of the reaction, fillers inert to the reaction [108], or an initial liquid phase, e.g., sugar solution or furfuryl alcohol precursor for carbon where this is a constituent of the reaction.

The second extension of carbothermal processing is to more complex compounds than just binary compounds, e.g., of ternary compounds TiZrC and $TiZrB_2$ by Mroz [126], where such processing of the end members at ~ 2000°C resulted in particle sizes of ~ 2–13 μm and various stoichiometries of ternary solid solution compounds with intermediate particle sizes. The third extension of reduction processing noted above is often used to directly produce ceramic composites (Sec. 8.2.3) without specifically producing a powder that is subsequently densified, but the latter route has also been pursued. Thus, for example Cutler and coworkers [127] showed that composite powders produced by the following reactions gave composite powders that could yield composite character and properties comparable to those obtained by making the composites from constituents oxide and nonoxide powders:

$$3TiO_2 + 4\,Al + 3C \Rightarrow 3TiC + 2Al_2O_3 \tag{2.5}$$

$$3SiO_2 + 4\,Al + 3C \Rightarrow 3SiC + 2Al_2O_3 \tag{2.6}$$

Though investigation of ternary nonoxide ceramics is substantially less extensive than of ternary oxide ceramics, it might be expected that typical solid-state reactions of constituent members of the desired nonoxide compounds would be a common route to preparation of powders of the ternary compounds. Though much of such reaction processing is conducted during densification rather than separately producing a powder, there is some literature data on separate powder preparation. Thus, for example Groen and coworkers [128,129] report formation of $CaSiN_2$ and $MgSiN_2$ from the respective end members of Ca_3N_2 or Mg_3N_2 and Si_3N_4, each producing (at ~ 1250°C with a N_2 atmosphere to limit volatilization) the desired compound powders of 1–2 μm that could be sintered to reasonable densities. Similarly, Yamane and coworkers [130] have prepared Li_3AlN_2 powder from Li_3N and AlN at ~ 700–900°C

Another class of reaction processes are those carried out in a molten media. While some nonoxide ceramic powders can be produced in molten salts in a similar fashion to preparation of some oxide powders, as discussed below, molten metal baths can provide suitable solvents for reactions to produce nonoxide ceramic powders. Kieffer and Jangg [131,132] discussed producing particles of various binary nonoxides such as typical carbides of Nb, Ta, Ti, and W, several silicides, and a few borides, nitrides or carbonitrides in various molten metal baths. Particles, often having single crystal character and morphology, up to ~ 1 mm in size can be extracted by dissolving the solidified metal. While this leaves many questions, especially regarding practicality, Bairamashvili and coworkers [133] reported making powders of α-AlB_{12} or $MgAlB_{14}$ by crystallization in aluminum melts and acid extraction that could be hot-pressed to give suitable bodies of these materials.

Nonoxide ceramic powders can also be produced from molten salt baths similar to processes for some oxides (see Sec. 2.4). Thus, Morgan and Koutsoutis [134] discovered in attempts to produce $CaLa_2S_4$ the preparation of almost spherical particles of $NaLaS_2$ a few microns in diameter, by reaction of Na_2S and $LaCl_3$ in an eutectic bath of 2 Na_2S + 3NaCl at ~ 900°C under an atmosphere of H_2S. Though not successful in their attempt to make $CaLa_2S_4$, they noted considerable potential for making various chalcogenide and related compounds by similar methods. More recently, Chan and Kauzlarich [135] reported preparation of carbides of either Nb or Ta by elemental reaction in molten YCl_3 or $LuCl_3$ at 1000–1150°C for a few days. Hooker and Klabunde [136] reported that evaporation of Ni metal in the presence of alkali acetate, formate, or nitrate salt melts at 170–220°C could yield nanoscale particles of Ni, NiO, or Ni_3C depending on processing parameters.

Next consider vapor-phase preparation of nonoxide ceramic powders starting with CVD of mainly binary compounds, by first noting the high tonnages of very low-cost carbon black powder produced each year by pyrolysis, mainly of methane, via gas-phase nucleation instead of surface nucleation and growth for CVD graphite. There has been substantial investigation of preparation of mainly binary nonoxide ceramic powders via CVD [8,9 and papers in Refs. 137,138], with some of this done at the pilot plant or production scale, for example, for SiC and Si_3N_4. Further, the broad applicability of CVD to binary ceramic compounds for bulk, coating, or thin film deposition shows broad applicability to powder preparation since the change from such surface deposition to gas-phase nucleation for powders is normally a reasonable controlled process. Similarly, the more limited demonstrations of CVD of mixed (composite) nonoxides or ternary nonoxide compound deposits (Sec. 6.6) implies good potential for CVD of powders of these and related bodies. Again commercial potential for CVD of ceramic powders is greater when materials costs are lower, which commonly favors uses of halide, especially chloride, sources of the metals or metalloids (e.g., B) and methane and ammonia or nitrogen for C and N. The often lower CVD temperatures of most organometallic sources generally do not compensate for their higher costs and frequent safety issues (and resultant added process costs). However, some alternatives may be feasible, such as trimethyl aluminum as a source of Al and SiS_2 as a silicon intermediary in making Si_3N_4 [139].

There are extensions of vapor-phase preparation of ceramic powders by stimulating vapor reactions via lasers or plasmas from either arcs or induction heating. Thus, laser stimulation of CVD has been investigated to produce very fine high-quality powders of Si_3N_4 and SiC [93,138,140], but which are projected to be of high cost (Sec. 1.4). Laboratory scale investigation of making nanoscale powders such as TiC and SiC by arcing electrodes, commonly of the metal carbide desired under a dielectric fluid, which in this case can be the source of carbon [141]. While this has some potential versatility, mainly for nonoxides, especially carbides, by limiting melting of the electrode and using various electrode-fluid reactions, this process is probably limited to specialized laboratory applications. Arc plasmas have been used to produce on at least a pilot plant scale fine, good quality powders of TiB_2 [142] and SiC, but with probable high costs. Induction plasmas have been used to produce a variety of ceramic powders on a laboratory scale [93,142], but again would probably be of high cost.

Finally consider briefly preparation of powders of compounds or composites of both oxygen and nonoxide anions, the most extensively investigated of the former being oxynitrides, especially SiAlONs and ALON. All of these materials are commonly prepared by reaction sintering from constituent compounds, as are many ternary and higher compounds as well as composites. However, separate preparation of constituent powders is also frequently done, usually via one or more common reaction processes such as carbothermal reduction. Corbin [143] has briefly reviewed this and other aspects of ALON.

2.6 POWDER PARTICLE COATING AND CHARACTERIZATION

It is increasingly being realized that coating of powder particles can be important from several standpoints. A narrower aspect of powder coating is for limiting interaction of the powder particles with the environment during storage and early stages of processing. This has been of interest in recent years to limit moisture effects on AlN powders, especially those for producing electronic substrates or packages of high thermal conductivity, and is apparently being done on some production powders. Such organic-based coatings would also suggest possible coating with organic binders for various fabrication processes since coating them on the particles would yield more uniform bodies, eliminating binder deficiencies and excesses, both of which reduce component quality. However, since each fabrication process use different types and amounts of binders and many ceramic manufacturers consider binder technology part proprietary. This may create serious "territorial" issues between (mainly the larger) ceramic processors and raw material suppliers for such binder coating, but may be an asset to smaller ceramic manufacturers.

More broadly being investigated are the possibilities of coating ceramic powder particles with either additives such as densification aids, such as for Si_3N_4, or composite phases, such as ZrO_2 for zirconia toughened composites. In these cases better uniformity of the distribution of the added phase should again result in more uniform, better quality components. Of even greater possible benefit is coating much or all of the matrix material on whiskers or platelets so that the resultant whisker or platelet composites can be freed of much of the constraints on pressureless sintering normally found in such composites (which are thus normally hot-pressed), besides possible benefits in resultant uniformity of the composite structure. Another important coating area for fiber composites is fiber coatings to prevent strong fiber-matrix bonding in order to have suitable toughening and noncatastrophic failure.

In response to these needs and opportunities there has been a fair amount of investigation and development in this area, examples of which are given below. Basically three coating techniques, two based on liquid processes, and one on CVD, have been used depending in part on the materials involved and the function of the coating. One liquid method of partial coating, for example of densification aids, on particles that will form the bulk of the resultant body is via colloidal techniques using surface charges to attract smaller particles of the additive(s) to the oppositely charged surfaces of larger particles of the main body composition. More extensive has been use of salt solutions, polymeric precursors, and especially of sols to coat particles as well as some whiskers, platelets, and fibers—the former two often heavily, but much of the fiber coating, as well as some particle coating is done by CVD. Somewhat heavier coatings, while also possibly used some for processing additives, are used more extensively for composites (e.g., cermets such as WC-Co or finer scale analogs of fiberous monoliths), and especially heavier coatings for whisker or platelet composites. Heavier coatings are also being used to fabricate bodies of ternary or more complex compounds. Finally, the third coating method is via vapor deposition, especially CVD for fiber coatings and coating nuclear fuel particles.

Han and coworkers [144] and Wang and Riley [145] used sol techniques to coat a few percent of alumina as a thin (e.g., 15 nm) coating on Si_3N_4 particles with resultant faster sintering than with conventional mixing of the alumina. Garg and De Jonghe [146] demonstrated similar, but also heavier, coating of Si_3N_4 particles with yttria or yttria-alumina precursors. Coating of YIG particles with nanoscale coatings of 0.5–2 w/o each of SiO_2 and MnO applied via respectively a sol and an acetate to yield good densification and uniform microstructure were reported by Cho and Amarakoon [147], while sol based coating of hydrous alumina on hematite, chromia, or titania were reported by Kratohvl and Matijevi [148]. Use of a polymeric precursor to successfully apply thin BN coatings of fine particles of alumina, magnesia, and titania (but not silica) was done by Borek and coworkers [149].

Turning to often heavier coatings, typically applied more explicitly for better fabrication of composites, Mitchell and De Jonghe [150] reported coating SiC whiskers or platelets with up to 20 v/o alumina via precipitation of a sulfate precursor. This allowed densification by pressureless sintering to closed porosity with at least 20 v/o SiC. Jang and Moon [151] reported more homogeneous ZTA composites by coating the zirconia particles on the alumina particles. Harmer and coworkers [152] reported sol coating of alumina particles with up to 50 v/o borosilicate markedly improved densification over mechanical mixtures of the ingredients. While the coating was moisture sensitive, this could be eliminated by a thin overcoat of silica. Huang coworkers [153] reported fabrication of improved aluminum titanate-25 v/o mullite composites by sol coating a mullite precursor on aluminum titanate particles.

Liquid-based particle coating can have other composite and noncomposite applications. Bartsch and coworkers [154] showing sol coating of amorphous silica on gamma alumina particles reduced the sintering temperature by ~ 300°C over that found for similar coated alpha-alumina particles or other alumina-silica mixtures. Composite applications of liquid-based particle coatings can also entail some metal phases; e.g., Ohtsuka and coworkers [155] reported solution-precipitation of nickel precursor coating on clay particles from nickel nitrate solutions followed by reduction. Alumina- nickel composites have been fabricated by electroless-nickel coating of alumina powder particles by Lin and Jiang [156].

Turning to CVD coating, more recent work has included considerable effort on coating finer ceramic (and metal) particles, with metals or ceramics. Thus, for example Franquin and coworkers [157] reported coating nanoscale Ni particles on γ-alumina for catalytic purposes and Chen and Chen [158] have used fluidized-bed CVD of Ni or Cu on Al_2O_3 or SiC particles to aid in their bonding in metal matrix composites. Itoh and coworkers [159] CVD coated TiN on Fe powder using $TiCl_4$, N_2, and H_2 at ~ 1000°C to significantly improve oxidation resistance. CVD coating of sintering aids on metal particles, e.g., Fe and Ni on W particles, has similar benefits as in ceramics, resulting in finer, more homogeneous microstructures. (A. Sherman, personal communication, 2000).

Turning to other ceramic particles, Li and coworkers [161] reported TiO_2 coatings on larger (10–30 μm) particles via CVD with a sol source of Ti in a rotary reactor. Itoh and coworkers [161] coated coarser (50 μm) Al_2O_3 particles with TiN, which upon hot-pressing gave composites with properties, such as electrical conductivity, tailorable with composition. Tsugeki and coworkers [162] also CVD coated Al_2O_3 particles (agglomerated to ~ 200 μm) with TiN (via $TiCl_4$, NH_3 at ~ 700°C) to similarly control properties, for example, give higher electrical conductivity. It should be noted, however, that there is a substantial background in CVD multilayer coatings, e.g., of ZrC or SiC and doped CVD graphite developed to extend the life of potential (sol-gel derived) oxide nuclear beads (~ $^1/_2$ mm dia.) for nuclear reactor fuels [163].

There has been substantial investigation and development of coatings for various ceramic fibers in various matrices, ranging from glass fibers in cement matrices to graphite and other ceramic fibers in metal or ceramic matrices. This is a large and specialized subject that cannot be fully treated here because of the diversity of matrix and fiber materials, needs, and processes. Instead, a summary of the most pertinent needs and results is presented for ceramic matrix composites. While protection of fibers from handling damage is desirable for all matrices, and a key need for glass fibers in cement is corrosion protection, a key need for SiC-based fibers in ceramic oxide matrices is coatings that limit fiber-matrix bonding. That such fiber coatings might be effective in improving fiber pullout and resultant toughness and noncatastrophic failure was suggested by mainly two sets of observations. First were those of Ysuda and Schlichting [164] showing that SiC coating of graphite fibers used in some ceramic matrices such as alumina improved strength and toughness. Second, Prewo and coworkers [165] showed that SiC fibers in crystallized glass matrices using TiO_2 as a nucleation agent for crystallization had greater fiber pullout than in matrices with ZrO_2 nucleating agent. This difference was associated with the TiO_2 nucleating agent reacting to form TiC along at least some of the fiber-matrix interface. This led Rice [166,167] to propose use of BN because of its inertness in many chemical interactions, and with many ceramic matrices, and related lack of bonding as well as its frequent cleavage-type failure like graphite, with which it is isostructural. This coating, originally applied by CVD using borazine (selected for its lower deposition temperatures), has become the standard for many ceramic fiber composites, and is now in commercial production (now using BCl_3 and NH_3 for lower costs). The rationale was originally given for two- and three-layer coatings, and some work has been done on SiC overcoats. Continuous SiC fiber tow coating with BN is available, along with some cloth coating, and bolt-to-bolt SiC cloth coating is expected soon (R. Engdahl, President of Synterials, Inc., Herndon, VA, personal communication, 2000).

The extensive continuing search for fiber coatings arises mainly due to the oxidative embrittlement problem encountered with composites with SiC (and probably other nonoxide) fibers composites upon high-temperature oxidizing exposure, which is not cured by BN or related coatings. The focus has thus turned to use of oxide fibers in oxide matrices with oxide fiber coatings, e.g., of rare earth phosphates [168,169], but whether these will prove both technically and economically viable is uncertain. There is also limited information on longer term limitations due to sintering or reaction of such all oxide composites constituents causing problems similar to and different from those of oxidation of fiber composites with nonoxide, fiber, constituents.

2.7 POWDER AND PARTICLE CHARACTERIZATION

Consider now characterization of powders and the powder particles, especially, starting with some overall characterizations that are of primary use for comparing one powder to another. A summary is given below with the reader referred to other sources (e.g., Refs. 1–9) for more detail.

An overall measure of powder flow is the *angle of repose*, that is, the included angle of a conical pile of powder poured onto a flat base. More flowable powders have higher angles (i.e., result in a shorter conical pile with a broader base). Two related parameters are the *pour density* and *tap density* (i.e., the apparent density of the powder mass respectively as poured and after tapping of the base on which the powder rests or the sides of the container into which the powder has been poured). The actual density of the particles is obtained from helium pycnometric density measurements of the powder.

Closely related to the above are the porosity in the powder mass. If the theoretical density of the powder material is known, then the ratio of the pycnometric to the theoretical density is the relative particle density and hence the volume fraction of solid (S_i), and one minus this is the volume fraction porosity in the particles (P_i), i.e.,

$$S_i = 1 - P_i \text{ and } P_i = 1 - S_i \tag{2.7}$$

The total volume fraction solid (S) and total volume fraction porosity (P) in the powder mass are similarly given by the same equations without the i subscripts. In principle, the volume fraction porosity between the powder particles (P_e) in the powder mass can then be obtained as $P - P_i$. However, such figures are only approximate since there is no clearly defined boundary between pores between and within the powder particles. Further, many agglomerated particles may undergo changes in their degree of agglomeration with powder handling and flow which will thus change these porosity values. The first of two other methods that can be useful in measuring porosity of both powder masses and green powder bodies is surface area. This requires assumption of a pore structure, usually of a single size of perfectly spherical pores, and thus typically is of primary use for relative comparison of powders. The second method of measuring porosity in powder particles is via transmission or scanning electron microscopy (TEM and SEM) where pores can often be directly seen. Porosity measurements are best done by stereological techniques, but this is dependent on "seeing" all of the pores in the particles which depends on both the technique, the porosity (size), and on the stability of agglomerated powder particles. As is true of all microstructural characterization, it is often of value or critical to compare measurements of parameters by different techniques, and sometimes under different conditions.

Consider now particle size measurement, which for agglomerated particles depends on their friability, reagglomeration, and aspects of their handling and measurement that may effect their attrition or agglomeration. Measurement methods depend in part on particle sizes. For larger particles with sizes of ~ 5, and especially ~ 40–50 μm in diameter, sieving is applicable and often used. Sedimentation is also used, covering much of the same particle size range as sieving, as well as particle about an order of magnitude finer. Both techniques also can give information on particle size distribution. Measurements via light scattering are quite rapid, versatile and extensively used, especially where dilute suspensions are available. Both x-ray and neutron scattering are also used, especially at fine particle sizes and are applicable to more concentrated suspensions.

Consider next grain or crystallite size measurement in polycrystalline powder particles. X-ray line broadening can be used, but does not distinguish between individual or agglomerated particles. Similarly x-ray and neutron scattering can be used. However, most common are SEM and TEM where individual crystallites (grains) can be seen. Stereological methods are of most value if sufficient grains can be observed. Where more than one phase is present, distinguishing the microstructures of both phases and their interrelation is important whether the second phase is another chemical or crystal phase or porosity of the intra- of inter granular type, or both can be important.

Both particle and crystallite shape and orientation can be important, with both often being interrelated for particles and for crystallites with some possible cross relations between the two. There is also some relation between size and shape of each. These factors are primarily determined by stereological measurements from TEM or SEM observations, which can also yield information on orientation, but much of this characterization depends on assumed size and shape parameters.

Finally, note that whether particles are coated or have gradients of composition can also be important and thus deserves characterization. This is true not only for particles that are already coated, but also for those that are going to be coated, for example, since porous particles with a coating of limited gas permeability may entrap some gases in internal pores. Similar entrapment may also occur for intragranular pores, especially if particle grain sizes are not particularly small. Also note that use of some characterization methods in controlling fabrication in manufacturing is discussed in Sec. 8.4.

2.8 DISCUSSION, SUMMARY, AND CONCLUSIONS

There is a substantial and growing diversity of powder preparation methods to address the broader range of powder compositions and character desired. However, many of the methods are of more limited investigation and leave considerable uncertainties about the uniformity, repeatability, scalability, and costs. Thus, traditional methods still dominate most commercial production, for example, salt precipitation and calcining for most binary oxides and solid-state reactions of binary oxide constituents of ternary oxides. An important exception is CVD preparation of some binary oxides such as Al_2O_3, SiO_2, and TiO_2; though residual Cl can inhibit densification from such typical processing, there are possible methods of addressing this. Where single-crystal particles (e.g., platelets or whiskers) are desired, hydrothermal or molten salt (or metal) methods are often appropriate, besides CVD, though these and other techniques are often influenced significantly by use of additives (see Sec. 3.2).

Traditional reaction processing still dominates for most binary nonoxide powder preparation. Thus, those compounds for which the metallic element is available at reasonable prices in suitable powder form and whose agglomeration due to melting is not an issue or can be controlled may be directly reacted with the appropriate cation powder to produce the desires nonoxide. This is most common for AlN and Si_3N_4. However, the dominant traditional reaction method of preparation of nonoxide powders remains carbothermal reduction for most binary nonoxide compounds, including commercial production of AlN and Si_3N_4. Other reaction processes, such as SHS and related processing, have shown some potential for some specific powders, but more demonstration of their practical aspects of uniformity, repeatability, scalability, and costs is needed. CVD and some plasma processing have been shown to have potential, but remain uncertain in their future roles. Morphological single-crystal particles, are often a product of CVD, often entailing use of additives (see Sec. 3.2) Newer techniques based on more diverse chemistry have considerable promise, but need much more evaluation and development.

Scale-up of powder processing from lab to pilot to production is an important and challenging transition as it is for other fabrication steps. For example, reactions that are either endothermic or exothermic provide challenges of keeping thermal uniformity as the scale of operation and sources of heat input and output increase and increasing masses of reactants and products often increase driving forces for agglomeration and back pressures limiting escape of gaseous reaction products. Such problems are often not documented, since successful scale-up, not difficulty along the way, is the goal. Thus, liquid media processing has resulted in powders of changing character on scaling from 10 to 100 to 1000 mL in the laboratory, changing raw materials has changed results due to previously unknown effects of modest levels of impurities, and scaling directions may be found incompatible with anticipations (J. Voight, D. Demos, R. Egan, Sandia National Lab., personal communication, 2001). For example, sol preparation of PZT powders required nonaqueous solvents, which in turn required explosion proof facilities, but such a peristaltic pump needed for the desired continuous process was not available, requiring reversion to a batch process in scale-up.

Further, three factors should be noted. First, ceramics is a diverse field and becoming more so, as both the number of ceramic compositions addressed and the diversity of product scale and microstructures increases. Thus, there may be more opportunities for speciality powders, for example, as shown for sol-gel derived abrasives and hydrothermal $BaTiO_3$. However, these were only achieved through substantial development to demonstrate quality, uniformity, scalability, repeatability, and acceptable cost for use in limited quantities. Larger volume applications require further cost reductions as demonstrated for sol-gel abrasives. Second, while many of the processes can produce micron- to nm-scale powder particles that are of interest for very fine microstructures, such particles pose important challenges for fabrication of bodies, especially dense ones. Third, many powders used for ceramic fabrications are not specifically made for such fabrication, but making tailored ceramic powders is becoming more common, as noted for alumina lamp envelopes. Such tailoring is also becoming more common even for special powder applications such as for thermal conducting ceramic particle-organic matrix composites for electronics and for plasma sprayed ceramic coating (H. Herman, personal communication, 2002) [170], as also discussed in Section 7.5.

REFERENCES

1. F. Singer, F.F. Singer, Industrial Ceramics. London: Chapman and Hall, 1984.
2. W. Ryan, Properties of Ceramic Raw Materials. New York: Pergamon Press, 1968.
3. M.N. Rahaman, ed. Handbook of Ceramic Engineering. New York: Marcel Dekker, Inc., 2001.
4. D.W. Richerson. Modern Ceramic Engineering. New York: Marcel Dekker, Inc., 1992.

5. P.E.D. Morgan, Chemical processing for ceramics (and polymers). In: H. Palmour III, R.F. Davis, T.M. Hare, eds. Processing of Crystalline Ceramics, Materials Science Research, Vol. 11. New York: Plenum Press, 67–77, 1978.
6. R.W. Rice. Ceramic Processing: An Overview. J. AiChE 36(4):481–510, 1990.
7. M.N. Rahaman. Ceramic Processing and Sintering. New York: Marcel Dekker, Inc., 1995.
8. C.R. Veale. Fine Powders Preparation, Properties and Uses. London: Applied Science Publishers LTD, 1972.
9. D. Segal. Chemical Synthesis of Advanced Ceramic Materials. Cambridge: Cambridge Univ. Press, 1989.
10. K.H. Stern, E.L. Weise. High temperature properties and decomposition of inorganic salts, part 1. Sulfates. NSRDS-NBS 7, U.S. Govt. Printing Office, 1966.
11. K.H. Stern, E.L. Weise. High temperature properties and decomposition of inorganic salts, part 2. Carbonates. NSRDS-NBS 30, U.S. Govt. Printing Office. 1969.
12. A.R. Miller. Possibility of error in limestone calcination. J. Am. Cer. Soc. 56(8):443, 1973.
13. P.J. Anderson, R.F. Horlock. Thermal decomposition of magnesium hydroxide. Trans. Faraday Soc. 58(478):1993–2004, 1962.
14. R.F. Horlock, P.L. Morgan, P.J. Anderson. Effects of water vapor on the decomposition of magnesium hydroxide. Trans. Faraday Soc. 59(483):721–728, 1963.
15. W.D. Caleister, Jr., I.B. Cutler, R.S. Gordon. Thermal decomposition kinetics of boehmite. J. Am. Cer. Soc. 49(8):419–22, 1966.
16. K.J.D. Mackenzie, P.J. Melling. The calcination of titania, II. Influence of atmosphere on crystal growth in anatase powders. Trans, J. Brit. Cer. Soc. 173(6):179–183, 1974.
17. J. Ewing, D. Beruto, A.W. Searcy. The nature of CaO produced by calcite powder decomposition in vacuum and in CO_2. J. Am. Cer. Soc. 62(11–12):580–584, 1979.
18. P.D. Garn, F. Freund. Variation of the thermal decomposition of magnesium hydroxide with water-vapor pressure. Trans J. Brit. Cer. Soc 74(1):23–28, 1975.
19. Ta.I. Zel'manovich, L.A. Reznitskii. Reversability of thermal decomposition of magnesium carbonate. Inorg. Matls. 18(12):1750–1753, 1982.
20. P.J. Anderson, R.F. Horlok, R.G. Avery. Some effects of water vapour during the preparation and calcination of oxide powders. Proc. Brit. Cer. Soc. No. 3, 33–42, 1965.
21. R.W. Rice. Fabrication of Dense MgO. Naval Research Lab. NRL Report 7334, 1971.
22. R.W. Rice. Characterization of Hot-Pressed MgO. Naval Research Lab. NRL Report 7335, 1971.
23. L.L. Musselman. Production processes, properties, and applications for aluminum-containing hydroxides. In: L.D. Hart, S. Lense, eds. Alumina Chemicals, Science and Technology Handbook. Westerville, OH: Am. Cer. Soc., 1990, pp. 75–92.
24. T.J. Carbone. Production processes, properties, and applications for calcined and high-purity aluminas. In: L.D. Hart, S. Lense, eds. Alumina Chemicals, Science and Technology Handbook. Westerville, OH: Am. Cer. Soc., 1990, pp. 281–294.
25. S.J. Wilson. Phase transformations and development of microstructure in boehmite-derived transition aluminas. Proc. Brit. Cer. Soc. No. 28, 1979, pp. 281–294.

26. J.L. Henry, H.J. Kelly. Preparation and properties of ultrafine high-purity alumina. J. Am. Cer. Soc. 48(4):217–218, 1965.
27. M.D. Sacks, T.-Y. Tseng, S.Y. Lee. Thermal decomposition of sperical hydrated basic aluminum sulfate. J. Am. Cer. Soc. 63(2):301–310, 1984.
28. E. Kato, K. Daimon, M. Nanbu. Decomposition of two aluminum sulfates and characterization of the resultant sluminas. J. Am. Cer. Soc. 64(8):436–443, 1981.
29. R.F. Horlock, P.J. Anderson. Molecular size pores in beryllium oxide powders. Trans. Faraday Soc. 63(531), (Part 3), 717–726, 1967.
30. B.A. Chandler, E.C. Duderstadt, J.F. White. Fabrication and Properties of Extruded and Sintered BeO. J. Nuc. Matls. 8(3):329–347, 1963.
31. R.E. Fryxell, B.A. Chandler. Creep, strength, expansion, and elastic moduli of sintered BeO as a function of grain size, porosity, and grain orientation. J. Am. Cer. Soc. 47(6):283–291, 1964.
32. R.W. Rice. CaO: I, fabrication and characterization. J. Am. Cer. Soc. 52(8): 1969.
33. G.V.S. Rao, M. Natarajan, C.N.R. Rao. Effect of impurities on the phase transformations and decomposition of $CaCO_3$. J. Am. Cer. Soc. 51(3):179–180, 1968.
34. D.T. Livey, B.M. Wanklyn, M. Hewitt, P. Murray. The properties of MgO powder prepared by decomposition of $Mg(OH)_2$. Trans. Brit. Cer. Soc. 56:217–236, 1957.
35. M.G. Kim, U. Dahmen, A.W. Searcy. Structural transformation in the decomposition of $Mg(OH)_2$ and $MgCO_3$. J. Am. Cer. Soc. 70(3):146–154, 1987.
36. S.M. Haile, D.W. Johnson, Jr., G.H. Wiseman, H.K. Bowen. Aqueous precipitationn of spherical zinc oxide powders for varistor applications. J. Am. Cer. Soc. 72(10):2004–2008, 1989.
37. R.K. Stringer, C.E. Warble, L.S. Williams. Phenomenological observations during solid state reactions. In: Kinetics of Reactions in Ionic Systems, Materials Science Research. Vol. 4. T.G. Gran, V.D. Frechette Eds. New York: Plenum Press, 1969, pp. 53–95.
38. A.W. Hey, D.T. Livey. Calcination processes in oxides. Proc. Brit. Cer. Soc. 12:13–31, 1969.
39. R.W. Rice. Processing induced sources of mechanical failure in ceramics. In: H. Palmour III, R.F. Davis, T.M. Hare, eds. Processing of Crystalline Ceramics, Materials Science Research. Vol. 11. New York: Plenum Press, 1978, pp. 303–319.
40. R.W. Rice. Failure analysis of ceramics. In: J. Varner, G. Quinn, eds. Fractography of Glasses and Ceramics IV.Westerville, OH: Am. Cer. Soc. 369–388, 2001.
41. R.W. Rice. The effect of gaseous impurities on the hot pressing and behavior of MgO, CaO, and Al_2O_3. Proc. Brit. Cer. Soc. Fabrication Science 2:99–123, 1969.
42. R. W. Rice. Fabrication and Characterization of Hot Pressed Al_2O_3. Naval Research Lab. NRL Report 7111, 1970.
43. R. Chandrasekhar, S.W. Charles, K. O'Grady, S. Mørup, J. Van Wonterghen. Preparation and characterization of barium hexaferrite powders produced by decomposition of organometallic complexes. Adv. Cer. Mtls. 2(1):65–68, 1987.
44. S. Kulkarni, J. Shrotri, C.E. Deshpande, S.K. Date. Synthesis of chemically coprecipitated hexagonal strontium-ferrite and its characterization. J. Mat. Sci. 24:3739–3744, 1989.

45. J. Takahashi, T. Toyda, T. Ito, M. Takatsu. Preparation of $LaNiO_3$ powder from coprecipitated lanthinum-nickel oxalates. J. Mat. Sci. 25:1557–1562, 1990.
46. K.-L. Lin, T.-H. Lin. Coprecipitatied Zirconia-Alumina Powders. Matls. Sci. and Eng. A111:211–216, 1989.
47. M. Déchamps, B. Djuričić, S. Pickering. Structure of zirconia prepared by homogeneous precipitation. J. Am. Cer. Soc. 78(11):2873–2880, 1995.
48. M. Akinc, D. Sordelet. Preparation of yttrium, lanthanum, cerium, and neodymium basic carbonate particles by homogeneous precipitation. Adv, Cer. Mtls. 2(3A): 232–238, 1987.
49. A. Pathak, D.K. Mukopadhyay, P. Pramanik. A new co-precipitation technique for the preparation of mixed-oxides. Mat. Res. Bul. 27:155–159, 1992.
50. I. Jaymes, A. Douy, D. Masssiot, J.-P. Busnel. Synthesis of a mullite precursor from aluminum nitrate and tetraethoxysilane via aqueous homogeneous precipitation: An ^{27}Al and ^{29}Si liquid- and solid-state NMR spectroscopic study. J. Am. Cer. Soc. 78(10):2648–2654, 1995.
51. S. Hokazono, K. Manako, A. Kato. The sintering behavior of spinel powders produced by a homogeneous precipitation technique. Br. Cer. Soc Trans. J. 91:77–79, 1992.
52. G.H. Maher, C.E. Hutchins, S.D. Ross. Preparation and characterization of ceramic fine powders produced by the emulsion process. Am. Cer. Soc. Bul. 72(5):72–76, 1993.
53. J.M. Fletcher, C.J. Hardy. Application of sol-gel processes to industrial oxides. Chem. and Ind. 75:48–51, 1968.
54. G. Wilson, R. Heathcote. Role of sol-gel powders in thermal-spraying processes. Am. Cer. Soc. Bul. 69(7):1137–1139, 1990.
55. H.K. Varma, T.V. Mani, A.D. Damodaran, K.G. Warrier, U. Balachandran. Characteristics of alumina powders prepared by spray-drying of boehmite sol. J. Am. Cer. Soc. 77(6):1597–1600, 1994.
56. J.H. Jean, D.M. Goy, T.A. Ring. Continuous production of narrow-sized and anagglomerated TiO_2 powders. Am Cer. Soc. Bul. 66(10):1517–1520, 1987.
57. B. Fegley, Jr., P. White, H.K. Bowen. Processing and characterization of ZrO_2 and Y-Doped ZrO_2 powder. Am. Cer. Soc. Bul. 64(8):1115–20, 1985.
58. A.J. Fanelli, J.V. Burlew. Preparation of fine alumina powder in alcohol. J. Am. Cer. Soc. 69(8):C-174–75, 1986.
59. K.S. Mazdiyasni. Powder synthesis from metal-organic precursors. Cer. Intl. 8(2):42–56, 1982.
60. P.F. Becher. Transient thermal stress behavior in ZrO_2-toughened Al_2O_3. J. Am. Cer. Soc. 64(1):37–39, 1981.
61. M.G. Schwabel, P.E. Kendall. Alumina abrasive grains produced by sol-gel technology. Am. Cer. Soc. Bul. 70(10):1596–1598, 1991.
62. M.A. Leitheiser, H.G. Sowman. Non-Fused Aluminum Oxide-Based Abrasive Mineral. U.S. Patent 4,314,827, 1982.
63. P.A. Lessing. Mixed-cation oxide powders via polymeric precursors. Am. Cer. Soc. Bul. 68(5):1002–1007,1989.
64. F.J. Schnettler, F.R. Monforte, W.W. Rhodes. A cryochemical method for preparing

ceramic materials. In:G. H. Stewart, ed. Science of Ceramics. V.l. 4. Brit. Cer. Soc, Stoke on Trent, 79–90, 1968.
65. R.E. Jaeger, T.J. Miller, J.C. Williams. Effects of ammonium hydroxide on phase separation in the cryochemical processing of salt solutions. Am. Cer. Soc. Bul. 53(12):850–852, 1974.
66. A.C.C. Tseung, H.L. Bevan. Preparation and characterisation of high surface area semiconduction oxides. J. Mat, Sci. 5:604–610, 1970.
67. J. Kelly, D.B. Hibbert, A.C.C. Tseung. A critical examination of a cryochemical method for the preparation of high surface area semiconducting powders. Part 1. Rate of freeze-drying. J. Mat. Sci. 13:1053–60, 1978.
68. D.B. Hibber, A.C.C. Tseung. A critical examination of a cryochemical method for the preparation of high surface area semiconducting powders. Part 2. Nonaqueous solvents. the behavior of solid solutions in freeze drying. J. Mat. Sci. 14:1665–1671, 1979.
69. D.B. Hibber, J. Lovegrove, A.C.C. Tseung. A critical examination of a cryochemical method for the preparation of high surface area semiconducting powders. Part 2. Factors which determine surface area. J. Mat. Sci. 22:3755–3761, 1987.
70. M.D. Rigterink. Advances in technology of the cryochemical process. Am. Cer. Soc. Bul. 51(2):158–161, 1972.
71. P.J. McGrath, R.M. Laine. Theoretical process development for freeze-drying spray-frozen aerosols. J. Am. Cer. Soc. 75(5):1223–1228, 1992.
72. M.P. O'Toole, R.J. Card. Y_2O_3-ZrO_2 powder synthesis via alcohol dehydration of aqueous salt solutions. Am Cer, Soc. Bul. 66(10):1486–1489, 1987.
73. K. Richardson, M. Akinc. Preparation of spherical yttrium oxide powders using emulsion evaporization. Cer. Intl. 13:253–261, 1987.
74. G.L. Messing, S.-C. Zhang, G.V. Jayanthi. Ceramic powder synthesis by spray roasting. J. Am. Cer. Soc. 76(11):1707–1726, 1993.
75. W.F. Kladnig, W. Karner. Pyrohydrolysis for the production of ceramic raw materials. Am Cer. Soc. Bul. 69(5):814–817, 1990.
76a. A.C. Sutorik, S.S. Neo, D.R. Treadwell, R.M. Laine. Synthesis of ultrafine $\ddot{\beta}$-alumina via flame spray pyrolysis of polymeric precursors. J. Am. Cer. Soc. 81(6):1477–1486, 1998.
76b. R. Baranwal, M.P. Villar, R. Garcia, R.M. Laine. Flame spray pyrolysis of precursors as a route to nano-mullite: powder characterization and sintering behavior. J. Am. Cer. Soc. 84(5):951–961, 2001.
77a. C.R. Bickmore, K.F. Waldner, R. Baranwal, T. Hinkin, D.R. Treadwell, R.M. Laine. Ultrafine titania by flame spray pyrolysis of a titanatrane complex. J. Eur. Cer. Soc. 18:287–297, 1998.
77b. R.M. Laine, T. Hinkin, G.Williams, S.C. Rand. Low-cost nanopowders for phosphor and laser applications by flame spray pyrolysis. Mats. Sci. Forum Vols. 343-346, J. Metastable and Nanocrystalline Mats. 8:500–510, 2000.
78a. J.J. Kingsley, K. Suresh, K.C. Patil. Combustion synthesis of fine particle rare earth orthoaluminates and yttrium aluminum garnet. J. Solid State Chem. 87:435–442, 1990.
78b. J.J. Kingsley, K. Suresh, K.C. Patil. Combustion synthesis of fine-particle metal aluminates. J. Mat. Sci. 25:1305–1312, 1990.

78c. J.J. Kingsley, N. Manickam, K.C. Patil. Combustion synthesis and properties of fine particle fluorescent aluminous oxides. Bull. Mater. Sci. 13(3):179–189, 1990.
79. S. Hirano, A. Kawabata, M. Yosinaka, K. Hirato, O. Yamaguchi. Formation and characterization of Ce_3ZrO_8 prepared by the hydrazine method. J. Am. Cer. Soc. 78(5):1414–1416, 1995.
80. R.H.G.A. Kiminami, M.R. Morelli, D.C. Folz, D.E. Clark. Microwave synthesis of alumina powders. Am. Cer. Bul. 79(3):63–67, 2000.
81. W.J. Dawson. Hydrothermal synthesis of advanced ceramic powders. Am. Cer. Soc. Bul. 67(10):1673–1678, 1988.
82a. N.S. Bell, S.-B Cho, J.H. Adair. Size control of α-alumina particles synthesized in 1,4-butanediol by α-alumina and α-hematiate seeding. J. Am. Cer. Soc. 81(6):1411–1420, 1998.
82b. J.H. Adair, S.-B Cho, N.S. Bell, H.A. Perotta. Recent developments in morphological control of α-Al_2O_3 particles synthesized in 1,4 butanediol solution. J. Dispersion Sci. Tech.; 22(2&3) 143–165, 2001.
83. Y. Hayashi, T. Kimura, T. Yamaguchi. Preparation of rod-shaped $BaTiO_3$ powder. J. Mat. Sci. 21:757–762, 1986.
84. R.H. Arendt, J.H. Rosolowski, J.W. Szymaszek. Lead Zirconate Titanate Ceramics from Molten Salt Solvent Synthesized Powders. Mat. Res. Bull. 14:703–709, 1979.
85. K.H. Yoon, Y.S. Cho, D.H. Kang. Review: molten salt synthesis of lead-based relaxors. J. Mat. Sci. 33:2977–2984, 1998.
86. T. Kimura, T. Kanazawa, T. Yamaguchi. Preparation of $Bi_4Ti_3O_{12}$ powders in the presence of molten salt containing LiCl. J. Am. Cer. Soc. 66(8):597–600, 1983.
87. Y. Hayashi, T. Kimura, T. Yamaguchi. Mechanism of Ni-Zn ferrite formation in the presence of molten Li_2SO_4-Na_2SO_4. J. Am. Cer. Soc. 69(4):322–325, 1986.
88. K. Nagata, K. Okazaki. One-directional grain-oriented lead metaniobate ceramics. Jap. J. Appl. Phy. 24(7):812–14, 1985.
89. G.G. Xanthopoulou. Self-propagating SHS of inorganic pigments. Am. Cer. Soc. Bul. 87–96, 1998.
90. J.D.F. Ramsay, R.G. Avery. Ultrafine oxide powders prepared by electron beam evaporization. Part 1. Evaporization and condensation processes. J. Mat. Sci. 9:1681–1688,1974.
91. J.D.F. Ramsay, R.G. Avery. Ultrafine oxide powders prepared by electron beam evaporization. Part 2. Powder Charcteristics. J. Mat. Sci. 9:1689–1695,1974.
92. J.D. Holmgren, J.O. Gibson, C. Sheer. Some characteristics of arc vaporized submicron particulates. J. Electrochem. Soc. 111(3):362–369, 1964.
93. L.M. Sheppard. Vapor phase synthesis of ceramics. Adv. Matls. & Proc. 53–58, 1987.
94. R.W. Rice. Hot Forming of Ceramics. In: J.J. Burke, N.L. Reed, V. Weiss, eds. Ultrafine-Grain Ceramics. Syracuse. NY: Syracuse Univ. Press, 1970, pp. 203–250.
95. I.V. Antipov, L.B. Kuz'min, L.M. Gofman, N.L. Opolchenova, V.S. Kutsev, V.S. Mirochnikov. Removal of chlorine-containing impurities from titanium dioxide obtained by the chlorine method. Inorg. Matls. 14(5):896–899, 1978.
96. Y. Suyama, M. Tanaka, and A. Kato. Submicron TiO_2-ZrO_2 powders produced by vapor phase reaction of $TiCl_4$-$ZrCl_4$-O_2 system. Cer. Intl. 5(2):84–88, 1979.

97. S. Hori, M. Yoshimura, S.S_miya. Al_2O_3-ZrO_2 ceramics prepared from CVD powders. In: N. Claussen, A.H. Heuer, eds. Science and Technology of Zirconia II. Adv. in Cer. Vol 12. Westerville, OH: Am. Cer. Soc., 794–805, 1984.
98. F.C. Gennari, J.J.A. Gamboa, D.P. Pasquevich. Formation of pseudobrookite through gaseous chlorides and by solid-state reaction. J. Mat. Sci. 33:1563–1569, 1998.
99. K. Itatani, T. Kubozono, F.S. Howell, A. Kishioa, M. Kinoshita. Some properties of mullite powders prepared by chemical vapor deposition. Part 1. Preparation of mullite powder. J. Mat. Sci. 30:1158–1165, 1995.
100. R.W. Rice. Processing of ceramic composites. In: J.G.P. Binner, ed. Advanced Ceramic Processing and Technology. Vol. 1. Park Ridge, NJ: Noyes Publications, 1990, pp. 123–213.
101. R.W. Rice, B.A. Bender, R.P. Ingcl, T.W. Coyle, J.R. Spann. Tougher ceramics using tetragonal ZrO_2 of HfO_2. In: L. L. Hench, D.R. Ulrich, eds. Ultrastructure Processing of Ceramics, Glasses, and Composites. New York: Wiley-Interscience,1984, pp. 507–523.
102. A.M. Wong, R. McPherson. The Structure of Plasma Dissociated Zircon. J. Mat. Sci. Let., 16:1732–1735, 1981.
103. J. Homeny, J.J. Nick. Microstructure-Property Relations of Alumina-Zirconia Eutectic Ceramics. Mat. Sci. & Eng. A127:123–133, 1990.
104. G.M. Crosbie, R.L. Predmesky, J.M. Nocholson, E.D. Stiles. Prepilot scale synthesis of silicon nitride under pressure. Am. Cer. Soc. Bul. 68(5):1010–1-14, 1989.
105. C.B, Ross, T. Wade, R.M. Crooks. Electrochemical synthesis of metal nitride ceramic precursors in liquid ammonia electrolyte solutions. Chem. Mater., 3:768–771, 1991.
106. J.J. Ritter, K.G. Frase. Low-temperature synthesis of ceramic powders for structural and electronic applications.In: L.L. Hench, D.R. Ulrich, eds. Science of Ceramic Chemical Processing. New York: John Wiley & Sons, 1986, pp. 497–503.
107. A. Celikkaya, M. Akinc. Preparation and mechanism of formation of spherical submicron zinc sulfide powder. J. Am. Cer. Soc. 73(8):2360–2365, 1990.
108. T.A. Ingles, P. Popper. The preparation and properties of boron nitride.In: P. Popper, ed. Special Ceramics. London: Academic Press, 1960, pp. 144–152.
109. D.A. White, S.M. Oleff, R.D. Boyer, P.A. Budinger, J.R. Fox. Preparation of silicon carbide from organosilicon gels: I, synthesis and characterization of precursor gels. Adv. Cer. Mats. 2(1):45–52, 1987.
110. D.A. White, S.M. Oleff, J.R. Fox. Preparation of silicon carbide from organosilicon gels: II, gel pyrolysis and SiC characterization. Adv. Cer. Mats. 2(1):53–59, 1987.
111. F. Hatakeyama, S. Kanaki. Synthesis of monodispersed spherical β-silicon carbide powder by a sol-gel process. J. Am. Cer. Soc. 73(7):2107–2119, 1990.
112. W.R. Schnidt, V. Sukumar, W.J. Hurly, Jr., R. Garcia, R.H. Doremus, L.V. Interante. Silicon nitride derived from an organometallic polymeric precursor: preparation and characterization. J. Am. Cer. Soc. 73(8):2412–2418, 1990.
113. D.A. Lindquist, T.T. Kodas, D.M. Smith, X. Xiu, S.L. Hietala, R.T. Paine. Boron nitride powders formed by aerosol decomposition of poly(borazinylamine) solutions. J. Am. Cer. Soc. 74(14): 3126–3128, 1991.

114. Z.A. Munir, J.B. Holt, eds. Combustion and Plasma Synthesis of High-Temperature Materials. New York: VCH Publishers, Inc. 1990.
115. K.A. Golubjatnikov, G.C. Stengal, R.M. Spriggs. The economics of advanced self-propogating, high-temperature synthesis materials fabrication. Am Cer. Soc. Bul. 72(12):96–102, 1993.
116. J. Lis, R. Pampuch, J. Piekarczyk, L. Stobierski. New Ceramics Based on Ti_3SiC_2. Cer. Intl. 19:219–222, 1993.
117. Y. Baik, K. Shanker. J. . McDermid, R.A.L. Drew. Carbothermal synthesis of aluminum nitride using sucrose. J. Am. Cer. Soc. 77(8):2165–2172, 1994.
118. B.G. Durham, M.J. Murtha, G. Burnet. Si_3N_4 by the carbothermal ammonolysis of silica. Adv, Cer. Matls. 3(1):45–48, 1988.
119. F.K. van Dijen. The design of a circulating fluid bed reactor for the carbothermal synthesis if silicon nitride. J. Eur. Cer. Soc. 14:397–401, 1994.
120. F.K. van Dijen, R. Metselaar. Chemical reaction engineering aspects of a rotary reactor for carbothermal synthesis of SiC. J. Eur. Cer. Soc. 5:55–61, 1989.
121. A. Tsuge, H. Inque, M. Kasori, K. Shinozaki. Raw material effects on AlN powder synthesis from Al_2O_3 carbothermal reduction. J. Mat. Sci. 25:2359–2361, 1990.
122. G.V. White, K.J.D. Mackenzie, J.H. Johnston. Carbothermal synthesis of titanium nitride. Part I. Influence of starting materials. J. Mat. Sci. 27:4287–4293, 1992
123. V.D. Krstic. Production of fine, high-purity beta silicon carbide powders. J. Am. Cer. Soc. 75(1):170–174, 1992.
124. H. Yoshimatsu, H. Kawasaki, Y. Miura, A. Osaka. Carbon-thermal reduction and nitridation of mixtures of SiO_2 and AlM->$2O_3$•$2H_2O$. J. Mat. Sci. 24:1280–1284, 1989.
125. Y. Sugahara, J. Miyamoto, K. Kuroda, C. Kato. Preparation of nitrides from 1:1 type clay minerals by carbothermal reduction. Appl. Clay Sci. 4:11–26, 1989.
126. C. Mroz. Processing TiZrC and $TiZrB_2$. Am Cer. Soc. 73(4):78–81, 1994.
127. R.A. Cutler, A.V. Virkar, J.B. Holt. Synthesis and densification of oxide-carbide composites. Cer. Eng. Sci. Proc. 6(7–8):715–728, 1985.
128. W.A. Groen, M.J. Krann, G. De With. New ternary nitride ceramics: $CaSiN_2$. J. Mat. Sci. 29:3161–3166, 1994.
129. W.A. Groen, M.J. Krann, G. De With. Preparation, microstructure and properties of $CaSiN_2$ Ceramics. J. Eur. Cer. Soc. 12:413–420, 1993.
130. H. Yamane, S. Kikawa, M. Koizumi. Lithium aluminum nitride, Li_3AlN_2, as a Lithium Solid Electrolyte. Solid State Ionics, 15:51–54, 1985.
131. R. Kieffer, G. Jangg. Production of hard compounds according to the menstruum process. Pwd. Met. Intl. 4(4):191–193, 1972.
132. R. Kieffer, G. Jangg. Production of hard compounds according to the menstruum process. Pwd. Met. Intl. 5(1):25–27, 1973.
133. L.A. Bairamashvili, L.I. Kekelidze, G.A. Golikova, V.M. Orlov,"The preparation of α-AlB_{12} and $AlMgB_{14}$ samples and an investigation of their electrothermal properties. J. Less-Common Metals 67:461–464, 1979.
134. P.E.D. Morgan, M.S. Koutsoutis. Fused salt synthesis of materials for Ir windows. Mat. Res. Bull. 22:617–621, 1987.
135. J.Y. Chan. S.M. Kauzlarich. Rare-earth halides as fluxes for the synthesis of tantalum and niobium carbide. Chem. Mater. 9:531–534, 1997.

136. P.D. Hooker, K.J. Klabunde. Reaction of nickel atoms with molten salts. A new approach to the synthesis of nanoscale metal, metal oxide, and metal carbide particles. Chem. Mater. 5:1089–1093, 1993.
137. H.O. Pierson, ed. Chemically Vapor Deposited Ceramics. Westerville, OH: Am Cer. Soc. 1981.
138. R.F. Davis, H. Palmour, III, R.L. Porter. Emergent process methods for high-technology ceramics, Mater Sci Res. Vol. 17. 1984.
139. P.E.D. Morgan, E.A. Pugar. Synthesis of Si_3N_4 with emphasis on Si-S-N chemistry. J. Am. Cer. Soc. 68(12):1985.
140. Y. Suyama, R.M. Mara, J.S. Haggerty, H.K. Bowen. Synthesis of ultrafine SiC powders by laser-driven gas phase reactions. Am. Cer. Soc. Bul. 64(10):1356–1359, 1985.
141. A. Kumar, R. Roy. Reactive-electrodc submerged-arc process for producing fine non-oxide powders. J. Am. Cer. Soc. 72(2):354–356, 1989.
142. H.R. Baumgartner, R.A. Steiger. Sintering and properties of titanium diboride made from powder synthesized in a plasma-arc heater. J. Am. Cer. Soc. 67(3):207–212, 1984.
143. N.D. Corbin,. Aluminum oxynitride spinel: a review. J. Eur. Cer. Soc. 5:143–154, 1989.
144. K.R. Han, C.S. Lim, M.J. Hong, S.K. Choi, S.H. Kwon. Surface modification of silicon nitride powder with aluminum. J. Am. Cer. Soc. 79(2):574–576, 1996.
145. C.-M. Wang, F.L. Riley. Alumina-coating of silicon nitride powder. J. Eur. Cer. Soc. 10:83–93, 1992.
146. A.K. Garg, L.C. De Jonghe. Microencapsulation of silicon nitride particles with yttria and yttria-alumina precursors. J. Mater. Res. 5(1):136–142, 1990.
147. Y.S. Cho, V.R.W. Amarakoon. Nanoscale Coating of Silicon and Manganese on Ferrimagnetic Yttrium Iron Garnet. J. Am. Cer. Soc. 79(10):2755–2758, 1996.
148. S. Kratovil, E. Matijvi_. Preparation and properties of coated, uniform, inorganic colloidal particles: I, aluminum (hydrous) oxide on hematite, chromia, and titania. Adv. Cer. Mats. 2(4):798–803, 1987.
149. T.T. Borek, X. Qiu, L.M. Rayfuse, A.K. Datye, R.T. Paine, L.F. Allard. Boron Nitride Coatings on Oxide Substrates: Role of Surface Modifications. J. Am. Cer. Soc. 74(10):2587–2591, 1991.
150. T.D. Mitchell, Jr., L.C. De Jonghe. Processing and properties of particulate composites from coated particles. J. Am. Cer. Soc. 78(1):199–204, 1998.
151. H.M. Jang, J.H. Moon. Homogeneous fabrication and densification of zirconia-toughened alumina (ZTA) composite by the surface-induced coating. J. Mater. Res. 5(3):615–, 1990.
152. M.A. Harmer, H. Bergna, M. Saltzberg, Y.H. Hu. Preparation and Properties of Borosilicate-Coated Alumina Particles from Alkoxides. J. Am. Cer. Soc. 79(6):1546–1552, 1996.
153. Y.X. Huang, A.M.R. Senos, J.L. Baptista. Preparation of an Aluminum Titanate-25 vol% mullite composite by sintering of gel-coated powders. J. Eur, Ser. Soc. 17:1239–1292, 1999.
154. M. Bartsch, B. Saruhan, H. Schneider. Novel low-temperature processing route of

dense mullite ceramics by reaction sintering of amorphous SiO_2- coated γ-Al_2O_3 particle nanocomposites. J. Am. Cer. Soc. 82(6):1388–1392, 1999.
155. K. Ohtsuka, J. Koga, M. Suda, M. Ono. Fabrication of metal-layer (nickel) silicate microcomposite particles by a surface-nucleated precipitation route. J. Am. Cer. Soc. 72(10):1924-, 1989.
156. Y-J. Lin, B.-F. Jiang. Sintering and phase evolution of electroless-nickel-coated alumina powder. J. Am. Cer. Soc. 81(9):2481–2484, 1998.
157. D. Franquinm S. Monteverdi, S. Molina, M.M. Bettahar, Y. Fort. Colloidal nanometric particles of nickel deposited on γ-alumina: characteristics and catalytic properties. J. Mat. Sci. 34:4481–4488, 1999.
158. C.-C. Chen, S.-W. Chen. Nickel and copper deposition on Al_2O_3 and SiC particulates by using the chemical vapour deposition-fluidized bed reactor technique. J. Mats. Sci. 32:4429–4435, 1997.
159. H. Itoh, K. Hattori, S. Naka. Rotary powder bed chemical vapour deposition of titanium nitride on spherical iron powder. J. Mat. Sci. 24:1641–1646, 1989.
160. C. Li, J. Han, Z. Zhang, H. Gu. Preparation of TiO_2-coated Al_2O_3 particles by chemical vapor deposition in a rotary reactor. J. Am. Cer. Soc. 82(8):2044–2048, 1999.
161. H. Itoh, H. Sugimoto, H. Iwahara, J. Otsuka. Microstructure and properties of the sintered composite prepared by hot pressing of TiN-coated alumina powder. J. Mat. Sci. 28:6761–6766, 1993.
162. K. Tsugeki, T. Kato, Y. Koyanagi, K. Kusakabe, S. Morooka. Electroconductivity of sintered bodies of α-Al_2O_3-TiN composite prepared by CVD reaction in a fluidized bed. J. Mat. Sci. 28:3168–3172, 1993.
163. J.L. Kaae, S.A. Sterling, L. Yang. Improvement in the performance of nuclear fuel particles offered by silicon-alloyed carbon coatings. Nuc. Tech. 35:536–547, 1977.
164. E. Yasuda, J. Schlicting. Carbon fiber reinforced Al_2O_3 and mullite. Z. Werkstofftech, 9:3110–3115, 1978.
165. K.M. Prewo, J.J. Brennan. Fiber Reinforced Glasses and Glass Ceramics for High Performance Applications. In: S.M. Lee, ed. Reference Book for Composites Technology. Lancaster, PA: Technomic Pub. Co. Inc.,1989, pp. 97–116.
166. R.W. Rice, D. Lewis, III. Ceramic fiber composites based upon refractory polycrystalline ceramic matrices. In: S.M. Lee, ed. Reference Book for Composites Technology. Vol. 1. Lancaster, PA: Technomic Pub. Co. Inc., 1989, pp. 117–142.
167. R.W. Rice. BN Coating of Ceramic Fibers for Ceramic Fiber Composites. U.S. Patent 4,642,271, 1987.
168. D.B. Marshall, P.E.D. Morgan, R.M. Housley, J.T. Cheung. High-temperature stability of the Al_2O_3-$LaPO_4$. J. Am. Cer. Soc. 81(4):951–956, 1998.
169. T.A. Parthasarathy, E. Boakye, M.K. Cinibulk, M.D. Petry. Fabrication and testing of oxide/oxide microcomposites with monazite and hibonite as interlayers. J. Am. Cer. Soc 82(12):3575–3583, 1999.
170. H. Herman. Powders for thermal spray technology. KONA, Powder and Particles, No. 9, 187–99, 1991.

3

Use of Additives in Powder Preparation and Other Raw Material and Nondensification Uses

3.1 INTRODUCTION

Use of additives plays a large role in ceramic technology, with two of the largest uses being to aid densification, as addressed in Chapter 5, and in modifying properties. The latter is a very large field, with various classes of properties and mechanisms of controlling properties ranging from various microstructural control, second phase, crystal structure, or lattice effects, and combinations of these. Many of these require considerable expertise in specific properties to be properly addressed and hence are not generally addressed in this book (but are illustrated some, e.g., in Sec. 3.3 and Chap. 5). However, there are other applications of additives in fabricating and processing of ceramics that are addressed here. These include processing of ceramic powders, whiskers, and platelets, (i.e., of some raw materials), and of enhancing, retarding, or eliminating formation of some crystalline phases, i.e. of structural changes that occur and are significant in some important ceramics. Such transformation control impacts some aspects of fabrication as well as some important applications, including some structural ones and especially the very important field of catalysis. Other uses of additives covered include nucleation of crystallization of glasses, solidifying melts, and of seeding and control of grain structure in and following sintering, including in situ growth of single crystals. Finally, additives also play a role in flux growth of ceramic crystals, and for various other miscellaneous uses such as surface effects. All of these uses are covered in this chapter, in the order listed. Besides being of importance to the specific goal for the additive, their individual uses can have implications for other uses, including for densification.

3.2 USE OF ADDITIVES IN PREPARING CERAMIC POWDERS

Most ceramic powder preparation is done without additives. However, there are some cases where additives are used for preparing ceramic powders for subsequent densification into solid ceramic bodies. This is generally more extensive for nonoxide versus oxide powders, but additives are also used in preparing oxide powders. Another use of additives is in preparing powders for specific applications, generally not used in making densified ceramic bodies. Both are discussed here, with oxide powder preparation discussed first.

Most powders of binary oxides are produced by thermal decomposition, that is, calcination, of salt precursors such as hydroxides, carbonates, sulfates, and so forth, as discussed in Chapter 2. Such powder preparation generally does not use additives, especially those that are typically introduced as solid phases (which may have their effect in the solid, liquid, or gaseous state). However, it is important to first note that gases in the calcination atmosphere during part of the thermal cycle of preparing binary (and some other) oxides, can be extremely important in the resultant powder character. Both the character and amount of gases given off in the salt decomposition, commonly H_2O, CO_2, and SO_4, and their extent and time of retention in the powder mass being calcined, can play an important role in the resultant powder character (Sec. 8.2.1). However, other gas species, e.g. variable amounts and types of species adsorbed on the precursor powder surfaces, as well as purposely introduced gases, including very reactive ones such as Cl_2 can be very important. An extension, and in part an example, of this is work of Shimbo and coworkers [1] on additions (2 m/o) of AlF_3 and especially MgF_2, with or without added moisture, to hydroxide powders for obtaining MgO and effects on resultant crystal growth and surface area. They showed that while moisture enhanced crystallite growth with or without MgF_2, MgF_2 additions resulted in finer MgO crystallite size in calcining from 600 to nearly 900°C (e.g., 15–30 nm) and larger and rapidly accelerating crystallite sizes at higher temperatures. In a similar study, they showed that addition of AlF_3 (hydrolyzed by steam) delayed decomposition of brucite ($Mg(OH)_2$), while MgF_2, which was less hydrolyzable, had little effect on decomposition [2].

Much of the above and subsequent effects of additives is via their presence as a separate phase in the solid, liquid, vapor, or mixed phase at the surface of solid particles of the material whose behavior is to be modified. However, some effects may also entail the additive in solid solution in the starting or final phase to be affected. Both effects are illustrated in effects of additives on the decomposition of $CaCO_3$ [3–5].

An important application for coarser, uniform powders of Al_2O_3 (and BN, as well as AlN and possibly MgO, all with particle sizes of ~ 20–100 μm) is as fillers in organic matrices, e.g., rubber, especially silicone, to give good thermal conductivity of the resultant composites for use in electronic systems. Single-crystal particles that are polyhedral to ~ spherical in shape in order to make good thermal contact with each other in the matrix are particularly effective. Crushed fused Al_2O_3 has been used, but is generally more variable in size and angular in shape than desired. An example of such Al_2O_3 powders was discussed in the open literature [6]. While preparation of such powders is generally proprietary and not known in detail, one probable approach is partial sintering of powder compacts with additives that both promote grain growth (which, as noted in Chap. 5, is common) and that remain in sufficient quantities and character at grain boundaries (and preferably only there) to allow subsequent dissolution to yield the individual polyhedral grains, which then may be milled to round sharp edges and corners of the grains. Firing of compacts in active, e.g., Cl_2, atmospheres may also be effective if crushing or other reduction to individual grains can be achieved in high yield [7–10].

Preparation of oxide powders with different stoichiometry is a fairly common need for compounds of anions allowing varying stoichiometry, e.g., CeO_2 versus Ce_2O_3, TiO_2 versus Ti_2O_3, and FeO versus Fe_2O_3 or Fe_3O_4. While carbothermal reduction of oxides to metals is common, less extreme reduction is needed to obtain lower degrees of oxidation, which may often be obtained by controlling the degree of atmospheric reduction. However, some use of solid (or liquid) reducing agents is made. Thus, Hauf and coworkers [11] showed that a range of Ti-O compositions could be obtained from TiO_2 using Si powder, e.g., with added $CaCl_2$, a static versus flowing atmosphere (at 800–1000°C), and use of anatase or rutile as the TiO_2 powder. Jallouli and Ajersch [12] reported effects of Al_2O_3, CaO, MgO, and SiO_2 on the hematite to magnetite (Fe_2O_3-Fe_3O_4) transformation, especially on swelling and resultant cracking.

Another important method of powder preparation of increasing interest is hydrothermal preparation, which can also benefit from use of additives as discussed in Section 2.4. As an additional example, McGarvey and Owen [13] showed that such preparation of magnetite resulted in different crystallite morphologies, whether their preparation was without or with CuO additions.

In processing mixed-oxide compounds, additives may be used to accelerate reaction of particles and the formation of phases. Thus, for example, Huang and coworkers [14] reported that reaction of Al_2O_3 and MgO powder particles to form magnesia alumina spinel was accelerated by addition of (2 w/o) LiF or LiF and $CaCO_3$ and that the additives could increase the alumina content and aid sintering of the spinel. This work corroborates earlier work of Kosti and coworkers

[15], which showed that the presence of AlF_3 or CaF_2 additions (e.g., 1.5%) significantly aided reaction of Al_2O_3 with $MgCO_3$. Kosti and coworkers also showed that AlF_3 accelerated reactions between Al_2O_3 and Dy_2O_3, Y_2O_3, or SiO_2, as well as lowed the Al_2O_3 γ-α transition by 140°C. Note similar effects of LiF on formation of $ZnAl_2O_4$ in Section 5.4. Kosti and coworkers [16] also showed enhanced reaction of NiO and Fe_2O_3 to form $NiFe_2O_4$ at lower temperatures (e.g., 800–900°C), but a retardation at higher temperatures.

Additives are used more frequently for preparation of nonoxide powders of binary compounds than for binary oxides. While a large amount of nonoxide powder is produced by carbothermal (or other elemental reactions) or by CVD (Chap. 2), preparation with additives is also common. Thus, for example Mansour and Hanna [17] showed that addition of Fe (as $FeSO_4$) to rice hulls heated to 1500°C substantially aided formation of β-SiC, with optimum results with a Fe/SiO_2 ratio of 0.075. Small amounts of Si_3N_4 and Fe_3Si also formed, but Si_3N_4 was the dominant product in a NH_4 atmosphere at 1400°C. They concluded that the controlling step was solubility of Si in molten Fe. Similarly Krishnarao [18] showed that small addition of $CoCl_2$ to burnt rice hulls resulted in production of primarily β-SiC powder at 1600°C. He extended these results showing some β-SiC whisker formation occurred under some processing conditions [19]. McCluskey and Jaccodine [20] showed that the depth and nitrogen content of nitrided layers on Si wafers heated in a 30% NH_4 70% N_2 atmosphere at 1000–1200°C were greatly increased by the addition of 200 ppm NF_3, especially at the lower temperature.

Diamond, an elemental carbide, is industrially synthesized from graphite using transition metal catalysts, particularly Fe, Co, and Ni [21]. However, these have also been used in combination with one another or other metals, e.g., Ni with Fe, Mn, Cr, Ti, or Zr [22], as well as ternary alloys, for example, $Ni_{70}Mn_{25}Co_5$ [23], and some other elements have been used by themselves (e.g., P [21]. At least one nonmetal has been used, $MnCO_3$ (but MnO was apparently not effective) [24]. Pressures of 5–8 GPa and temperatures of 1500–2000°C, usually intermediate values, are common.

Carbothermic preparation of AlN powder is important, but so is CVD, and especially direct nitridation of Al powder. While, direct nitriding can be done without additives, this often requires higher oxygen content to keep the Al particles from coalescing [25]. Thus, Komeya and coworkers [26] reported that CaF_2 additions were an excellent promoter of Al nitridation. Li (e.g., 2.3 w/o) alloying of the Al is also reported to promote nitrdation and limit oxygen contamination [27]. LiF additions have been used, apparently aiding nitidation by attacking oxide coating on the Al particles. Regardless of how AlN powder is prepared, it is of value to stabilize its surface from reaction with water and oxygen; carboxylic acids as coatings on particles have been reported to be useful [28].

Turning to cubic boron nitride (c-BN), this is formed from hexagonal BN using catalysts at high temperatures and especially high pressures analogous to diamond production from graphite, with the catalysts being in each case similar to those for sintering polycrystalline compacts of the cubic phase. However, for both cubic powder preparation and sintering, BN processing typically uses ceramic catalysts versus typical transition metal catalysts for diamond. Mg_3N_2 is a common catalyst for forming (and sintering) BN, as discussed by Endo and colleagues [29–31] and Lorenz and colleagues [32,33]. Formation of $Mg_3B_2N_4$ occurs, which apparently plays an important role in eutectic formation and solution-precipitation of BN and also reacts with B_2O_3, typically present, to produce MgO, which forms as a layer around the c-BN particles, with both the oxygen (i.e., borate) content and resultant MgO impacting the process. Typical pressures and temperatures used were 5–6 GPa and 1300–1400°C, and decreasing oxygen content and adding Zr metal powder increased c-BN yields [29]. $Mg_3B_2N_4$ can be formed directly at normal pressures and then used for c-BN synthesis at ~ 4 GPa and temperatures to ~ 1550°C [34]. Other ternary nitrides such as Li_3BN_2 [31,35] and $Ca_3B_2N_4$ [35] have also been used and effects of B_2O_3 studied further [36], along with effects of ammonium borate [37], which lead to the discovery of using B with urea or ammonium nitrate [38]. Fluorides, especially NH_4F, have been used, the latter yielding very fine particle size (0.2–1 μm) due to the lower temperatures [39].

Other catalysts have been used to form c-BN, for example, Al metal [40]. More refractory ceramic catalysts such as MgB_2 [41] and AlN [42] have also been used. Besides the above static synthesis of c-BN, it has been shock synthesized from wurtzite BN, again with catalysts, with B enhancing conversion and TiB_2 retarding conversion [43].

Some study of additives on converting Si powder particles to Si_3N_4 powder has been made, but much of the effect of additives on nitriding Si is in Si compacts to form RSSN, so the reader is referred to effects of additions of metal such as Fe or their oxides in Section 5.5. Other studies include Jennings' study of effects of Fe on nitriding [44a], and of Cofer and Lewis on effects of Cr [44b], the latter showing extensive nitridation with 5 a/o Cr at 1150–1200°C versus 1300–1400°C for Fe additions. Bhatt and Palczer [45] reported that addition of ~ 0.5 w/o Fe or Ni (the latter as NiO) allows complete nitridation at 1250°C. Other studies show that small amounts Na [46,47], Al [47], Ca [47], or Ba [48] fluorides also can accelerate, at least earlier stages of nitriding Si, attributed mainly to disruption of oxide coatings on the Si. However, higher melting fluorides, such as CaF_2, appear better and some of them, e.g., CaF_2 increase α-Si_3N_4 content. In carbothermal reduction of SiO_2 and simultaneous nitridation, the fine nature of sol-derived SiO_2 can be an advantage, as can addition of chemical sources of C or N, or both, (e.g., dimethyl formamide or pan fibers) [49]. Pavarajarn and Kimura [50] have recently reported some more comprehensive evaluation of catalysts for nitriding Si.

3.3 ADDITIVE EFFECTS ON CRYSTALLOGRAPHIC-PHASE TRANSFORMATIONS

Additives play an important role in effecting phase transformations, as shown by effects in preparing diamond or cBN powders discussed above. Some effects of additives on phase transformations involving chemical changes, in the decomposition of hydroxides to oxides, were also noted in the previous section. This section focuses primarily on effects of additives on crystal structure transformations, i.e., from one crystal structure to another without any change in composition. Such transformations have important applications, mainly catalytic or structural integrity ones, the former with the material mainly in the form of lightly- or unsintered powder, the latter in a bulk polycrystalline solid. Catalytic applications, especially of alumina and titania materials, are commonly sought for their high surface area as powder (wash) coats on a ceramic (e.g., extruded, porous cordierite) support with limited or no powder sintering to support the actual catalyst; for example γ-Al_2O_3 to support noble metal catalyst for control of automotive exhaust emissions. Crystallographic transformations in a bulk solid are of interest to control strength reductions that often accompany such structural changes, and especially transformation toughening achieved in ZrO_2 with some metastable tetragonal content. These and other miscellaneous applications, briefly noted later, generally require either changing the temperature at which a transformation occurs, or eliminating any transformation of the body.

The issue of influencing crystallographic transformations arises in those materials which can exist in more than one crystal structure, especially those that do so as a function of temperature, as opposed to those that do so as a function of pressure, since the former are encountered more than the latter in normal practice. This section addresses additive effects on transformations, primarily in oxide materials, since a number of them have more than one potential crystal structure as a function of processing, with some having known additives effecting the transformations with important application consequences. Important transformations occurring in nonoxide materials, for example, hexagonal C (graphite) or BN to cubic C or BN (Sec. 3.2), are used to produce diamond or c-BN with additives generally not encountered in normal processing and use of hexagonal C or c-BN. The α-β transformations in SiC and Si_3N_4 occur to varying extents during processing, and may affect resultant microstructures and properties. The focus of this section is on those additive transformation effects in binary oxides having important applications—e.g., neglecting transformations such as those occurring in BeO at 2050–2150°C [51] since this typically has no impact on the processing or use of BeO because it occurs at high temperature. The reader is referred to other reviews of ceramic crystal transformations, such as those of Kriven and colleagues [52–54].

Consider Al_2O_3, which typically exists in the thermodynamically stable α-phase of rhombohedral (trigonal) structure, but can exist in about six other structures depending on processing route and history, as reviewed by Levin and Brandon [55]. Thus, the only precursor to directly yield α-Al_2O_3 is diaspore (α-AlOOH), which does so at the lowest temperatures of ~ 700–800°C. Other aluminum hydrates of different structure, composition, or both decompose first to other Al_2O_3 crystal structures, usually at lower temperatures (e.g., 150–700°C), then convert to α-Al_2O_3 at 900–1100°C, after existing in one or two other intermediate phases. Such initial forming of other crystal phases, with subsequent conversion at higher temperatures to α-Al_2O_3 occurs not only with other chemical precursors, but also with other processing methods, including anodic films, vapor deposited (e.g., CVD), as well as via melting. The latter is mainly in rapidly quenched, especially melt-sprayed material, and quenching also plays a role in some of the deposited non-α phases.

The existence of Al_2O_3 in the above phases has various consequences that are the driving forces behind the use of additives to impact transformation. Thus, conversion of non-α-Al_2O_3 phases in melt-sprayed or vapor-deposited bodies, such as coatings, to α-Al_2O_3 leads to increased porosity and possibly microcracking due to the α-phase having the highest density, hence the smallest volume, of the Al_2O_3 phases. However, the greatest interest in control of Al_2O_3 phase arises from its use for catalysis. It has long been known that γAl_2O_3, which forms at 300–500°C from boehmite (γ-AlOOH), is much more effective for catalysis than other phases, especially α-Al_2O_3. It was uncertain whether this was the result of intrinsic or extrinsic causes; the latter possibly stemming from the much lower temperatures for obtaining γ- versus α-Al_2O_3 and the resultant much greater surface area of γ- versus α-Al_2O_3, thus greatly favoring catalytic effects of γ- versus α-Al_2O_3. However, Tsuchida [56] reports that γ-Al_2O_3 is intrinsically better for catalysis based on tests of high surface area α-Al_2O_3 from diaspore. Thus, an important factor in catalysis is avoiding additives or impurities that enhance γ- to α-Al_2O_3 transformation, and finding and using additives that retard it.

Consider use of additives to effect the γ- to α-Al_2O_3 transformation. Both Xue and Chen [57] and Ozawa and coworkers [58] investigated such effects of mainly oxide additives introduced via solutions. Though their transition temperature for undoped Al_2O_3 differed by 64°C (due to different precursors and processing conditions) and their relative trends differ due to different levels of additives (respectively 1 and 10 m/o), they both show most additives lowering rather than raising the transition temperature, with greater lowering than increasing of the transition temperature (Fig. 3.1). Xue and Chen showed only B, Si, and Zr oxide additions increasing the transition temperature by ~ 40–75°C. They also showed that increasing the additive level to 5 m/o had negligible effect on changes due to B additions, but substantially further lowered transformation

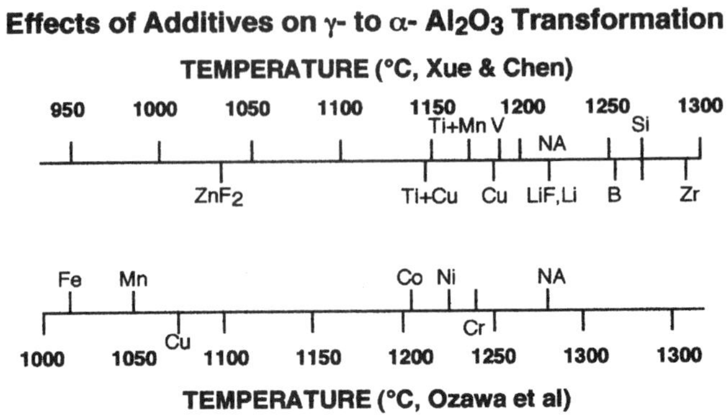

FIGURE 3.1 Effects of additives on the γ to α-transformation temperature of Al_2O_3 powders after Xue and Chen [57] and Ozawa and coworkers [58]. Note that: (1) since the transition temperatures with no additives (NA) differed between the two studies, the two temperature scales have been shifted so these two transition temperatures align vertically; (2) oxide additions are shown by the metal only (the two fluoride additives are fully designated), and (3) data of Xue and Chen is for 1m/o additions and that of Ozawa and coworkers is for 10 m/o additions, which is a factor in their larger transition temperature shifts.

temperatures with CuO/CuO_2 or ZnF_2 additions respectively by ~ 5 and 10%. Note that both investigations showed reductions of up to 150–260°C and that Ozawa and coworkers' results showed a trend for substantially lower surface areas (hence increased particle size, sintering, or both) with additives giving lower transition temperatures. Earlier work of Bye and Simpkin [59] showing increased α Al_2O_3 formation with Fe versus Cr additions is generally consistent with Ozawa and coworkers' results.

Besides effects of additive precursor and processing noted above, there are other effects of additives on the γ to α-Al_2O_3 transformation. Thus, Saito and coworkers' study [60] showed that the form of SiO_2 additions is important with crystalline additions, such as quartz or cristobalite additions enhancing transformation to α-Al_2O_3, while amorphous SiO_2 retards it, as also shown by Xue and Chen [57]. Similarly, Messing and colleagues [61–63] showed that while addition of Fe_2O_3 via solution significantly reduced the θ to α-Al_2O_3 transformation, so did seeding with fine α-Fe_2O_3 particles, which also enhanced microstructural development and aided densification at ~ 1200°C. The success with seeding was attributed to epitaxial growth of α-Al_2O_3 on the α-Fe_2O_3 seeds. Similarly, they showed that seeding γ-Al_2O_3 with α-Al_2O_3 significantly enhanced the γ- to α-Al_2O_3 transformation, which was also further aided by a wet versus a dry firing atmosphere, again showing the substantial effects that atmosphere can have on microstructural development. Consistent with this, Lopasso and coworkers [64] showed that firing in a Cl_2 atmosphere accelerates the conversion to α Al_2O_3 and cited earlier work on this as well as other work showing similar acceleration of transformations to rutile TiO_2 and monoclinic ZrO_2.

Transformations of TiO_2 powders can occur between its 3 forms: brookite (orthorhombic), anatase (tetragonal) or rutile (tetragonal), with the latter being the stable phase, and anatase or anatase/rutile mixtures being of particular interest for catalysis. Eppler, in an earlier brief review, noted that the anatase to rutile transformation was inhibited by WO_3, as well as by chloride, sulfate, fluoride, or phosphate ions and accelerated by alkalies and transition metal ions [65]. He also showed that Sb_2O_3 (a common impurity in TiO_2) accelerates this transformation. Similarly, Debnath and Chaudhuri [66] briefly reviewed additive effects and showed that $AlPO_4$ and (fumed) SiO_2 both inhibited transformation to rutile, which commonly occurs at 550–800°C. Al_2O_3 additions retard the transformation, e.g., to ~ 1060°C with Al/Ti ratios of 0.2, as shown by Yang and coworkers [67], as do rare earth oxide additions (e.g., Hishita and coworkers [68], but is lowered by Fe_2O_3 additions as shown by Gennari and Pasquevich [69]. Again, other factors such as precursors and processing history need to be considered, such as atmosphere effects of Cl_2 noted earlier and in conjunction with Fe_2O_3 additions [70]. Other factors affecting phase transitions are grain size [71], and mechanical treatment of the powder, typically by milling [72], which also can cause transformations in other materials (e.g., Fe_2O_3) [73].

The above use of additives is to impact, commonly delay, the crystallographic transformation of powders or very porous bodies to higher temperatures or longer times at temperature at or beyond the range where a high surface area can be retained. However, there are also important needs to suppress such transformations in materials where densification, use temperatures, or both, require heating a body of at least limited porosity to a reasonable fraction of its melting point. Usually the need and approach is to suppress transformation by forming solid solutions of the material to be stabilized and a stabilizing agent such that the resultant solution has a crystal structure, usually a natural high temperature structure of the material of interest, that is stable over the temperature range of interest. There are a number of such stabilizations, but only a few are summarized here. A more recent example of this is the stabilization of Dy_2O_3 in its high temperature monoclinic (not its higher temperature cubic) structure via ~ 8 m/o CaO by Kim and Kriven [74], but with some microstructural complications.

Another example is Bi_2O_3, which is of interest for neutron, fuel cell, and other electrical and optical applications, but transitions through four polymorphic structures before melting at ~ 825°C, which limits its utility without stabilization in a suitable high temperature structure [75]. Thus, Bi_2O_3 has been stabilized in a high temperature cubic fluoride structure with either 25 m/o Y_2O_3 or 15 m/o Nb_2O_5, which have differing effects on both the grain structure and electrical properties [76]. It has also been stabilized in the metastable high-temperature tetragonal form with 4–10 a/o Sb_2O_3, which again affects its electrical properties [77–79].

Consider what is probably the most investigated and used stabilization to prevent crystal-phase transformation, i.e., of ZrO_2 (and secondarily that of its closely related compound HfO_2), which have been the subject of a number of (mostly earlier) reviews [80–87]. The primary phases, other than possible metastable phases, are monoclinic, tetragonal, and cubic structures which exist respectively to ~ 1000, 2370, and 2700°C for ZrO_2 and ~ 1700, 2600, 2900°C for HfO_2, where the highest temperatures are for melting. The cubic phase in both cases is that of the CaF_2 structure. The transformation temperatures are affected by other factors, with composition (including oxygen stoichiometry) being the most important as discussed below, but also include factors such as microstructure and rate and especially direction of temperature change. The latter arises due to hysteresis in the transformations, especially that for the monoclinic to tetragonal transformation of ZrO_2 which is ~ 150–300°C, but is much less for HfO_2, i.e. ~ 20–30°C.

Though ZrO_2 has been studied much more, data for HfO_2, as well as it being isostructural with ZrO_2, indicates that the following phase stabilization for ZrO_2 applies to HfO_2. As with most solid-phase stabilization, this is accomplished by using additives in solid solution that yield the high-temperature structure across the temperature range of interest, that is, the fluorite structure. The primary stabilizers for this for ZrO_2 are either alkaline earth oxides, especially MgO or CaO, and rare earth and related oxides, especially Y_2O_3 or CeO_2, usually with more than a few m/o for complete stabilization of the cubic structure. There is also considerable use of combinations of two or more stabilizers, especially various combinations of rare earth oxides. These combinations are commonly either naturally occurring mixtures or such mixtures after removal of selected, more expensive individual rare earth oxides, thus lowering stabilizer costs, which are typically a measurable cost factor in the bodies. The first of three things to note about stabilization of ZrO_2 is that much investigation and considerable use has been made of partial stabilization to retain some metastable tetragonal ZrO_2 and the resultant transformation toughening [82]. Second, oxygen deficiency can also stabilize ZrO_2 [86,87]. Thus, processing, such as heat treating to achieve complete solid solution of the stabilizer under high-temperature reducing conditions aids stabilization, which can then be lost, usually via subsequent precipitation under more oxidizing conditions. Such stabilization effects of reducing atmospheres have been neglected in some studies of partially stabilized ZrO_2. Third, the diversity of stabilization additives for ZrO_2 and their combinations allow opportunities to balance electrical property effects where these are of importance in ZrO_2 use, for example, as sensors or fuel cell components.

Finally, while the above examples have been for binary oxides, additives are also used for stabilizing desired crystal structures for ternary and more complex oxides, as well as for binary and more complex nonoxides, though there is less data on both. In particular, note that ternary and more complex oxide phases have a number of important electrical and other applications which depend critically on the crystal phase in which they exist and the temperature, compositional, etc., ranges over which the desired phase(s) exist. An example of this is the well-known solid solution of lead titanate and lead zirconate (PZT) for sonar transducers.

3.4 USE OF ADDITIVES IN THE GROWTH OF CERAMIC AND RELATED WHISKERS AND PLATELETS

Though generally constrained in their use by health issues, ceramic whiskers are of interest for use in ceramic composites, which have found some industrial use as speciality wear parts and especially in cutting tools, and are also of interest for some metal and possibly other composites. Whiskers, which are small filamentary single crystals, are also of importance as a tool in understanding both the crystal physics of their properties and the crystal chemistry of their growth. Platelets, which are an extension of whisker growth processes since they basically reflect a change from filamentary to plate growth of small single crystals, are also similarly of interest and have also been investigated for use in ceramic and other composites.

There are various growth mechanisms for whiskers, all of which can be affected some by additives or impurities, but the vapor-liquid-solid (VLS) mechanism of whisker growth is a major method of producing excellent whiskers. VLS whisker growth is inherently dependent on use of additives, which are the source of the liquid phase, as well as possible modifications of it and is thus the focus of this section. In the VLS growth mechanism, gaseous sources of the whisker constituents are dissolved in liquid droplets of the "additive phase," which at the start of whisker growth are on a substrate supporting them and the subsequent whiskers. Growth proceeds with the whisker material precipitating from the liquid droplet such that the whisker is attached to the substrate and the liquid droplet is carried along on the tip of the whisker for further growth. Thus, a general characteristic of VLS whisker growth is the solidified droplet ball on top of the whisker (see Fig. 3.2), unless the droplet is lost, due to subsequent evaporation or by being broken off.

While whisker growth has been extensively studied over a number of years, with much attention to VLS growth, much less is published on the specifics of platelet growth, though some specific aspects of their growth is noted from the literature. Also note that the morphology of single-crystal particles growing in a liquid is often significantly impacted by minor amounts of materials in the the growth liquid, so growth of platelets is often a modification of

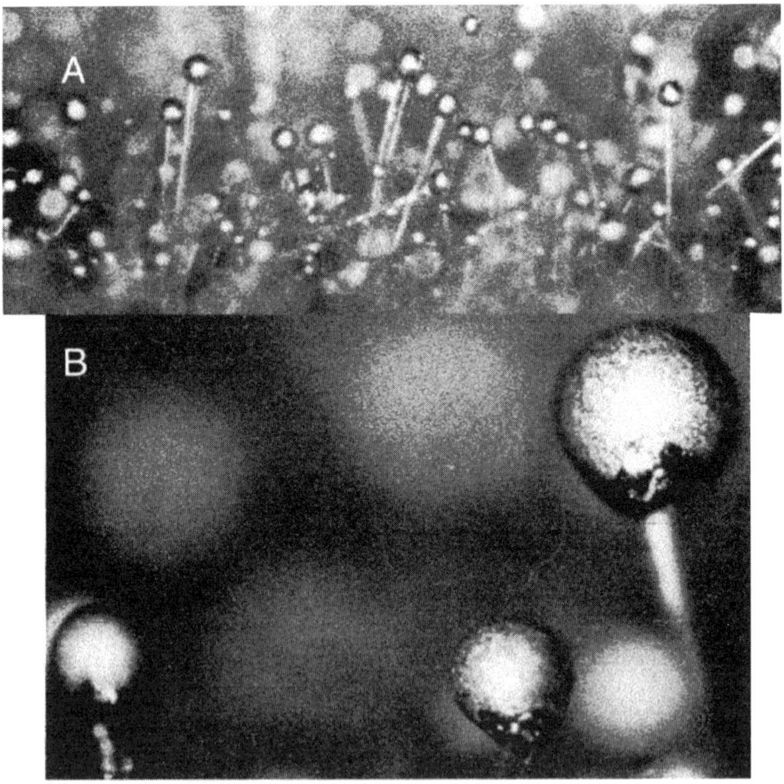

FIGURE 3.2 Examples of solidified balls of liquid Al or Al-Si on the tips of sapphire whiskers. (A) Lower magnification showing a ball for every whisker; (B) higher magnification showing more detail. Note the darker material on the bottom of the balls (identified as precipitated Si) near their intersection with the whisker. (From I. Ahmed [88].)

whisker growth. Thus, though the focus of this section is on refractory materials, especially ceramics, there is substantial information on the morphological, i.e., single-crystal shape aspects, of growth of less refractory whiskers, platelets and other particles with varying crystalline morphologies. Thus, for example Genk [89] has reviewed the morphological aspects of growth of many inorganic materials such as salts from water or other solutions, noting the pronounced effects that additives can have on growth habits.

While the liquid phase for VLS whisker growth varies some with the source of the vapor phase for a given material of whisker growth, there is a substantial commonality of liquid phases used, e.g., as shown in Givargizov's extensive review [90]. Thus, noble metals such as Au, Ag, Pt, and Pd, as well as metals such as Cu and Ni are common liquid phases for growth of elemental whiskers of Si, Ge, and B, while metals such as Ni, Fe, and Mn are cited for growth of C whiskers (including diamond, for which these metals are also used for making diamond powder and compacts, Sec. 3.2). Metal whiskers are also often grown with metal liquid droplets.

VLS growth of whiskers of binary compounds of borides (e.g., of Nb, Ti, or Zr), carbides (e.g., of B or Si), nitrides (e.g., of Al, Si, Ti, or Zr), and phosphides (e.g., of B, Nb, Ti, or Zr) also commonly use liquids of the same or related metals, especially noble or transition metals as noted for growth of elemental whiskers noted above. Some other metals such as Al, Si, Cr, Fe, or Ta were used for SiC whisker growth. Such usage of liquid metals has continued, e.g., the excellent SiC whiskers grown by Milewski et al. [91] were grown using particles of 304 stainless steel as the source of the liquid droplets (with the size ~ 30 μm and spatial dispersion of the particles an important factor in the size and density of whisker growth). Subsequently, boric acid, e.g., ~ 5 w/o, has been found to influence SiC whisker growth, but it is uncertain whether the growth is still VLS growth and what is the resultant chemical form of the boric acid addition [92,93]. On the other hand, use of SiO_2-CH_4-Na_3AlF_3 as raw materials is reported to give SiC whiskers via VLS growth from Al/Si droplets formed on the graphite boats used [94]. Also replacing CH_4 with N_2 and addition of some Fe_2O_3 results in VLS growth of Si_3N_4 whiskers [95]. AlN whiskers have also been grown by the VLS method using carbothermal reduction of Al_2O_3 in a N_2 atmosphere using 2 w/o addition of CaF_2 and B_2O_3 for the liquid phase [96], but the nature of the resultant composition of the additive phase during actual whisker growth appears uncertain.

3.5 USE OF ADDITIVES IN OTHER CERAMIC PROCESSING, ESPECIALLY MELT PROCESSING

While the above VLS growth of whiskers involves a liquid droplet of an additive out of which a whisker grows, there is also some use of additives in melt processing of ceramics. Consider first flux growth of single crystals, wherein the material for the desired crystals is dissolved in a suitable flux and the crystals grown essentially by precipitation from the flux-based solution using various crystal seeding and growth mechanisms, including traveling solvent zones [97]. Flux crystal growth has been used most extensively for oxides where commonly no atmosphere protection is required. It is particularly advantageous for growing refractory crystals of materials that do not melt congruently, undergo little or no melting due to high vapor pressure, or have possibly destructive high temperature phase transformations (e.g., BeO). These uses balance out the limitations of slower growth rate and smaller faceted crystals (but whose growth habit can be changed by changes in the flux, e.g., Fig. 3.3). As discussed in the substantial review by Tolksdorf [97], fluxes for oxide crystal growth commonly consist of two or three of basic oxides and fluorides: PbO, PbF_2, BaO, BaF, Bi_2O_3, Li_2O, Na_2O, K_2O, KF; and acidic oxides: B_2O_3, SiO_2,P_2O_5, V_2O_5, MoO_3. Again, as with hydrothermal (as well as some vapor phase) growth of crystallites, modest changes or additives to the flux can readily change the growth habit of the crystals.

The above solvent method of crystal growth can be extended to a variety of ceramics via electrochemical deposition of crystals (or coatings) on an electrode in a conductive mixture of compounds containing the atomic constituents of the ceramic sought, as reviewed by Elwell [99]. Thus, for example solutions of borates in alkali have been used with a source of the cation to produce many (more than three dozen) refractory borides, and of phosphates to produce a similar number of phosphides, e.g., $NaPO_3$ and NaCl or NaF with WO_3 to produce W_2P or W_4P. Similarly some sulfides, e.g., of W and Mo, have been produced from the oxide dissolved in Na_2SO_4, $Na_2B_4O_7$, and NaF, and some carbides and silicides have also been formed electrochemically, though formation of carbides (and of nitrides) is hindered by the ease of decomposing carbonates (and nitrates). A number of oxides can also be produced by such molten salt electrolysis, especially tungsten bronzes, as well as some spinels.

Consider now the use of additives to refine the microstructures in fusion casting of oxide ceramics, for example, in the production of refractories, where such fusion casting produces some of the largest ceramic components. Thus, as reviewed by McNally and Beall [100], ZrO_2-Al_2O_3-SiO_2 refractories are arc-skull melted and cast for construction of glass-melting tanks. For higher temperature melting (e.g., 1550–1650°C), high ZrO_2 contents (68–82.5 w/o) are used with 10–20 w/o (preferably > 15 w/o) SiO_2, 0.5–2.5 w/o Na_2O, ≤ 1 w/o (preferable < 0.4 w/o) Fe_2O_3 + TiO_2, and Al_2O_3 so the Al_2O_3/SiO_2 ratio is 0.3–0.65, with the Na_2O content and the Al_2O_3/SiO_2 ratio being particularly important. Fusion casting of pure Al_2O_3 results in weak bodies due to growth of large columnar grains, but addition of 1.2–1.8 m/o CaO breaks up the coarse grains and improves room temperature strengths. The addition of ~ 0.9 m/o of metal fluoride aids manufacturing and further improves strengths. They noted that additions of a few percent of Li_2O to MgO-chrome fusion cast refractory compositions give bodies with more favorable phase composition as well as moisture stability. They also noted some effects of composition and additives in controlling microstructures of fusion cast compositions in the Ti-B-C and ZrC-based systems.

The more extensive field of using nucleating agents to obtain volume crystallization of glasses is widley used to produce such oxide glasses, commonly referred to as glass-ceramics. Thakur [101] lists a number of such additives grouped as (1) metals and compounds inducing phase separation (Pt, Ag, Au, Cu, and sulfides, fluorides, etc.); (2) oxides such as TiO_2, SnO_2, MoO_3, WO_3; (3) ox-

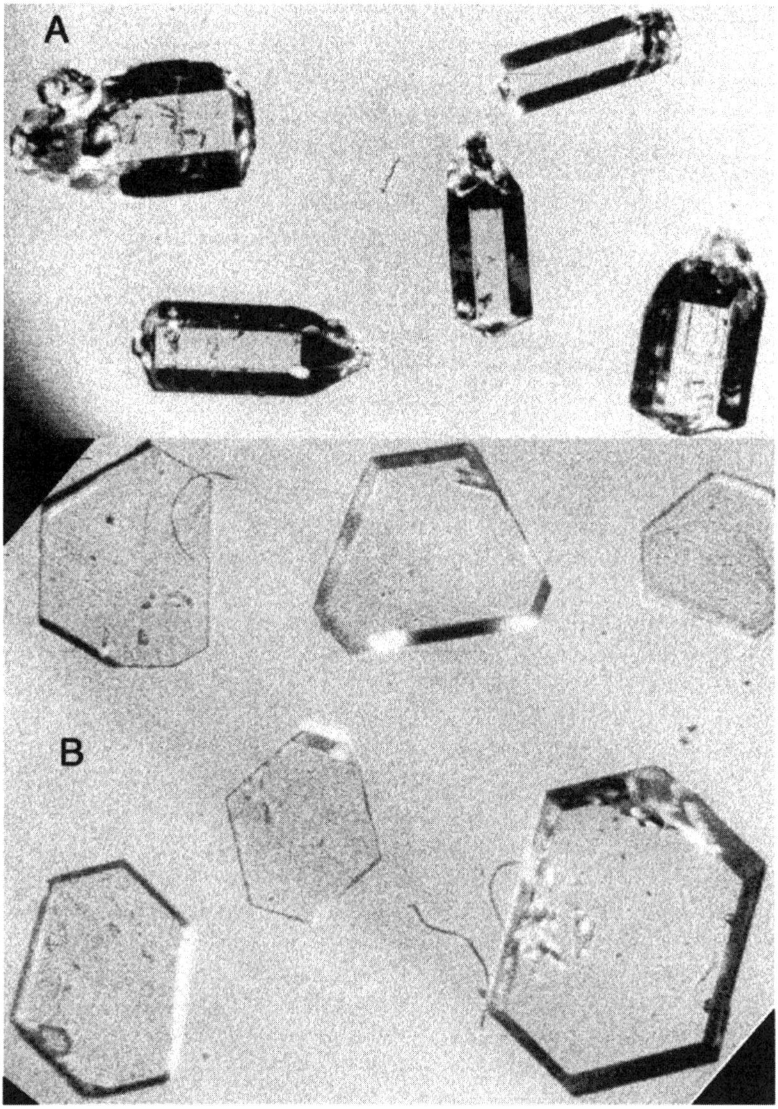

FIGURE 3.3 BeO crystals grown from fluxes based on Li_2MoO_4 with additions of MoO_3, after Newkirk and Smith [98]. Maximum crystal dimensions are of the order of 0.6 cm. Note the different growth habits due to changes in flux composition and temperature and that small (e.g., 0.5 w/o) additions of flux additives such as PbO, SnO_2, MnO_2, Al_2O_3, CaO, MgO, TiO_2, ZrO_2, LiF, and Li_2SO_4 can improve crystal quality with limited effects on growth rates and habit.

ides which can exist in two valence states in glass melts (V_2O_5, Fe_2O_3, Cr_2O_3); and (4) oxides with high coordination for the cations (ZrO_2,ThO_2, Ta_2O_5). Gutzow and Toschev [102] studied noble metal nucleation of crystallization of model Na phosphate and borate glasses, while Hammel [103], Tomozawa [104], Nielson [105], and Stewart [106] reported on the use of one or more of the common oxide nucleating agents of P_2O_5, TiO_2, or ZrO_2. McNally and Beall [100] also discussed nucleation of SiO_2-Al_2O_3-Li_2O glasses with 4 m/o TiO_2, noting that the glass phase separates on quenching, and that on subsequent heating to ~ 825°C aluminum titanate, crystallites form to start crystallization, with subsequent typical complex crystallization as a function of thermal exposure. They note even greater crystallization complexity in SiO_2-Al_2O_3-MgO glasses with ~ 10 w/o of TiO_2, ZrO_2, or combinations of them, and use of Cr_2O_3 as a nucleating agent in commercial production of fused basalt products.

As indicated above, some additives affect phase separation in glasses. Other studies also show this; for example, Markis and coworkers [107] reported that fluoride additions via NaF to Na_2O-SiO_2 glasses raised the temperature at which phase separation commenced, e.g., by ~ 10% at 1.2 m/o addition, but with no change in the scale of the phase separated microstructure, in marked contrast to large increases in water containing glasses. Similarly, additives can alter phase seperation in crystalline materials; Takahashi and coworkers [108] tested effects of 14 oxide additions to the SnO_2-TiO_2 system, noting substantial effects of ZrO_2 additions, more with Ta_2O_5, and especially, WO_3 or Sb_2O3.

Consider now use of additives in a specialized reaction process that is often referred to as the Lanxide™ process, which was originally and most extensively developed for making Al_2O_3-based bodies via controlled oxidative growth from molten Al held in an open refractory container compatible with the process. While this process may be used to form large monolithic bodies of Al_2O_3 with limited residual Al, it is more commonly used to form composite bodies, again of potential large size, by growth of such a matrix through a preform of particulate (e.g., of larger grains of Al_2O_3 or SiC) or fiber (e.g., SiC) reinforcement. The extent of such oxidative growth is sensitive to limited quantities of additives, especially Mg as well as a second addition of Si, Ge, Sn, Pb [109–111], or Zn [112] Such additions are typically added as alloying agents in the Al used, typically at 2–10 w/o (usually 2–5 w/o), which greatly increase the extent of oxidative growth (e.g., 2–4 cm of thickness per day), particularly in the oxidation temperature range 1150–1300°C. Apparently some additives or impurities inhibit the process, and can be used to define at least the approximate shape of the product body by placement in the particulate or fiber preform, in addition to some product body shaping via shaping of the preform.

The above method of oxidative growth from molten metal in a refractory container open to the furnace environment has been extended to composites of a nitride ceramic: AlN, TiN, or ZrN with some residue of the respective metal of

Al, Ti, or Zr [113]. Such growth can again occur through an open preform, e.g., of compatible grains of other compatible refractory ceramic phases to make various composites. Use of additives in the growth of such nitride-based bodies has not been discussed in the literature, but growth of Al-based materials has been reported using commercial Al alloys such as A380.1 (8.5% Si, 3.5% Cu, 3% Zn, 1% Fe, 0.5% Mn, and 0.1% Mg), that is, with constituents similar to important additives for growing Al_2O_3-based bodies [114].

Consider next use of additives in the in situ growth of macroscopic single crystals within a dense polycrystalline body. Researchers at General Electric in the last few years accidently discovered that part or all of Lucalox Al_2O_3 bodies or portions of them could be converted to sapphire single crystals by first firing in H_2 at 1850°C to boil out the MgO, reducing the level of Mg^{++} to at least or below the solubility limit of 30–50 ppm (C. Scott, General Electric, Cleveland OH, personal communication, 1999, [115]). Then firing at 1880–1900°C will convert the area of sufficiently reduced MgO content of such Al_2O_3 bodies to sapphire in situ in the body, essentially via greatly exaggerated grain growth, provided the body is suitably free of porosity, microcracks (i.e., has a grain size below that for microcracking or has not been cooled between the two firings), and other second phases impeding grain boundary migration (e.g., La_2O_3 or other rare earths present as second phases, used instead of MgO to obtain full density). The second heating to bring about the polycrystalline to single-crystal conversion, which occurs in minutes, even over substantial lengths, for example, in thin wall tubes > 60 cm long, can be via laser heating of one end. This conversion, besides being impeded by some additives or impurities, as noted above, can also be accelerated by some other materials in solid solution, e.g., 50–100 ppm of Ga_2O_3 (discovered since alumina powders used having this as an impurity showed easier conversion), Cr_2O_3 (i.e., resulting in ruby), and to some extent TiO_2. Thus, for example, rods on which a spiral pattern of Ga_2O_3 powder was painted, than appropriately fired resulted in a corresponding spiral of conversion to sapphire. While limited by the depths from which MgO can be sufficiently diffused out of the body, such growth of single-crystal parts from a polycrystalline body has potential for producing at least thin, shaped single-crystal parts. However, such growth appears limited to lower quality crystals due to residual porosity and second phases incorporated from the polycrystalline material, which may often more seriously limit optical performance. Thus, use of the in situ crystal growth generally presents trade-offs between lower quality and costs versus higher quality and costs for comparable parts machined from conventionally grown bulk crystals.

Limited work has also been conducted to grow single crystals of other materials via grain growth within polycrystalline bodies. Earlier work investigating effects of limited excesses of TiO_2 added to $BaTiO_3$, such as the work of Hennings and coworkers [116], showed exaggerated growth of isolated grains—e.g., ~ 50 μm versus a matrix grain size of a few microns, attributed to liquid-phase effects on firing above 1312°C is in part a precursor to such poly- to single-crystal conversion. More recently Yoo and coworkers [117] reported that adding a small amount of a SiO_2 slurry on top of a green $BaTiO_3$ compact subsequently sintered at 1370°C for up to 80 hrs resulted in in situ growth of grains to sizes of a few centimeters. SiO_2 was seen as a catalyst in forming twinned seed grains that were seen as central to the subsequent grain growth. Combination of this SiO_2 additive approach with other additives such as LiF that aid densification and enhance grain growth (Sec. 5.4) may be worthy of exploration. Recently Kahn and coworkers and others [118] have reported growth of various lead titanate or niobate-based crystals for electrical functions via seeded polycrystal conversion. Thus, conversion of polycrystalline areas or bodies to single crystal areas or parts shows some promise, but is limited by diffusion requirements and distances and by incorporation of microstructural variations such as residual grain boundary pores or second phases in the polycrystalline precursor into the resultant single area.

Another use of additives, primarily SiO_2, is to increase the surface smoothness of bodies with as fired surfaces, which often also improves the as-fired strengths. Yamada and coworkers [119] reported that mixing additions of up to 0.1 w/o of fine, high purity SiO_2 powder with a high purity Bayer Al_2O_3 with 0.2 w/o of MgO and 0.03 w/o of Cr_2O_3 increased surface roughness (as measured by surface gloss) as the SiO_2 content and firing temperatures (1500–1650°C in H_2) increased despite some reduction in grain size, such as from 1.6 to 1 μm without and with 0.1 w/o SiO_2, respectively. However, forming a SiO_2 coating on dense Al_2O_3 fibers commonly increases strengths, e.g., by up to a maximum of ~ $^1/_3$ at ~10% addition, despite some elastic moduli decrease as SiO_2 content increases [120–122]. Reduced grain sizes of the Al_2O_3 phase, regardless of its crystal structure, with increased SiO_2 is clearly an important factor in the increased strength despite the decrease in elastic moduli. However, increased surface smoothness with increased SiO_2 addition is a factor since improved strengths were obtained in FP Al_2O_3 fibers in manufacturing them with an added surface SiO_2 coating rather than SiO_2 additions to the bulk of the fiber [120,121]. The differences between these fiber observations and those of Yamada and coworkers for bulk Al_2O_3 bodies must reflect differences primarily in processing and possibly some in composition (though both Yamada and coworkers' bulk bodies and FP fibers contained MgO as an additive). (See also use of glass coatings on sapphire windows to eliminate the need for polishing the windows, as discussed in Sec. 8.3.1.)

3.6 DISCUSSION, SUMMARY, AND CONCLUSIONS

This chapter, which is mainly on use of additives in preparation of powders and other raw materials, illustrates the diversity of ceramic fabrication and related processing steps by both the diversity of approaches, additives, materials, uses and mechanisms noted or outlined. Thus, for example, additives are critical in producing materials such as C and BN in their high-pressure cubic diamond phase which, while not the normal stable phase, will be stable at atmospheric pressure and to do so while also providing the powder character needed for subsequent densification. It should also be noted that a variety of factors can be interactive with the use of additives, such as temperature, pressure, atmosphere, and particle and grain size, as well as impurities. In this spirit, note the overlap or complementary character of treatment of topics in this chapter with that of Chapters 4–8, especially Chapter 5. Thus, the use of silica coatings to give smoother surfaces on alumina fibers was noted in this chapter, while other coatings of mostly other fibers with other materials and methods, for example, with BN by CVD, are discussed in various subsequent chapters. Again note that besides the extensive use of additives to aid densification (Chap. 5), they are extensively used to modify properties (addressed some in Sec. 3.3 and illustrated some in other chapters, especially Chap. 5, such as making black alumina microelectronic packages, Sec. 5.3).

REFERENCES

1. M. Shimbo, Z.-e Nakagawa, Y. Ohya, K. Hamano. Effect of fluoride addition on crystal growth of MgO. J. Cer. Soc. Jap. Intl. Ed. 97:844–850, 1989.
2. M. Shimbo, Z.-e Nakagawa, Y. Ohya, K. Hamano. Effect of HF addition on decomposition of $MgO(OH)_2$. J. Cer. Soc. Jap. Intl. Ed. 97:629–633, 1989.
3. G.V.S. Rao, M. Natarajan, C.N.R. Rao. Effect of impurities on the phase transformations and decomposition of $CaCO_3$. J. Am. Cer. Soc. 51(3):179–180, 1968.
4. G.F. Knutsen, A.W. Searcy, D. Beruto. Effect of LiCl on the rate of calcite decomposition. J. Am. Cer. Soc. 65(4):219–222, 1982.
5. D. Beruto, G. Belleri, L. Barco, V. Longo. Interactions of LiBr with calcite and calcium oxide powders. Cer. Intl. 9(2):53–, 1983.
6. A spherical alumina filler. Material Innovation. Am. Cer. Soc. Bul. 70(6):963–964, 1991.
7. M.J. Readey, D.W. Readey. Sintering of ZrO_2 in HCl atmospheres. J. Am, Cer. Soc. 69(7):580–582, 1986.
8. M.J. Readey, D.W. Readey. Sintering of TiO_2 in HCl atmospheres. J. Am, Cer. Soc. 70(12):C–358–361, 1987.
9. D.W. Readey, D.J. Aldrich, M.A. Ritland. Vapor transport and sintering. In: R.A. German, G.L. Messing, R.G. Cornwall, eds. Sintering Technology. New York: Marcel Dekker, Inc., 1996, pp. 53–60.
10. D.J. Aldrich, D.W. Readey. Microstructures in multicomponent oxides by vapor phase sintering. In: T.S. Srivatsanm, J.J. More, eds. Processing and Fabrication of Advanced Materials V. TMS,1996, 545–555.
11. C. Hauf, R. Kniep, G. Pfaff. Preparation of various titanium suboxide powders by reduction of TiO_2 with silicon. J. Mat. Sci. 34:1287–1292, 1999.

12. M. Jallouli, F. Ajersch. Analytical model for the swelling of sintered iron oxide pellets during the haematite-magnetite transformation. J. Mat. Sci. 21:3528–3538, 1986.
13. G.B. McGarvey, D.G. Owen. Copper (II) oxide as a morphology directing agent in the hydrothermal crystallization of magnetite. J. Mat. Sci. 31:49–53, 1996.
14. J.-L. Huang, si-Y. Sun, Y.-C. Ko. Investigation of high-alumina spinel: effect of LiF and $CaCO_3$ addition. J. Am. Cer. Soc. 80(12):3237–3241, 1997.
15. E. Kostić, S. Bošovi, Š. Kiš. Influence of fluoride ion on the spinel synthesis. J. Mat. Sci. Let. 1:507–510, 1982.
16. S. Bosovi, E. Kosti, S. J. Kiss. Synthesis of $NiFe_2O_4$ in the presence of fluorine. J. Mat. Sci. Let. 4:200–202, 1985.
17. N.A.L. Mansour, S.B. Hanna. Silicon carbide and nitride from rice hulls II. Effect of iron on the formation of silicon carbide. Trans. & J. Brit. Cer. Soc. 78(6):132–136, 1979.
18. R.V. Krishnarao. Effect of cobalt chloride treatment on the formation of SiC from burnt rice husks. J. Eur. Cer. Soc. 12:395–401, 1993.
19. R.V. Krishnarao. Effect of cobalt catalyst on the formation of SiC from rice husk silica-carbon mixture. J. Mat. Sci. 3645–3651, 1995.
20. F.P. McCluskey, R.J. Jaccodine. Effect of NF_3 on the direct thermal nitridation of silicon. J. Electrochem. Soc. 136(8):2328–2331, 1989.
21. M. Akaishi, H. Kanada, S. Yamaoka. Phosphorous: an elemental catalyst for dimond synthesis and growth. Science 259:1592–1593, 1993.
22. H. Kanada, O. Fukunaga. Growth of large diamond crystals. In: S. Akimoto, M. H. Manghnani, eds. Advances in Earth and Planetary Sciences Center for Academic Publications, Tokyo, Japan, 1982.
23. L. Gou, S. Hong, Q. Gou. Investigation of the process of diamond formation from SiC under high pressure and high temperature. J. Mat. Sci. 30:5687–5690, 1995.
24. V. Srikanth, M. Akasishi, S. Yamaoka, H. Yamada, T. Taniguchi. Diamond synthesis from graphite in the presence of $MnCO_3$. J. Am. Cer. Soc. 80(3):786–790, 1997.
25. S. Ito, I. Ebato, H. Fukui, N. Koura, N. Yoneda. Perparation of aluminum nitride using slightly oxidized Al powders. J. Cer. Soc. Jap. 100:622, 1992.
26. K. Komeya, E. Mitsuhashi, T. Mrguro. Synthesis of AlN powder by carbothermal reduction-nitridation method-effect of additives on reaction rate. J. Cer. Soc. Jap. 101:366, 1993.
27. H. Scholz, P. Greil. Synthesis of high purity AlN by nitridation of Li-doped Almelt. J. Eur. Cer. Soc. 6:237–642, 1990.
28. M. Egashira, Y. Shimizu, Y. Tako, R. Yamaguchi, Y. Isikawa. Effect of carboxylic acid adsorption on the hydrolysis and Ssintered properties of aluminum nitride powder. J. Am. Cer. Soc. 77(7):1793–1798, 1994.
29. T. Endo, O. Fukunaga, M. Iwata. Growth pressure-temperature region of cubic BN in the system BN-Mg. J. Mat. Sci. 14:1375–1380, 1979.
30. T. Sato, H. Hiraoka, T. Endo, O. Fukunaga, M. Iwata. Effect of oxygen on the growth of cubic boron nitride using Mg_3N_2 as catalyst, J. Mat. Sci. 16:1829–1834, 1981.
31. T. Endo, O. Fukunaga, M. Iwata. Precipitation mechanism of BN in the ternary system B-Mg-N. J. Mat. Sci. 14:1676–1680, 1979.

32. H. Lorenz, U. Kühne, C. Hohlfeld, K. Flegel. Influence of MgO on the growth of cubic boron nitride using the catalyst Mg_3N_2. J. Mat. Sci. Let. 7:23–24, 1988.
33. H. Lorenz, B. Lorenz, U. Kühne, C. Hohlfeld. The kinetics of cubic boron nitride in the system BN- Mg_3N_2. J. Mat. Sci. Let. 7:23–24, 1988.
34. S. Nakano, H. Ikawa, O. Fukunaga. Synthesis of cubic boron nitride by decomposition of magnesium boron nitride. J. Am. Cer. Soc. 75(1):240–243, 1992.
35. T. Endo, O. Fukunaga, M. Iwata. The synthesis of cBN using $Ca_3B_2N_4$. J. Mat. Sci. 16:2227–2232, 1981.
36. J.-Y. Choi, S.-J. L. Kang, O. Fukunaga, J.-Ku Park, K.Y. Eun. Effect of B_2O_3 and hBN crystallinity on cBN synthesis. J. Am. Cer. Soc. 76(10):2525–2528, 1993.
37. T. Kobayashi, K. Susa, S. Taniguchi. Pressure and temperature stability region of cubic BN in the presence of ammonium borate. Mat. Res. Bul. 12:847–852, 1977.
38. T. Kobayashi, K. Susa, S. Taniguchi. New catalyst for the high pressure synthesis of cubic BN. Mat. Res. Bul. 10:1231–1236, 1975.
39. T. Kobayashi. Solvent effects of fluorides in cubic BN high pressure synthesis. Mat. Res. Bul. 14:1541–1552, 1979.
40. N.M. Olekhnovich, O.I. Pashkovskii, I.M. Starchenko, V.T. Sharai, V. B. Shipilo. Mechanism of the phase transition in BN during its reaction with Al at high pressures and temperatures. Iorg. Mats. 16:1209–1212, 1980
41. T.I. Serebryakova, V.A. Ponomarenko, A.I. Karasev, V.I. Shemanin, E.V. Marek. Magnesium diboride and its application in synthesis reactions of cubic boron nitride. Sov. Pwd. Met. & Met. Cer. 19:791–792, 1981.
42. S.-I. Hirano, T. Yamaguchi, S. Naka. Effects of AlN additions and atmosphere on the synthesis of cubic boron nitride. J. Am. Cer. Soc. 64(12):734–736, 1981.
43. T. Akashi, A. Sawaoka, S. Saito. Effect of TiB_2 and boron additions on the stability of wurtzite-type boron nitride at high temperatures and pressures. J. Am. Cer. Soc. 61(5–6):245–246, 1978.

44a. H.M. Jennings. On the influence of iron on the reaction between silicon and nitrogen. J. Mater. Sci. 3(5):907–909, 1988.

44b. C.G. Cofer, J.A. Lewis. Chromium catalysed silicon nitridation. J. Mat. Sci. 29:5880–5886, 1994.

45. R.T. Bhatt, A.R. Palczer. Effects of surface area, polymer char, oxidation, and NiO additive on nitridation kinetics of silicon powder compacts. J. Mat. Sci. 34:1483–1492, 1999.
46. T. Hayashi, H. Ushida, H. Saito, S. Hirano. Effects of NaF and NH_3 on preparation of Si_3N_4 powders from SiO_2. J. Cer. Soc. Jap. 95(2):278–, 1987.
47. D. Campos-Loriz, S.P. Howlett, F.L. Riley, F. Yusaf. Fluoride accelerated nitridation of silicon. J. Mat. Sci. 14:2325–2334, 1979.
48. S.K. Biswas, J. Mukerji. Effect of BaF_2 on the nitridation of commercial silicon. J. Am. Cer. Soc. 63(3–4):232, 1980.
49. V. Raman, V.K. Parashar, O.P. Bahl. Effect of additives on the thermal nitridation of sol-gel derived silica gel. J. Mat. Sci. 28:4159–4162, 1993.
50. V. Pavarajarn, S. Kimura. Catalytic effects of metals on direct nitridation of silicon. J. Am. Cer. Soc. 81(8):669–674, 2001.

51. D.K. Smith, C.F. Cline. The crystal structure of β-beryllia. Acta Cryst. 18:393–397, 1965.
52. W.M. Kriven. Possible alternative transformation tougheners to zirconia: crystallographic aspects. J. Am. Cer. Soc. 71(12):1021–1030, 1988.
53. J.L. Shull, W.M. Kriven, W.D. Porter, C.R. Hubbard. High temperature phase transformations in $Y_4Al_2O_9$, $Gd_4Al_2O_9$, and $Dy_4Al_2O_9$. In: W.C. Johnson, J.M. Howe, D.E. Laughlin, W.A. Soffa, eds. Solid→Solid Phase Transformations. The Minerals, Metals, and Materials Soc., 1994, 95–100.
54. W.M. Kriven. Displacive transformations and their application in structural ceramics. J. de Physique IV, Colloque C8, Supplement au J. de Physique III. 5:C8–101–110, 1995.
55. I. Levin, D. Brandon. Metastable alumina polymorphs: crystal structures and transition sequences. J. Am. Cer. Soc. 81(8):1995–2012, 1998.
56. T. Tsuchida. Preparation of high surface area α-Al_2O_3 and its surface properties. Appl. Cat. A:General 105:L141–46, 1993.
57. L.A. Xue, I.-W. Chen. Influence of additives on the γ-to-α transformation of alumina. J. Mat. Sci. Let. 11:443–445, 1992.
58. M. Ozawa, O. Kato, S. Suzuki, Y. Hattori, M. Yamamura. Sintering and phase evolution of γ-Al_2O_3 with transition-metals addition at around α-transition temperature. J. Mat. Sci. Let. 15:564–567, 1996.
59. G.C. Bye, G.T. Simpkin. Influence of Cr and Fe formation of α-Al_2O_3 from γ-Al_2O_3. J. Am. Cer. Soc. 57(8):367–, 1974.
60. Y. Saito, T. Takei, S. Hayashi, A. Yasumori, K. Okada. Effects of amorphous and crystalline SiO_2 additives on γ-Al_2O_3-to-—Al_2O_3 phase transition, J. Am. Cer. Soc. 81(8):2197–2200, 1999.
61. J.L. McArdle, G.L. Messing. Transformation and microstructure control in boehmite-derived alumina by ferric oxide seeding. Adv. Cer. Mat. 3(4):387–392, 1988.
62. J. Tartaj, G.L. Messing. Effect of the Addition of α-Fe_2O_3 on the microstructural development of boehmite-derived alumina. J. Mat. Sci. Let. 16:68–70, 1997.
63. R.B. Bagwell, G.L. Messing. Effect of seeding and water vapor on the nucleation and growth of α-Al_2O_3 from γ-Al_2O_3. J. Am. Cer. Soc. 82(4):825–832, 1999.
64. E.M. Lopasso, J.J. A. Gamboa, J.M. Astigueta. Enhancing effect of Cl_2 atmosphere on transition aluminas transformation. J. Mat. Sci. 32:3299–3304, 1997.
65. R.A. Eppler. Effect of antimony oxide on the anatase-rutile transformation in titanium dioxide. J. Am. Cer. Soc. 70(4):C 64–66, 1987.
66. R. Debnath, J. Chaudhuri. Inhibiting effect of $AlPO_4$ and SiO_2 on the anatase → rutile transformation reaction: an x-ray and laser raman study. J. Mater. Res. 7(12):3348–, 1992.
67. J. Yang, Y.X. Huang, J.M.F. Ferreira. Inhibitory effect of alumina additive on the titania phase transformation of a sol-gel-derived powder. J. Mat. Sci. Let. 16:1933–1935, 1997.
68. S. Hishita, I. Mutoh, K. Koumoto, H. Yanagida. Inhibition mechanism of the anatase-rutile phase transformation by rare earth oxides. Cer. Intl. 9(2):6–, 1983.

69. F.C. Gennari, D.M. Pasquevich. Kinetics of the anatase-rutile transformation in TiO_2 in the presence of Fe_2O_3. J. Mat. Sci. 33:1571–1578, 1998.
70. F.C. Gennari, D.M. Pasquevich. Enhancing effect of iron chlorides on the anatase-rutile transition in titanium dioxide. J. Am. Cer. Soc. 82(7):1915–1921, 1999.
71. X.-Z. Ding, X.-H. Liu, Y.-Z. He. Grain size dependence of anatase- to-rutile structural transformation in gel-derived nanocrystalline titania powders. J. Mat. Sci. Let. 5:1789–1791, 1996.
72. E.M. Kosti, S.J. Kiss, S.B. Boskovi, S.P. Zec. Mechanically activated transition of anatase to rutile. Am. Cer. Soc. Bul. 76(6):60–61, 1997.
73. R.M. Torres Sanchez. Phase transformation of γ to α-Fe_2O_3 by grinding. J. Mat. Sci. Let. 13:461–462, 1996.
74. Y.J. Kim, W.M. Kriven. Crystallography and microstructural studies of phase transformations in the Dy_2O_3 system. J. Mater. Res. 13(10):2920–2931, 1998.
75. J.R. Keski. Bismuth oxide ceramics. Am, Cer. Soc. Bul. 51(6):527–531, 1972.
76. A.V. Joshi, S. Kulkarni, J. Naclas, J. Diamond, N. Weber, A.N. Virkar. Phase stability and oxygen transport characteristics of yttria- and niobia-stabilized bismuth oxide. J. Mat. Sci. 25:1237–1245, 1990.
77. A.M. Azad, S. Larose, S.A. Akbar. Review: bismuth oxide-based solid electrolytes for fuel cells. J. Mat. Sci. 29:4135–4151, 1994.
78. M. Miyayama, H. Terada, H. Yanagida. Stabilization of β-Bi_2O_3 by Sb_2O_3 doping. J. Am. Cer. Soc. 64(1):C-19, 1981.
79. M. Mayayama, S. Katinichi, Y. Suenaga, H. Yanagida. Electrical conduction in β-Bi_2O_3 doped with Sb_2O_3. J. Am. Cer. Soc. 66(8):585–588, 1983.
80. E.C. Subbarao, H.S. Maiti, K.K. Srivastava. Martensitic transformation in zirconia. Phys. Stat. Sol. (a) 21 (9):9–40, 1977.
81. M. Yoshimura. Phase stability of Zirconia. Am. Cer. Soc. Bul. 67(12):1950–1955, 1988.
82. S. Somiya, N. Yamamoto, H. Hanagida, eds. Science and Technology of Zirconia III, Advances in Ceramics. Vol. 24. Westerville, OH: Am. Cer. Soc., 1988.
83. R.C. Garvie. Zirconium dioxide and some of its binary systems. In: A. M. Alper, ed. High Temperature Oxides. Part II. Oxides of Rare Earths, Titanium, Zirconium, Hafnium, Niobium, and Tantalum. New York: Academic Press, 1970, 117–166.
84. C.T. Lynch. Hafnium oxide. In: A.M. Alper, ed. High temperature Oxides. Part II. Oxides of Rare Earths, Titanium, Zirconium, Hafnium, Niobium, and Tantalum. New York: Academic Press, 1970, pp. 193–216.
85. J. Wang, H.P. Li, R. Stevens. Review hafnia and hafnia-toughened ceramics. J. Mat. Sci. 27:5397–5430, 1992.
86. R. Ruh, H.J. Garrett. Nonstoichiometry of ZrO_2 and its relation to tetragonal-cubic inversion in ZrO_2. J. Am Cer. Soc. 50(5):257–261, 1967.
87. R.J. Ackermann, S.P. Garg, E.G. Rauh. The lower phase boundary of ZrO_{2-x}. J. Am Cer. Soc. 61(5–6):275–276, 1978.
88. I. Ahmad. Contributions to the Development of the Whisker Reinforced Composites. Presentation from U. S. Army Watervliet Arsnel, for the 13^{th}Refractory Composites Working Group in Seattle, 1967.

89. W.J. Genck. Crystal habit and size distribution: the influence of additives and impurities. Chem Process. 78–82, 1990.
90. E.I. Givargizov. Growth of whiskers by the vapor-liquid-solid mechanism. In: E. Kaldis, ed. Current Topics in Materials Science, 1. New York: North-Holland Pub. Co., 1978, 79–145.
91. J.V. Milewski, F.D. Gac, J.J. Petrovic, S.R. Skaggs. Growth of beta-silicon carbide whiskers by the VLS process. J. Mat. Sci. 20:1160–1166, 1985.
92. L.J. Serovi, S.K. Milonji, N.M. Bibi. Influence of boric acid concentration on silicon carbide morphology. J. Mat. Sci. Let. 14:1052–1054, 1995.
93. V. Raman, V.K. Parashar, O.P. Bahl. Influence of boric acid on the synthesis of silicon carbide whiskers from rice husks and polyacrylonitrile. J. Mat. Sci. Let. 16:1252–1254, 1997.
94. T. Hashishin, Y. Kaneko, H. Iwanaga, Y. Yamamoto. Silicon carbide whisker synthesis from SiO_2-CH_4-Na_3AlF_6 system. J. Mat. Sci. 34:2189–2192, 1999.
95. T. Hashishin, Y. Kaneko, H. Iwanaga, Y. Yamamoto. The synthesis of silicon nitride whiskers from SiO_2-N_2-Na_3AlF_6 system. J. Mat. Sci. 34:2193–2197, 1999.
96. He-P. Zhou, H. Chen, Y. Wu, W.-G. Miao, X. Liu. Structure characteristics of AlN whiskers fabricated by the carbo-thermal reduction method. J. Mat. Sci. 33:4249–4253, 1998.
97. W. Tolksdorf. Flux growth. In: D.T.J. Hurle, ed. Handbook of Crystal Growth, Vol. 2. New York: Elsevier Science B.V. 1994, 563–611.
98. H.W. Newkirk, D.K. Smith. Studies on the Formation of Beryllium Oxide Macrocrystals. U. Ca. Lawrence Radiation Laboratory report for Contract No. W-7405-eng-48, 1963.
99. D. Elwell. Electrolytic growth from high-temperature solutions. In: E. Kaldis and H. J. Scheel, eds. Current Topics in Materials Science, 1976 Crystal Growth and Materials, 2. New York: North-Holland Pub. Co., 1977, 605–637.
100. R.N. McNally, G.H. Beall. Crystallization of fusion cast ceramics and glass-ceramics. J. Mat. Sci. 14:2596–2604, 1979.
101. R.L. Thakur. Determining the suitability of nucleating agents for glass-ceramics. In: L.L. Hench, S.W. Freiman, eds. Advances in Nucleation and Crystallization in Glasses, Special Pub. No. 5. Westerville, OH: Am. Cer. Soc., 1971, p. 166.
102. I. Gutzow, S. Toschev. The Kinetics of Nucleation and the Formation of Glass-Ceramic Materials. In: L.L. Hench, S.W. Freimans, eds. Advances in Nucleation and Crystalization in Glasses, Special Pub. No. 5. Westerville, OH: Am. Cer. Soc., 1971, pp. 10–21.
103. J.J. Hammel,"Nucleation in Glass-A Review. In: L.L. Hench, S.W. Freimans, eds. Advances in Nucleation and Crystalization in Glasses, Special Pub. No. 5. Westerville, OH: Am. Cer. Soc., 1971, pp. 1–9.
104. M. Tomozawa. Effects of oxide nucleating agents on phase seperation of simple glass systems. In: L.L. Hench, S.W. Freimans, eds. Advances in Nucleation and Crystalization in Glasses, Special Pub. No. 5. Westerville, OH: Am. Cer. Soc., 1971, pp. 41–48.
105. G.F. Neilson. Nucleation and crystallization in ZrO_2-nucleated glass-ceramic systems. In: L.L. Hench, S.W. Freimans, eds. Advances in Nucleation and Crystaliza-

tion in Glasses, Special Pub. No. 5. Westerville, OH: Am. Cer. Soc., 1971, pp. 73–82.

106. D.R. Stewart. TiO_2 and ZrO_2 as nucleates in a lithia aluminosilicate glass-ceramic. In: L.L. Hench, S.W. Freimans, eds. Advances in Nucleation and Crystalization in Glasses, Special Pub. No. 5. Westerville, OH: Am. Cer. Soc., 1971, ibid. pp. 83–89.
107. J.H. Markis, K. Clemens, M. Tomozawa. Effect of fluorine on the phase seperation of Na_2O-SiO_2 glasses. J. Am. Cer. Soc. 64(1):C-20, 1981.
108. J. Takahashi, M. Kuwayama, H. Kamiya, M. Takatsu, T. Oota, I. Yamai. Decomposition behaviors of dopant-free and doped solutions in the TiO_2-SnO_2 System. J. Mat. Sci. 23:337–342, 1988.
109. M.S. Newkirk, A.W. Urquhart, H.R. Zwicker, E. Breval. Formation of Lanxide™ ceramic composite materials. J. Mater. Res. 1:81–89, 1986.
110. S. Antolin, A.S. Nagelberg, D.K. Creber. Formation of Al_2O_3/metal composites by the directed oxidation of molten aluminum-magnesium-silicon alloys: Part I, microstructural development. J. Am. Cer. Soc. 75(2):447–454, 1992.
111. S. Antolin, A.S. Nagelberg, D.K. Creber. Formation of Al_2O_3/metal composites by the directed oxidation of molten aluminum-magnesium-silicon alloys: Part II, growth kinetics", J. Am. Cer. Soc. 75(2):455–462, 1992.
112. P. Xiao, B. Derby. Al_2O_3/Al composites formed by the directed oxidation of an Al-Mg-Zn alloy. J. Eur. Cer. Soc. 12:185–195, 1993.
113. M.S. Newkirk, H.D. Lesher, D.R. White, C.R. Kennedy, A.W. Urquhart, T.D. Claar. Preparation of lanxide™ ceramic matrix composites: matrix formation by the directed oxidation of molten metals. Cer. Eng. & Sci. Proc. 8(7–8):879–885, 1987.
114. D.K. Creber, S.D. Poste, M.K. Aghajanian, T.D. Claar. AlN composite growth by nitridation of aluminum alloys. Cer. Eng. & Sci. Proc. 9(7–8):975–982, 1988.
115. C. Scott, M. Kaliszewski, C. Greskovich, L. Levinson. "Conversion of Polycrystalline Al_2O_3 into Single-Crystal Sapphire by Abnormal Gram Growth," J. Am. Cer. Inc. 85(5):1275–1280, 2002.
116. D.F.H. Hennings, R. Janssen, P.J.L. Reynen. Control of liquid-phase-enhanced discontinuous grain growth in barium titanate. J. Am. Cer. Soc. 70(1):23–27, 1987.
117. Y.-S. Yoo, M.-K. Kang, J.-H. Han, H. Kim, D.-Y. Kim. Fabrication of $BaTiO_3$ single crystals by using the exaggerated grain growth method. J. Eur. Cer. Soc. 17:1725–1727, 1997.
118a. A. Khan, F.A. Meschke, T. Li, A.M. Scotch, H.M. Chan, M.P. Harmer. Growth of Pb(Mg1/3Nb2/3)O3-35 mol% PbTiO3 single crystals from (111) substrates by seeded polycrystal conversion, J. Am. Cer. Soc. 82(11):2958–2962, 1999.
118b. J.A. Horn, S.C. Zhang, U. Selvaraj, S. Tolier-McKinstry. Templated grain growth of textured bismuth titanate. J. Am. Cer. Soc. 82(4):921–926, 1999.
118c. Y. Narendar, G.L. Messing. Seeding of perovskite lead magnesium niobate crystallization from Pb-Mg-Nb-EDTA Gels. J. Am. Cer. Soc. 82(7):1999.
118d. P.W. Rehig, G.L. Messing, S. Tolier-McKinstry. Templated grain growth of barium titanate single crystals. J. Am. Cer. Soc. 83(11):2654–2660, 2000.
119. S. Yamada, N. Kamehara, K. Murakawa. Effects of SiO_2 on surface smoothness and densification of alumina. Jpn. J. Appl. Phy. 17(1):73–78, 1978.

120. A.K. Dhingra. Advances in inorganic fiber developments. In: E. J. Vandenberg, ed. Contemporary Topics in Polymer Science. New York: Plenum Press, 1984, pp. 227–260.
121. J.D. Birchall, J.A.A. Bradbury, J. Dinwoodie. Alumina Fibers: Preparation, Properties and Applications. In: W. Watt, B.V. Perov eds. Handbook of Composites. Vol. 1: Strong Fibers. New York: Elsevier Sci. Pub. B. V., 1985, pp. 115–153.
122. M.H. Stacey. Development in continuous alumina-based fibers. Brit. Cer. Soc. Trans. J. 87:168–172, 1988

4

Forming and Pressureless Sintering of Powder-Derived Bodies

4.1 INTRODUCTION

The dominant method of fabricating polycrystalline ceramic bodies, especially monolithic ones, is via various methods of powder consolidation followed by pressureless sintering. The domination of such combinations arises from their advantages often outweighing their limitations. Advantages include versatility of such fabrication methods over a considerable range of materials, component sizes and shapes, often using techniques amenable to automation, and with moderate costs. There are some limitations of materials that can be processed, the individual consolidation methods, and the microstructures and hence properties achievable. Some of these limitations are reduced or removed by use of additives, which can also enhance results with materials amenable to such processing, as discussed in Chapter 5. There are also generally some other potentially competing methods, such as pressure sintering, CVD, and melt forming discussed in Chapter 6, that have some applicability and considerable potential for more, as well as some other processes for specialized fabrication discussed in Chapter 7. However, powder-based fabrication discussed in this chapter and some in Chapters 5 and 7 is expected to continue to be the dominant method of fabrication of ceramics as well as many ceramic composites.

The versatility of powder-based fabrication is due to the suitability of mixing processes and especially the versatility and variety of powder consolidation methods and their compatibility with pressureless sintering, and the often transparent nature of both consolidation and sintering to different materials. Various mixing procedures widely used in industry such as milling, are not addressed in detail here; readers are referred to other sources [1–4]. Powder consolidation is addressed first, starting with pressure consolidation with limited binder content and plasticity, that is, die then isopressing, followed by pressure consolidation with substantial binder content and plasticity, primarily extrusion and injection molding. Subsequently colloidal consolidation techniques, mainly slip, tape, and pressure casting, and electrophoretic deposition, as well as other miscellaneous fabrication methods are addressed. Finally, aspects of binders, drying, green machining, binder burnout, and bisque firing are briefly noted; then sintering is addressed. Again in keeping with the thrust of this book, these topics are treated less extensively than in other references more focused on these techniques. The focus here is on practical parameters, needs, and issues. Readers are referred to other sources for more detailed discussion of the techniques of this chapter, especially underlying principles [1–9].

4.2 POWDER CONSOLIDATION UNDER PRESSURE WITH LITTLE BINDER AND PLASTIC FLOW

4.2.1 Die Pressing

Powder can be consolidated by applying mechanical pressure, which is most simply done under uniaxial compression of powder in a die, hence the term die pressing, also referred to as cold pressing to distinguish it from hot pressing. Such pressing can be done with no or limited, e.g., < 12%, binder/lubricant content, though some is generally used, for example, water in simple cases, leading to terms such as dry or damp pressing with respectively ~ < 4% and > 4% water. Often better (i.e., greater and more uniform) consolidation can be done with hydrostatic compression, but with some limitations of speed, tolerances, and costs, as discussed in the following section.

Basically, die pressing is typically done in steel or other more wear-resistant dies, that is, those with WC-Co liners. Such dies have one (or more) vertical tubular through holes whose shape and dimensions are based on those of the ceramic piece desired allowing for sintered shrinkage and machining. Pressure is applied to the powder in the die cavity by opposing rams with limited die cavity clearances, e.g., ≤ 25 μm for fine micron sized powders and up to two to four times this for progressively coarser particles. An important issue is the relative motion of the top versus the bottom ram, with the simplest (and common laboratory) situation being motion of only the top ram, with a stationary bottom ram, i.e., typically with the die bottom and the bottom die ram sitting on the bottom press platen. However, due to pressing gradients discussed below such single-action pressing may be less desirable. The alternative is to allow motion of the bottom ram, usually so each ram provides equal compaction, which reduces, but does not eliminate, the resultant gradients in the pressed body as discussed below. Such improvements via double action pressing, which are more important for longer, high-aspect ratio parts, must be balanced against increased costs and some complication of production, especially with more complex dies. Complexity of dies can be considerably increased to allow more complex parts to be formed, e.g., use of dual rams to press parts with different heights on the top or bottom of the part, or both, and use of axial pins to form terminated or through holes. A basic requirement is that the part must have a constant overall cross section (including any axial holes), but external shaping can be done by practical contour grinding of green pressed parts, as for dry bag isopressing discussed in the next section. Dies may also have some limited taper to aid part ejection. Also, since die production life is typically limited by wear and most ceramic powders are quite abrasive, dies commonly have a cobalt bonded WC or other ceramic liners (e.g., ZTA, TZP, and PSZ are used in pressing battery parts which corrode WC liners). Such die pressing is summarized in Table 4.1 and in more detail below.

Pressing operations basically consists of a three-stage cycle: (1) die filling, (2) powder compaction, and (3) part ejection, with factors controlling each of these stages being driven by practical considerations, especially overall time and cost with good yield. Cycle times can be a fraction of a second for small parts, increasing substantially for larger parts, for example, to several minutes, with a common component time being of the order of half a minute. However, use of automated presses with multicavity dies, multistation presses, or both can increase production rates to over 5000 parts/min, but commonly in the range of 10–150/min. On application of the pressing pressure, initially compaction is most rapid, e.g., up to 5–10 MPa where it slows, with progressively diminishing returns beyond, which along with time factors and green body quality (discussed below), typically limit production pressing to at most 50 to < 100 MPa (where die wear also increases substantially). One problem is entrapment of air in the powder, which can be limited by higher starting densities and slowing the rate of ram motion as a function of ram travel and hence densification. Overall densification to > 50% of theoretical density is typically obtained, for example, for commercial alumina bodies, and substantially more than this for some traditional ceramics, but often much less for very fine powders (Fig. 4.1).

Critical to the above results are the preparation and character of the powder used, since they depend on both rapid and repeatable die fill and a high pour density (e.g., 25–30% of theoretical density). Very fine powders present particularly severe problems since they may not flow uniformly (e.g., due to greater and more variable moisture absorption) and compress poorly (Fig. 4.1). Powder

TABLE 4.1 Comparison Summary of Primary Green Body

Fabrication method	Binder: Type[a]	Binder: Amount	Powder prep.	Shape versatility	Size capability[c]	Microstructural effects
Die pressing	Often A	Low[b]	SD	Moderate	Large l, w, more limited t	Density variation, laminar defects at higher pressure, limited particle orientation
Isopressing	Often A	Low[b]	SD	Moderate	Long l, more limited w and t	More uniform
Roll compact.	A/NA	Low–med	SD	Limited	Sub. l, w, more limited t	Not well characterized
Extrusion	Often A	High	DA	Moderate	Long l, more limited w and t	Axial elongated pores, varying grain orientation
Injection molding	Often TP	High	DA	High	Moderate	Complex, depends of die shape/filling, needs more characterization
Slip casting	A/NA	Low–med	CS	Moderate	Moderate–large l, w, limited t	Some possible pore orientation
Tape casting	A/NA	Low–med	CS	Limited	Large l, intermediate w, limited t	Particle–thickness gradients ⇒ interlayer pores on lamination
Pressure casting	A/NA	Low–med	CS	Moderate	moderate[d]–large l, w, limited t	Pores from bends, i.e., exsolution of gas dissolved in slips on pressure release
EPD	A/NA	Low	CS	Moderate	Moderate–large l and w, limited t	Not well characterized

[a]A, aqueous; NA, nonaqueous; TP, thermoplastic; SD, spray dried; DA, deaired mix; CS, colliodal suspension.
[b]Higher contents needed for green machining.
[c]Using l = length, w = width, and t = thickness, as the larger, intermediate, and smaller dimensions respectively.
[d]Especially for centrifugal casting versus pressurized slip

FIGURE 4.1 Consolidation of fine (~ 0.1 μm) MgO powder. Top, as-poured column of powder of ~ 15 cm long and ~ 10% of theoretical density (in a plastic tube between partly visible rubber corks). Lower left, the powder consolidated nearly fourfold to ~ $^1/_3$ of theoretical density by die pressing at ~ 35 MPa. Lower right, the die pressed powder consolidated ~ threefold (via hot pressing) to ~99% of theoretical density. Such extensive consolidations typical of very fine powders present serious practical problems and increased costs in comparison to common production use of powders that can be poured and die pressed to respectfully ~ 25 and 50% of theoretical densities. (From Ref.6. Published with permission of AiChE.)

compaction typically requires some limited quantities of lubricants to aid both sliding of powder particles past one another and along the die wall, e.g. as stearates of a cation of the ceramic body or just stearic acid, and binders to enhance green, that is, compact, strength for part ejection from the die or mold and subsequent handling. Binder is also needed for forming spherical agglomerates, primarily by spray-drying, that give uniform powder flow and die fill needed for automated die pressing. Spray-drying ideally produces agglomerates that are (1) fairly spherical for good flow of the spray-dried powder, (2) sufficiently dense to give good powder pour density, (3) strong enough to retain their integrity under the loads and conditions of manufacturing (e.g., of hundreds of pounds of weight of overburden on powder in the bottom of powder containers and abrasion in handling large amounts of powder), and (4) otherwise totally crushed or loose most or all of their identity during the pressing operation. Agglomerate strength and some of its deformation capability comes from binders used, e.g., 2–4 w/o of polyvinyl alcohol (PVA) with a polyethylene glycol plasticizer. While problems of some significantly distorted versus spherical agglomerates occur, the most severe problems occur in failures to meet the challenges of balancing traits 3 and 4 above, usually too hard an agglomerate (see Fig. 4.2 and [3]). Some of the problem is related to spray drying binders, but some of this is due to environmental effects (e.g., storage and humidity), since other suitable binder systems or agglomeration methods have not been found, often making spray-drying a necessary evil in die pressing (as well as some isopressing, Sec. 4.2). Thus, for example, PVA absorbs moisture, lowering its glass transition temperature and acting as a plasticizer [10,11], and this combined with effects of temperature variations can give a seasonal variation to density and properties. Thus, higher humidities and temperatures in summer versus winter can increase the density achieved in alumina by ~ 0.5% of theoretical density, altering shrinkage control,with corresponding changes in microstructure and properties [11,12]. Common process controls on spray-dried powders are the sphericity of the agglomerates (e.g., versus donuts) and the angle of repose.

Consider now some of the technical problems and limitations of die pressing. The occurrence and extent of the problems may depend on powder and pressing parameters, e.g. powder particle shape and size distribution, agglomeration, die shape, dimensions, fill and pressing parameters. However, there are basic trends, for example, in pressing simple shapes, such as cylinders or discs as commonly used in both component production and test specimen preparation, that are important to keep in mind as a guide to testing and characterization. Such pressings typically have both radial and axial density gradients [13–18], which tend to increase with increasing length of ram travel, i.e., with single-action versus double-action pressing, taller pressed parts, and lower pour density as

FIGURE 4.2 Example of substantial identity retention of spray-dried agglomerates in trial industrial die pressing. See also Figs. 4.6 and 4.8 for other examples of inadequate deformation and loss of spray-dried agglomerate identity and resultant defects.

commonly encountered with finer particles. The powder compacted nearest to a moving die ram commonly reaches the highest density at the compact periphery and decreases by a few or more percent of theoretical density with distance from the die wall, often approaching a nearly constant density over the central part, for example, one-half, of the die diameter (Fig. 4.3). Progressing further from a moving ram, the axial density values decrease, while the radial density variation first decreases, often nearly disappearing over about the middle one-third of the compact height for a single moving ram, but then progressively increases in about the bottom one-third of the compact height. However, the radial density gradient in the bottom portion of the die with only the top ram moving is typically the reverse of that in the top of the compact,that is, the lowest radial density is at the bottom of the compact periphery, with the pressed density increasing with radial distance in from the compact periphery. This radial density change is commonly nearly a mirror image of that at the top, both in relative and absolute values. Use of double-acting rams results in reduced axial density gradients be-

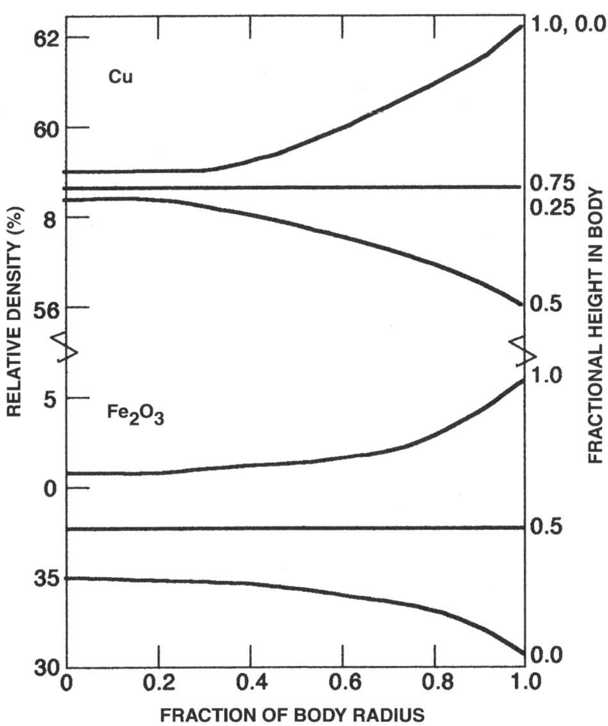

FIGURE 4.3 Plots of the of the cross-sectional density gradients in die-pressed cylinders. (Data from Refs. 13 and 14.)

tween the top and bottom of pressed cylinders since the bottom half becomes approximately a mirror image of the top, which may yield more density uniformity in the central portion, due to radial and axial gradients that may be introduced there [14]. Basically such double-action pressing results in the density gradients in the bottom half of the pressing being a mirror image of the top half of a pressing with a single-acting press, rather than the density trends in the bottom half of a single-action pressed body. As die wall friction increases, as in pressing cylindrical shells of limited wall thickness, or in using finer, more voluminous powder, these density gradients may be more severe, as indicated by gradients observed in commercially die-pressed sonar transducer rings (Fig. 4.4).

Other pressing defects that often occur individually, are defects of a general laminar character collectively, that is, nominally normal to the pressing axis. These are often areas of limited or no bonding or contact of parts of adjacent powder layers in a nominally vertical direction. Such laminations may be fairly diffuse or continuous in a given layer, and may occur approximately periodically over much of the length, especially of longer pressed parts. Springback of the compacted powder on release of pressing pressure is a major factor in the occurrence and extent of such laminar defects. Near the end(s) of pressed parts in contact with moving die ram(s) less laminar defects may occur, consisting of partial to complete separation of part or all of various shaped end sections, referred to by such terms as *end-capping* [3,19] or *ring capping* [3], all from the same underlying causes. However, air entrapment can also be an important factor in the forming of some laminar defects [18,20] in which case they are more discrete and produce greater vertical separation of the opposing powder layers, e.g., commonly forming lenticular pores (Fig. 4.5), but may be more varied [10].

The occurrence and severity of the above laminar defects are increased by greater springback of the powder compact upon release of the pressing pressure. Thus, they are limited or prevented by better particle flow and less die wall fric-

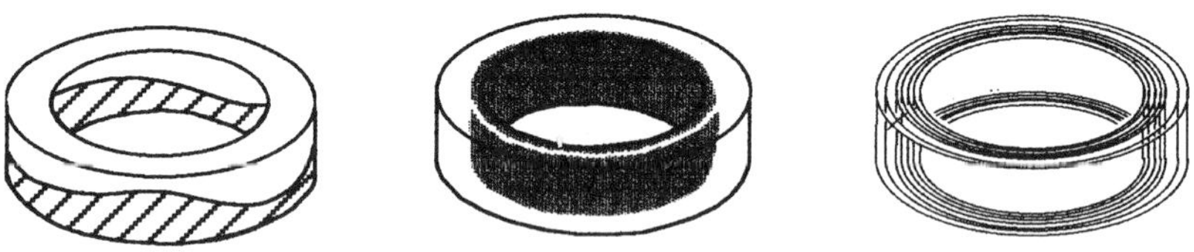

FIGURE 4.4 Schematics of indicated density variations in PZT sonar transducer rings (inner dia. ~ 5, wall thickness ~ 1, and height ~2 cm) commercially produced by (A) Die (probably double-action) pressing; (B) isopressing; and (C) slip casting. Note shaded areas are somewhat more porous (i.e., lower density) areas revealed by etching. (From Ref. 17, published with permission of the American Ceramic Soc.)

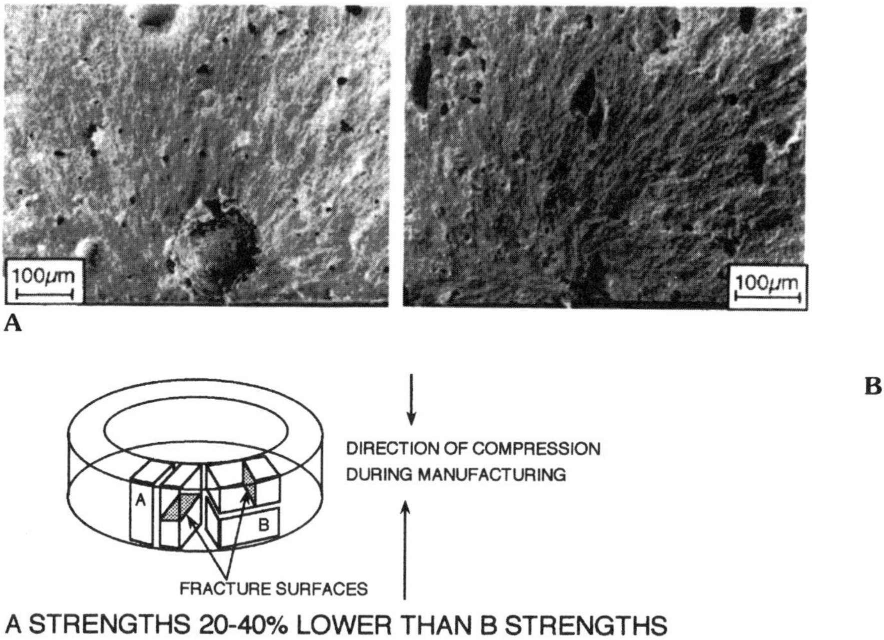

FIGURE 4.5 Fracture initiation from large laminar pores in a commercial die-pressed PZT sonar transducer ring showing the orientation of lens-shaped laminar pores normal to the pressing direction and resultant strength anisotropy, i.e., lower strengths of A versus B specimens. (From Ref. 17, published with permission of the American Ceramic Soc.)

tion. that is, better lubrication, greater green strength (e.g., stronger binders), and less ram travel (e.g., denser packing powder and lower pressing pressures as well as lower part aspect ratios). Note that sometimes isopressing following die pressing is used because of potential advantages of better part shape definition of die pressing and greater and more uniform compaction of isopressing, including possible reduction of die pressing lamination problems. However, besides requiring an added operation and cost, more recent evaluation of such dual pressing indicates that while laminar and related defects from die pressing may be reduced by subsequent isopressing, some lamination defects often remain [21,22], and once they occur, they are difficult to reduce or remove.

Besides powder, die, and pressing parameters impacting the above density variations and possible laminar defects, powder agglomerates can have considerable effect depending on their number and character. Agglomerates that do not adequately lose their identity will leave a fairly uniform pattern of their remnant effects, for example, larger pores in the remaining interstices between the agglomerates (e.g., Fig. 4.2). However, a limited population of spray-dried or other hard agglomerates that undergo limited or no collapse during pressing can result in other serious defects. One is simply a larger pore in the interstices between parts or all of three or four contacting agglomerates that underwent limited or no deformation (Fig. 4.6), the occurrence and character of which may depend on differences in the extent and timing of the collapse of surrounding agglomerates. Often more serious is the formation of a partial peripheral crack/pore (Fig. 4.7), e.g., about halfway around an isolated hard agglomerate,

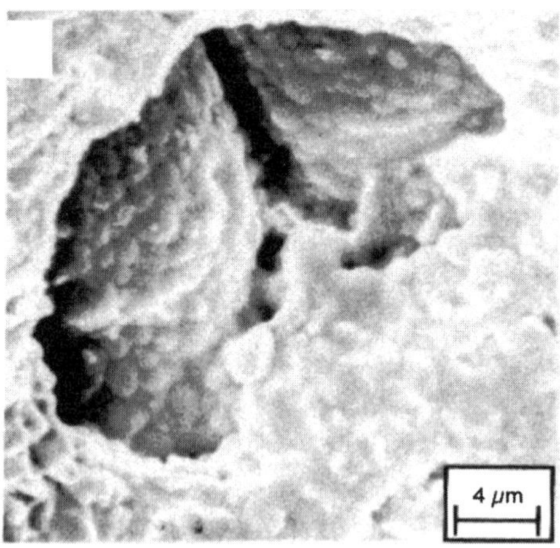

FIGURE 4.6 Pores between remnant agglomerates near the sintered surface in sintered commercial TZP.

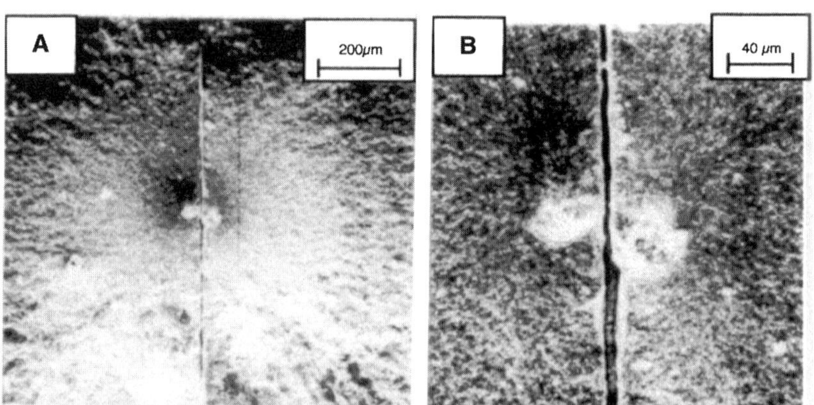

FIGURE 4.7 Example of fracture occurring from an isolated hard agglomerate in testing a sintered die-pressed alumina.

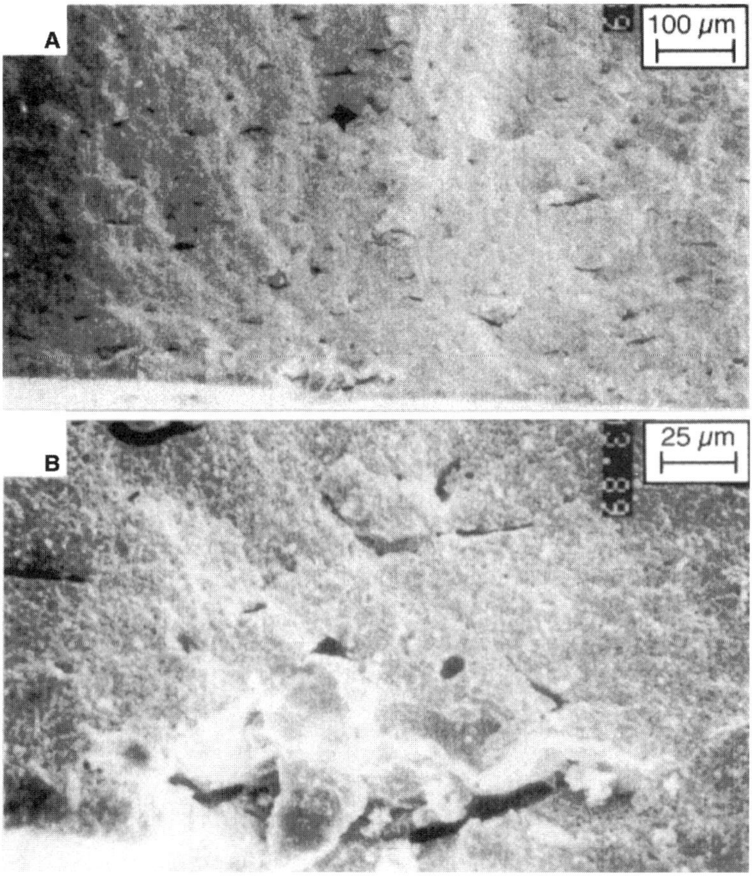

Figure 4.8 SEM example of laminar cracks due to inadequate deformation of spray dried agglomerates in die pressing test bars of commercial TZP.

due to factors such as poor bonding of surrounding powder and differential matrix-agglomerate springback on release of the pressing pressure. Use of spray-dried agglomerates can change the picture substantially, introducing local laminar cracks/pores when they do not deform and lose much of their identity, at least in some pressed part sizes and configurations. Increasing relative humidity in the powder compact can substantially exacerbate green body density gradients, which are typically accentuated on sintering.

Die pressing can also result in considerable preferred orientation of platy or elongated particles, particularly in pressing bodies of low aspect ratio. Further, the above differences in die pressed density reflect differences in relative motion of the powder, and thus should also have some correlation with varying orientation of platy or elongated particles during pressing. Again such preferred orientation is often increased on sintering [23]. Also note that some of the density gradients of (cold) die pressing can occur in hot pressing, unless theoretical density is achieved, and some laminar defects may still occur in bodies hot pressed to translucency, that is, near theoretical density (Sec. 8.2.1), corroborate such hot pressing effects. Further, greater interparticle motion in hot pressing versus (cold) die pressing enhances the opportunity for orientation of platy or elongated particles.

Die pressing is widely used in industry, especially for smaller parts (e.g., a few to several cm in dimensions), mainly with lower aspect ratios (e.g., < 3). Larger bodies, for example, for refractory and wear uses are pressed with larger, especially lateral, dimensions (e.g., of tens of cm), but die pressing must compete with other fabrication methods. Some pressing variations or defects such as laminar defects may be partially eliminated or ameliorated by sintering, but they can be exacerbated as density gradients typically are, and while variations in preferred orientation of grains crystallographic structure can be diminished by sintering, it is often increased, especially with grain growth [23]. The limitations of such larger bodies of consolidated powder lies as much or more in their sintering than in their green formation and die pressing is generally a reasonable option for fabrication of large bodies (Table 4.1). However, powder-based fabrication generally has increasing costs and decreasing component capabilities as component size increases. Increasing costs of larger presses, powder loading and green part handling, as well as much longer pressing cycles all contribute to increasing pressed part costs as their size increases. However, die pressing is used for making many larger parts, especially those that have two or all three sizeable dimensions.

4.2.2 Hydrostatic/Isostatic Pressing

Isostatic pressing, i.e., using hydrostatic pressure, is commonly done on both laboratory and industrial scales by placing powder in plastic, or more commonly rubber or urethane, bags that are sealed (and often evacuated), then placed in a hydrostatic pressure (typically cylindrical) vessel with a suitable pressing fluid, which upon closing and sealing the vessel is pressurized. This process, referred to as *wet bag isopressing*, which is discussed further below, has considerable industrial use, especially for larger, specialty components. There is, however an approximately hydrostatic pressing process referred to as *dry bag isopressing* or *dry bag pressing* that is widely used industrially because of its practicality.

Basically dry bag pressing can be envisioned as a hybrid between fully isostatic pressing and die pressing. It essentially consists of an oversized "metal die" as for die pressing, but with a thick rubber bladder wall in the metal "die" or housing. The bladder and housing are designed so that upon filling the central volume formed by the rubber bladder and sealing the bladder to the interior of the metal housing, the area between the bladder and the metal housing can be pressurized with hydraulic fluid to provide radial (e.g., biaxial) pressing on the powder. The top of the bladder is typically sealed by an applied axial mechanical pressure (that does not apply pressure to the powder), while the bottom is either permanently sealed or pressure sealed in conjunction with sealing the top. Such pressing is limited to substantially lower pressures than much wet bag isopressing, e.g., to 20–60 MPa. However, such pressures encompass the range where most of the compaction occurs in pressing, and the radial nature of the pressure is more effective for compaction, as is limited or no "die wall" friction. Dry bag pressing, while having some shape limitations, also has some shaping advantages, in particular the rubber bladder can be shaped some, to produce bodies of tapering or varying diameter. Thus, the process has been used for production of spark plug insulators, where partial shaping of the taper and shoulders reduces the amount of contour grinding needed. (Some trade-off of better pressed body quality with broader particle size distribution, e.g., 5–40 μm particles, was required since contour grinding costs increase with larger particle sizes [24].) Other shapes produced include both cylindrical and various shaped "spherical" alumina pieces for all but the smaller sizes (e.g., > 1 cm) of ball-milling media (Fig. 4.9), and tubes, for exam-

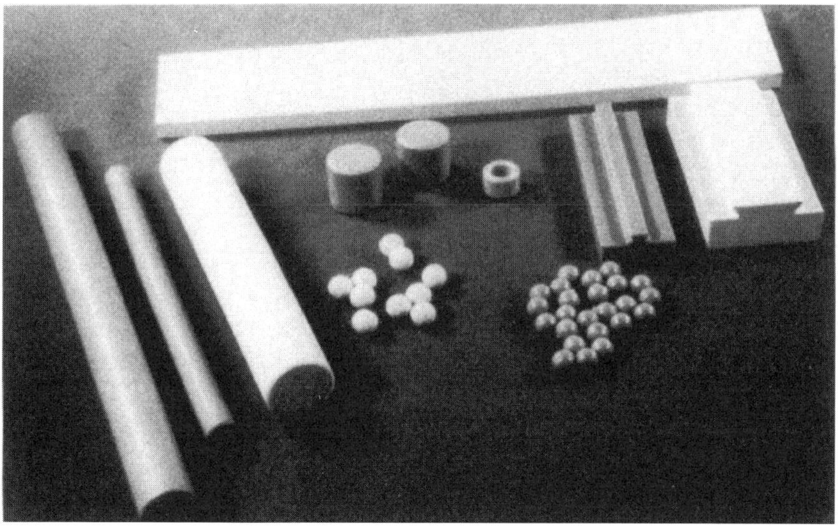

FIGURE 4.9 Examples of industrially produced ceramic parts, e.g., milling media via dry bag isopressing (central area) and rod- and bar-shaped parts via extrusion (and tubes that are not shown). (Photo courtesy of Diamonite Products.)

ple, high quality ones for high-pressure, high-intensity lamps [25]. Such fabrication is practical in view of reasonable pressing cycles of minutes, substantially less than normal isopressing, due to the small volumes of fluid to be pressurized and the rapid mechanical sealing. Again since "die" fill is automated and depends on good and uniform powder flow, spray-dried powder is commonly used, but it is less likely to produce the extent of defects as often occur in die pressing due to higher and hydrostatic pressures of isopressing. Moderate equipment cost, reasonable production rates, and generally more uniform, less defective bodies often make this a suitable production process.

Wet bag isopressing is carried out in pressure vessels that are produced up to a few meters in height and a meter or more in diameter, but at substantial cost for such larger sizes. These costs may be justified with sufficient volume since larger volumes in one press save on the substantial cycle time (minutes), when averaged over more parts in a run. Pressure capabilities commonly range from ~ 100 MPa to ~ 700 MPa, with substantial cost increases for high pressures, especially as size increases. It is used more for special fabrication, such as longer tubes or rods where extrusion is not available or suitable, for example, for tubes or rods with flanges, simple, limited change of diameter(s) or shape (including hemispherical sections and ogives, e.g., for IR domes or radomes), or closed-end tubes (such as for some batteries [26].) It can also produce very large pieces, for example, of cross sectional areas of 0.1–0.3 m^2 and lengths to $\geq$ 2.5 m [27]. Another important use is for prototyping components. Due to their overall good uniformity, logs or blocks of isopressed powders are often used, for example, by ceramic manufacturers, for prototyping components by green machining them from isopressed bodies, then firing.

As noted earlier, powder (which is sometimes spray-dried for better fill) is filled into plastic or rubber bags, which are sealed (then often evacuated) and loaded in the press, pressed, then unloaded and unbagged. Most or all of these steps being manual with limited or no automation and pressing cycles being of the order of several minutes, and shapes often being limited have constrained use of isopressing. However, its use is aided first by higher green densities, e.g., commonly 55–65% of theoretical densities [28] to as high as ~ 75% [29]. Shape versatility is another plus, and it, and especially accuracy, can be improved by the common practice of loading bags in perforated metal or wire-mesh holders, thicker shaped bags, and especially use of mandrels, for example, for cylinders and other tooling designs. Thus, PZT sonar rings have been produced by isopressing cylindrical tubes on a mandrel [Fig. 4.4B], with the rings then being gang sliced from the tubes. While direct feeding of raw powder is typical for laboratory and some production processing, use of agglomerated, especially, spray-dried, powder is used to aid uniformity of bag fill, which is likely to be needed for increased automation to make the process more cost-effective.

The nature of the pressing behavior and results has many similarities to die (uniaxial) pressing, but with less heterogeneity problems and frequently higher density. Thus, compaction is most rapid at low pressures, for example, at ≤ 100 MPa where alumina green density may reach ~ 60% of theoretical density, then increase more slowly, reaching ~ 65% at 700 MPa [28–30] (often with less increase in fired density [28]). (Note, however, that use of cyclically applied hydrostatic pressure has been reported to allow achieving densities normally obtained at higher pressures, though at some increase in cycle time [29-33].) There is also some density inhomogeneity, e.g., near mandrels (e.g., Fig. 4.4C), especially their ends, and near the bag surface [30], that is, similar to, but less extreme than higher density near moving die-pressing rams. One important difference between iso- and diepressing is that die pressing often results in varying anisotropy due to preferred orientation of anisotropically shaped particles, formation of laminar defects or pores, or both as noted earlier. On the other hand, isopressing results in little or none of this, even with fairly anisotropically shaped powder particles [23]. However, isopressing costs are often higher and allowable lateral dimensions more constrained than in die pressing (Table 4.1).

4.3 PLASTIC FORMING

4.3.1 Extrusion

There are a variety of plastic forming processes that are made feasible by mixing ceramic powder with additives that allow the resultant mix to be plastically formed [1–4]. Originally this was accomplished by using traditional ceramic raw materials, primarily clays that can be made quite plastic with considerable addition of water. This allowed such clay-derived bodies as pottery and bricks to be formed by forcing a mass of a plastic mix by hand into a mold. Later, the potter's wheel was invented for more versatile pottery forming and is still used today in much the same fashion as originally developed. Technology of the potter's wheel has also evolved to use rotating wheels for highly automated plastic forming of cookware and china, as well as some for institutional and electrical porcelain bodies. This also includes making large (e.g., ~1 m diameter and > 1 m high) stoneware vessels. The first of two other important derivatives of this original plastic forming technology are injection molding, which is discussed in the following section; the second is extrusion, i.e., forcing a plastic mass through a die to shape it. Extrusion is still used today not only for production of bricks, but also for other clay-derived bodies, such as sizeable sections of ceramic pipe, as well as other ceramics.

The focus of this section is extrusion as a fabrication method for technical ceramics, which first requires use of additives (binders) to provide the plasticity that is provided by clays in traditional ceramics, as well as selection of the type of extruder, which can also impact the selection of binders. One type of extruder is a ram extruder, which uses a hydraulically activated ram to force a plastic ceramic mass at controlled rates down the encompassing metal barrel and through the product forming die at the opposite end of the barrel. Such extrusion is typically done at room temperature, usually with water-based binders, but other liquid-based binders can be used (Table 4.1). An important requirement is that the plastic ceramic mass, which is mixed elsewhere, be de-aired after being placed in the barrel—this makes ram extrusion a batch process, though this constraint can be reduced with a two barrel extruder where one barrel is loaded and evacuated while the other is being used for extrusion. Typical binder systems consist of a binder/flocculant (e.g., methylcellulose or PVA), a coagulant (e.g., $CaCl_2$ or $MgCl_2$), and a lubricant (e.g., stearates, silicones, oils, fine talc or graphite, needed mainly with high die and extrudate surface areas) [3,34]. However, other systems such as sol-based systems may be feasible as noted below.

The other type of extruder is one that uses a screw or auger to force the plastic mass through the product forming die. Such screw extruders typically also have a mixing and a de-airing and evacuation section prior to the extrusion section, which in principal makes screw extruders potentially capable of continuous operation. Screw extruders are generally operated at room temperature with the same or similar binder systems as used with ram extruders. In such cases, while there are various differences between this and ram extrusion, there are substantial similarities; and while both are used for ceramics, ram extrusion may be more widely used. However, many screw extruders, which are used extensively for extrusion of thermal plastics, can be operated at elevated temperatures at which such plastics are sufficiently plastic to be extruded. Thus, such extruders can use ceramic mixes made plastic by use of thermoplastic binders, that is. which extrude essentially as a highly filled plastic [35]. This is very similar to injection molding discussed in the next section, including binder systems mixing and operation. While used some for ceramic extrusion, such thermoplastic binder extrusion may have greater future potential use with preceramic polymers as the binder.

Choice of extruder type involves a variety of factors, such as initial and operating costs (e.g., wear, especially of screws), availability, operation and resultant extrudate character, including size and complexity, and yield. However, other than in some special cases discussed below, the net balance is often similar, for example, since die costs are similar (and can be several to many tens of thousands of dollars, e.g., for extruding cellular bodies). Thus, there is often not a large difference between the two; while ram extrusion is very common for technical ceramics, screw extruders are widely used for clay-derived bodies and have been used for at least some extruded exhaust catalyst supports, which is a very demanding application (Fig 4.10). There is some advantage of screw extruders for continuous production, and they may be somewhat more scalable to larger extrudate sizes, which are currently of the order of 0.1 m^2 in

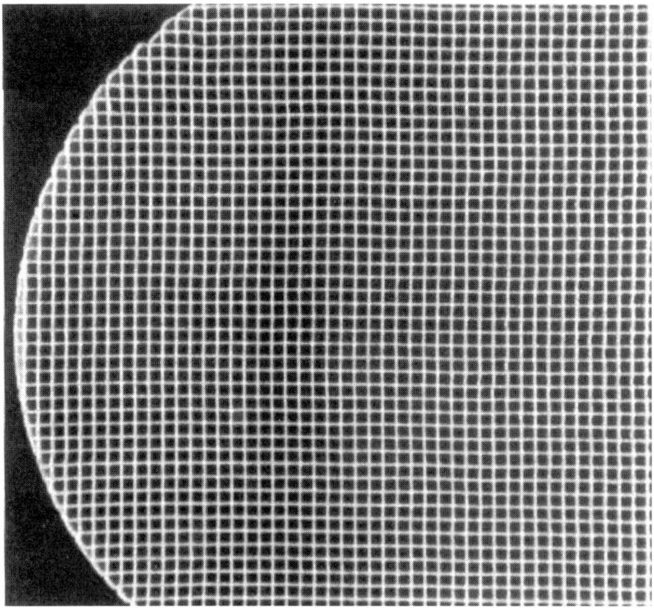

FIGURE 4.10 Section of a cordierite-based honeycomb automobile exhaust catalyst support. The complete honeycomb body is typically 5-cm thick, 10-cm wide, and 10-cm long, but may be larger for larger engines. While ~ 60 cells/cm^2 with wall thicknesses of 0.15 mm are typical, these can be changed to at least ~ 90 cells/cm^2 with cell wall thicknesses of 0.1 mm

cross-sectional area. One area where ram extrusion is necessary is where a cross-sectional aspect of the mass to be extruded must be maintained. The first of two examples of this is extrusion of fibrous monolith bodies, where a structure of aligned, coated pseudo-fibers is made (e.g., Si_3N_4 "fibers" coated with BN are made by initial extrusion of a cylindrical billet of Si_3N_4 with a thick coating of BN), then sections of the resultant extrudate to be re-extruded are aligned, and the process is repeated several times until the scale of coated fibrous structure desired is obtained. While the initial extrusion can be done with a screw extruder, it will not maintain the desired fiberous structure on re-extrusion [36]. Second, fine scale ceramic parts, for example, on a mm or finer scale such as may be needed for some advanced actuators, are also reported to be fabricatable by repeated extrusion, which can only be done with ram extruders [37]. Also, note that the barrel of ram extruders can generally be varied from horizontal to vertical operation, with angles sometimes being an advantage in handling the exiting extrudate.

Note some factors for either type of extruder; Extrusion of many objects, especially those with axial holes, such as for tubes and honeycomb structures, requires die designs that entail the extruded material to be flowing around and cut by parts of the die, then be knitted back together, that is, healed or rebonded, beyond this change of flow. This requires more complex and expensive dies, but whose engineering is well established. The presence of a considerable content of fine, submicron, particles generally aids extrusion. Extrusion pressures are commonly a few to several, e.g., 15, MPa, with lower values for bodies with increasing clay content. Extrudate formation rates can be up to ~ 1 m/sec and with large extruded bodies can result in productions of up to the order of a hundred tons per day for bodies with substantial clay content.

Typical applications of extrusion are rods and tubes, the latter ranging from thermocouple insulators, larger insulators, lamp envelopes, and furnace tubes, but including other shapes (Fig. 4.9). One of the largest markets is cellular ceramics for various uses, (see Fig. 4.10), the largest of which is for auto exhaust catalyst supports, a market that is probably in excess of a hundred million dollars per year and growing. However, there are a variety of other environmental uses of such cellular materials for catalytic uses and filtering, as well as for rotary heat exchangers, some of which are likely to grow substantially with increasing constraints on pollution.

Turning to the character of extruded ceramics, there are two key aspects that, while often not given adequate attention, are important. The first is the nature of the porosity. Theoretically, the axial distribution of plastic binder in the extrudate, though being complex due to its three-dimensional connections around the ceramic particles, should also have a basically tubular character in the axial direction, which probably varies with the range and absolute scale of the particle size. An overall axial tubular character of much of the binder must result from the streamline flow necessary for producing a coherent, uniform extrusion. Thus, while the binder acts as a lubricant for the ceramic particles to move to accommodate the deformation of the extrusion, the plastic flow of the binder itself is also important in the extrusion and any locally higher binder content will be elongated axially. While sintering will substantially reduce the amount of the porosity left from binder burnout and reduce its degree of connectivity, considerable remnants of the tubular character probably will remain. Though little study has been made of such axial tubular character of porosity in extruded ceramics (or metals), recent evaluation of the porosity dependence of extruded ceramics is consistent with such axial tubular character [21,22] (Table 4.1). A basic consequence of such axial porosity character in extruded bodies is a substantial anisotropy of porosity dependent properties, which, while not studied much, is supported by limited data showing anisotropy in extruded bodies. Some of this results from preferred orientation of the ceramic structure as discussed below, but some most likely arises from some tubular pore character and its preferred axial orientation. The substantial anisotropy of properties that can result from such axial tubular pore character are both a strong reason to evaluate porosity character in extruded bodies, and can be an important tool in demonstrating and defining it. Even less attention has been given to the possibility of radial gradients of porosity that may occur in extrusion and of the effects of knitting sections of extrudate back together after being temporarily separated in some extrusions, for example, in tubes and cellular structures.

The other key microstructural factor in extruded ceramics that has received insufficient attention is preferred orientation of the crystal structure of the grains in extruded ceramics. Varying degrees of this are expected as a function of the degree and range of anisotropy of particle/crystallite shape and its relation to the crystal structure, particle sizes and size distributions, and extrusion lateral dimensions and pressures. While quantification of effects of these factors is not available, some demonstration of their occurrence and their substantial impact is shown by the following two examples. The first is substantial study of BeO by Fryxell and Chandler [23] using both AOX and UOX powders, which respectively have equiaxed particles and considerable content of fine morphological particles with the long axis of elongated particles being the c-axis (Sec. 2.2). They showed that while both powders gave isotropic bodies of sintered BeO when first consolidated by isopressing, as did the equiaxed AOX powder when consolidated by extrusion, extrusion of the the UOX powder with some elongated particles resulted in a substantial c-axis texture along the extrusion axis. Thus, extrusion resulted in substantial axial orientation of the elongated powder particles, and this orientation increased with increasing sintering and grain growth (hence being an important forerunner to the substantial microstructural seeding that has been of more recent interest). The resultant anisotropy of the BeO from UOX powder, e.g., up to a hundred-fold higher x-ray intensity of preferred peaks than in a randomly oriented body, translated into anisotropy of bulk properties. Thus, the axial thermal expansion of the UOX derived BeO bodies decreased up to 7% below that of isotropic BeO, consistent with lower expansion along the c-axis. Similarly, Young's and shear moduli were up to 7% higher along the extrusion axis of UOX derived BeO, and both room- and higher temperature strengths of larger grain bodies were higher in the axial direction.

The second example of the occurrence and impacts of preferred crystal orientation from extrusion, which also shows other important engineering benefits is extrusion of cordierite auto exhaust catalyst supports by Lachman and coworkers [38,39]. They used clay as a substantial ingredient for forming the resultant cordierite, which not only significantly lowered raw materials costs, e.g., possibly by as much as an order of magnitude, but also greatly aided the extrudability of the body. Extrusion of the green honeycomb body resulted in significant preferred crystal orientation of the clay particles as expected. However, rather than being destroyed in the reaction of the clay with the other constituents to form cordierite, the preferred orientation of the clay particles resulted in a preferred orientation of the resultant cordierite grains. This unexpected cordierite orientation resulted in another advantage of using clay, namely a beneficial anisotropy of thermal expansion, that is, the highest cordierite expansion direction normal to the honeycomb cell walls, which is the most benign, and thus beneficial to thermal stress and shock performance of the catalyst support. Thus, preferred orientation of even reactive ingredients that lose their chemical identity en route to the final product can have an impact on the resultant extruded and fired body in the intervening reactions if orientation of the reactive ingredient particles carries over to the fired product (Table 4.1).

Now consider recent developments and future possibilities, focusing on three promising areas, the first being binders. Chen and Cawley [40] have extruded alpha alumina using a colloidal boehmite, AlO(OH), sol with 0.3 w/o PVA as binder, and Kumar and coworkers [41] reported similar extrusion of both α- and γ-alumina. Further demonstration of this, as well as of the second and third areas of orientation and composites, is work of Blackburn and Tawson [42] forming composites of chopped, oriented alumina fibers in a mullite matrix by extruding mixes of fine silica and the alumina fibers using a boehmite sol. Another example of successful alignment of anisotropically shaped particles is Muskat and coworkers' [43] extrusion of 15% β-Si_3N_4 whiskers in a predominately α-Si_3N_4 (using a conventional organic-based binder system) with subsequent sintering to 95% of theoretical density, preserving the high degree of orientation of β-Si_3N_4 whiskers. While this latter Si_3N_4 extrusion used a conventional fugitive organic binder, use of preceramic polymer binder systems should be feasible, and though being a cost factor, may offer potential in both compositions, densities, and qualities achievable.

4.3.2 Injection Molding

Injection molding is the other major plastic forming method besides extrusion, of which it is basically a derivative, since it essentially consists of extruding a spaghetti-size stream of plastic ceramic mix into a mold of the component to be formed. Typically the binder-ceramic plastic mix is heated above the glass transition temperature of the binder, for example, by 125–150°C, and the mold cavity is cooled, with the mix injected into the cavity ideally becoming rigid just as cavity filling is completed. This process, which was developed about 75 years ago, has been attractive for forming modest size parts, from dimensions < 1 cm up to a few tens of cm (Table 4.1), with larger bodies being fairly open, since massive bodies present various problems, such as slow temperature changes. The attraction of injection molding is rapid forming (cycles from ~1 min) with considerable intricacy, for example, nominal thicknesses of 0.5–5 mm and (closed or through) holes > 1.5 mm diameter, with tight tolerances, < 0.1 mm.

Injection molding became the production process for spark plug insulators in the late 1930s (but was replaced by dry bag isopressing and contour grinding as noted in Sec. 4.2.2), has had a variety of industrial uses since then, (see Fig 4.11), and has been of considerable interest for fabricating turbocharger and turbine engine rotors and other components, such as individual vanes or complete vane assemblies. (However, note that recent tests of pressure-cast tensile test specimens has shown that they have much better strength and reliability than those made by injection molding, Sec. 4.4.1).

Keys to successful injection molding are several interactive factors as

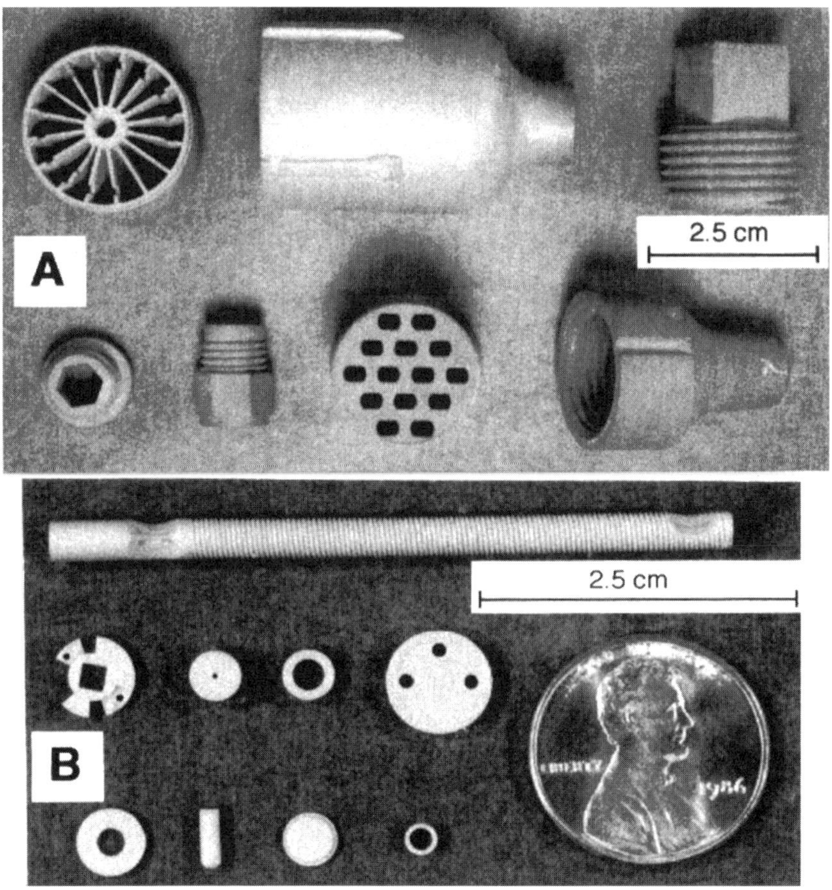

FIGURE 4.11 Examples of injection molded ceramics: (A) intermediate-sized parts and (B) smaller parts. (Photos courtesy of Diamonite Products.)

discussed in a number of reviews of the subject [3,44–50]. These include the binder system, mixing uniformity, particle size and distribution, mold design, and operational parameters (e.g., temperatures, pressures and material flow rates). Thermoplastic binders are most common, but other binders are used or being explored as discussed below. Usually a major and minor binder constituent is used so that the latter is burned off at lower temperatures, aiding in removal of the major binder content, which is one of the challenges for injection molding. Particle size effects manifest themselves mainly via their effect on viscosity, which impacts both mixing and actual molding, with finer particles substantially increasing viscosity above ~ 50 v/o, but with some reductions of the rates of viscosity increases with increasing particle size distribution. Finer particles also increase difficulties of binder removal. Thus, typically used powder particles are one to several microns in diameter. Obviously a challenge is to have the mold fill completed just before the body "freezes," which depends on how the extrudate is entered into the mold, the mold-material temperature differences, and the rate of material flow into the mold.

Failure to adequately meet the above challenges results in various problems which are often interactive and commonly fall into four categories. The first is inadequate molding—e.g., pores from inadequate de-airing, incomplete filling of the mold, and incomplete knitting of adjacent strands of the extrudate [51,52]. The second is inhomogeneous mixing from agglomerates of the binder or the ceramic, as well as less extreme but larger scale composition variations. Third is residual stresses from both composition variations and, especially, "frozen in" stress from thermal gradients and frequent resultant cracking as molding is completed or during early stages of burnout. Fourth is binder burnout, which is challenging because of the volume of binder to be burned out, combined with its expansion and plasticity during burnout, hence allowing distortion and exacerbation of molding problems [51–53]. Possible issues that have received little attention are probable tubular pore character and alignment of anisotropic particles within extrudate strands expected (with some demostration, Ref. [54], and resultant effects of probable anisotropy of properties of the strands causing sections of them to act as pseudo large grains with anisotropic properties and hence sources of stress concentration.

A variety of changes are in varying degrees of demonstration to improve the basic process or make it more versatile. One of broader development is use of other binders and molding parameters to allow much lower pressure molding, that is, by ~ 10–100-fold from those typically used in normal (high, 7–70 MPa) pressure molding [3,55]. Other efforts include other novel binders, such as those based on water—e.g., use of water-soluble organic-based gelling agents in water-based binders [55] or water as the main binder constituent, with freezing of the water via a cooled mold the method or rigidization [56]—as well as preceramic polymer-based binder systems [57]. All of these significantly change one or more aspects of injection molding, especially binder "burnout." Finally, some investigation of injection molding for possible advantageous forming of ceramic composites—e.g., with possible favorable preferred orientation of the dispersed phase—has been made. For example, Tsao and Danforth [58] showed that while average lengths of SiC whiskers were approximately halved, there was considerable preferred alignment in injection molding them in a silicon nitride matrix.

4.4 COLLOIDAL PROCESSING

4.4.1 Slip, Tape, and Pressure Casting

Forming ceramic bodies by slip casting is a long standing process consisting of making a slurry of the green body constituents, then casting this in a porous mold, typically of (naturally porous) plaster, and more recently porous plastics. As much of the slurry liquid at or near the mold surface is absorbed in the pores in the mold, a layer of solid—the cast layer—is formed by the interlocking solid particles in the region near the mold surface. As the process continues, this solid layer increases so long as the mold pores continue to absorb slurry liquid. While thick deposits, including solid bodies without any cavity, can be made, slip casting is particularly suited for making hollow bodies, often with thin to modest wall thickness (Table 4.1). Such hollow bodies, which can be quite sizable and versatile in shape, are made by simply pouring excess slip out of the mold. The first of two key factors is the slurry, that is, the slip. For environmental, cost, practical, and historical reasons water is the primary liquid used, but some other liquids are used. For reasonable shelf life and uniformity of the resultant green body, stable slips are important, for which there is substantial background on the colloidal chemistry necessary for such slip stability [59]. (Increasingly, such colloidal concepts are also being used to guide development of other colloidal systems, such as binders for injection molding and extrusion.) A major application and driving force for further development of slip casting has been the sanitary ware industry, since ceramic toilets, urinals, and washbasins would be difficult and very costly to make by other means. They being large, hollow shapes with modest wall thickness make them amenable to slip casting [60]. These products are also a natural for the process, since clays are a major raw material for them, and clays are very amenable to forming good slips and bridging over die pores (clays were thus the origin of the technology, much as they were for extrusion). Today the process and its derivatives are used for a variety of other, mostly nonclay derived, ceramics and diverse applications, with potential for more use.

A key step in the process following the actual slip casting is drying, part of which occurs in the mold, causing shrinkage of the part that is important for its removal from the mold (typically made in two or more pieces to facilitate part removal), as well as subsequent drying. Common drying shrinkages are of the order of 3% or more, and can be anisotropic, the combination of which often requires process adjustment to avoid cracking. Another important step for production is drying of the molds so they can be reused.

Slip casting, as traditionally practiced, like any process, has its advantages and disadvantages/limitations (Table 4.1). The biggest advantage is its versatility, first in terms of shape and second in terms of size and materials applicability. It can also accommodate a range of particle sizes, working with typical particle sizes from a fraction of a micron (green densities of ~ 40–50% for ~ 0.5 micron particles [61] to several microns. It can also work fairly well with finer, nanoscale particles, though with challenges of particle flocks adequately bridging mold pores (mainly in newer plastic molds which have larger pores) and lower particle volume fractions and resultant green densities, for example, 32–36% for Al_2O_3-ZrO_2 composites [62] while most common for oxides, it has some applicability to non-oxides, e.g., SiC [63]63]. Traditional slip casting is done with low capital cost—standard mixing equipment, plaster molds, moderate drying facilities, and standard binder burnout and sintering facilities. Because of these factors, it has been a handy process to make a limited number of pieces, for example, prototypes.

However, slip casting has limitations, which, from an operational standpoint, are primarily its slow casting rates, with thickness proportional to the square root of casting time, and hence increased cost of casting and of drying to avoid cracking, as well as costs for preparing and maintaining a large mold inventory and facilities for mold storage and drying. Drying times of one to a few hours per cm of thickness are often required [3], though they may be reduced some for bodies by impinging of slip-cast deposits from adjacent or opposing sides of a mold. There are also some technical issues such as segregation of larger particles and resultant gradients of particle grains and oriented/laminar arrays of porosity (Fig. 4.4C), that, while not widely addressed, should be considered. However, flexure bars cut from slip-cast sonar transducers normal to the lamellar porosity had strengths comparable to those for bars from isopressed and the strong direction of die-pressed transducer rings [17]. These limitations result in slip casting being used in areas such as sanitary ware, since nothing else is really practical, but slipcasting being displaced by other processes for other products, such as sonar transducer rings.

While slip casting is often applied to modest size components, it can be used for larger bodies. Some of these are larger crucibles or other vessels where the wall thickness is modest, e.g., <1 cm. Again, sanitary ware such as toilets, urinals, and washbasins were made only by slip casting for years before improvements by pressure casting were introduced. Beyond this, large refractory bodies have been made. Thus, large, ~ 2.5 m long alumina refractories with lateral dimensions of ~ 0.3–0.4 m whose isopressing was successfully shown by Werenberg and coworkers [27], were typically made by slip casting.

The first of four developments that have significantly improved and extended the use of slip casting is the use of pressure in casting. This entails some form of pressure or stress to enhance casting rates and may often also include use of vacuum to increase fluid infiltration into the mold. One method of enhancing casting has been centrifugal casting, that is, casting in a mold in a centrifuge, which though studied for the dinnerware industry [60], has been primarily, if not exclusively, a laboratory tool. Such casting, which uses greater centrifugal stress on the colloidal particles versus on the fluid, can increase green density, e.g., to ~ 57% of theoretical density for TZP [64], as well as align reinforcing particles [65]. However, higher green densities (e.g., 60–70% of theoretical) and shorter casting times are an attraction [66]. Recent developments show promise for making larger ceramic tubes, due to lower rotating speeds, and hence possible increased safety [67].

Of much broader use is the application of pressure (e.g., from < 1 to a few MPa) on the slip to increase the fluid flow through the deposit and the mold and thus decrease the casting time, for example by three- to sixfold for a thickness of ~ 1 cm, increase green density, and reduce drying shrinkage, by two- to fourfold. Such pressure casting is now extensively used in the sanitary ware industry, where along with other advances, namely microwave drying of parts and molds and the automation of the process, has reduced both process energy needs and production time, for example, from 2–3 days to 9–10 hrs [68]. Polymeric molds have replaced plaster molds because of large increases in mold life, e.g. ~ 10,000 versus ~ 100 cycles respectively for plastic and plaster molds (but also with much larger pores in plastic versus plaster molds, hence added challenges for slip casting fine powders).There has been considerable interest in application of pressure casting to ceramics for high-performance applications such as turbine and piston engines, since it shows promise of substantially increasing component reliability by reducing processing defects [69]. While this is promising, issues remain, in particular whether components with such improved performance can be produced cost-effectively. Thus, for example, residual stresses can be a factor in such casting, and use of higher pressures to further reduce casting times and increase green density can be limited not only by diminishing returns, as well as bubble formation by gases dissolved in the slip liquid under pressure, being released to form bubbles upon pressure release [70], that is, as in the "bends" in the blood of divers (Table 4.1).

While sanitary ware manufactured by pressure casting represents sizable bodies made by the process, additional perspective on the potential capabilities of the process is given by earlier work (under direction of Professor Smoke, results reported to this author by Professors McLaren and Haber, Rutgers University, New Brunswick, New Jersey). The goal of making 90% alumina seal rings ~ 1 m diameter, ~ 2.5 cm thick with a wall thickness of ~ 7.5 cm for a fusion Tokomac reactor was met by pressure casting, achieving a cast density of ~ 54% of theoretical. The ~ 1% drying shrinkage, though low, represented a substantial radial shrinkage of ~ 6 mm, which could cause cracking given the limited green strength. To accommodate this radial shrinkage, individual rings were placed on an array of radially oriented pie-shaped pieces of Teflon sheet such that radial shrinkage could be accommodated by either slippage of the rings on the Teflon pieces or by these pieces sliding on their support. (Similar novel engineering solution to the much greater firing shrinkage of these rings is described in Sec. 4.7).

Interest in the technical capabilities of pressure casting were significantly heightened by more recent demonstrations of tensile strengths and reliability with silicon nitride compositions known to be capable of high room temperature strengths with limited processing defects. Pujari and coworkers [71] showed that pressure casting of specimens for true tensile testing of such compositions resulted in remarkably high tensile strengths (e.g. ~ 1 GPa) and high Weibull moduli (~ 20), which are both much higher than have been obtained with injection molding. Lyckfeldt and coworkers [69] showed in a detailed study of applying pressure casting to making prototype silicon nitride engine rotors that many of the challenges of casting thicker (e.g., > 50mm) and more complex cross sections could be adequately met, but casting times to achieve this were long, ~ 3.5 hrs. They noted that use of a partially flocculated slip could reduce casting times by two- to three-fold, by bridging mold pore openings with smaller particles, but use of such slips required more development to avoid casting problems. Thus, pressure casting has substantial promise, but casting issues of quality versus faster casting (and lower costs) remain.

The fourth development that significantly extended the use, versatility, and productivity of slip casting is tape casting. This consists of casting a thin layer of slip onto an impervious surface, commonly a Mylar sheet that allows it and the cast sheet to be subsequently rolled up for storage and handling [72]. Casting is typically via a doctor blade system, that is, a slip-containing reservoir that sits on the Mylar tape and slides under the doctor blade reservoir so tape is formed by the Mylar sheet and the reservoir moving relative to one another, with slip flowing out of the bottom of one side of the reservoir, where there is an adjustable narrow gap between the reservoir bottom and the Mylar. Tapes of various thickness, e.g., < 10 μm to several tens of microns, are cast by adjusting the extent of this gap to various heights above the Mylar, as a function of the gap, the Mylar speed, and the slip character and drying shrinkage. Drying of such tapes occurs into the air above the top side of the tape rather into the surface (Mylar) on which it is cast. While there is increasing interest in aqueous-based slips for tape casting, many are nonaqueous based systems. Tapes up to a meter or more in width are produced in very large quantities for ceramic, especially alumina, substrates and multilayer packages, as well as barium titanates for multilayer capacitors. These, along with thicker colloidal pastes of metals and of other ceramics for screen printing (horizontal, i.e., *x-y*) electrical conductors and electrical components on the tapes, are basic to the large ceramic electronics industry (Fig. 4.12).

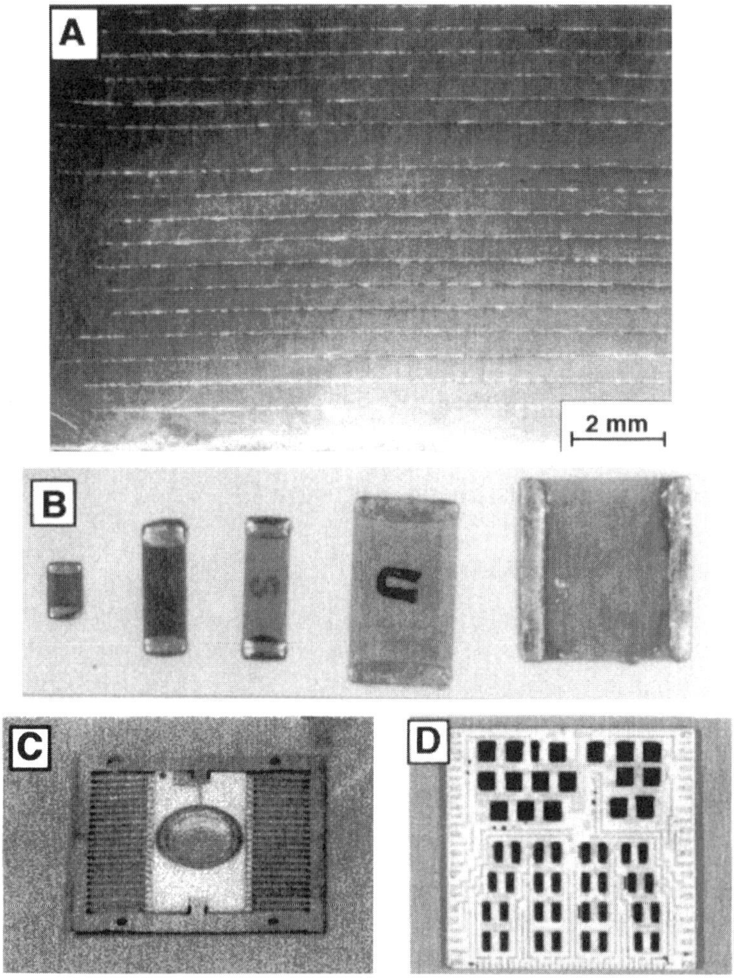

FIGURE 4.12 Examples of important electronic ceramic products made with tape casting. (A) Cross section of a multilayer capacitor showing interleaved metal electrodes; (B) Examples of finished capacitors, some with dimensions > 1 cm; (C) A simple multilayer package (with lead frame, about 2-cm square); and (D) a similar sized substrate with passive components.

For these (and other) applications punching of holes in the tapes (for "vias" to be filled with metal pastes for vertical, i.e., *z* direction conductors) and especially lamination of tapes carefully aligned on top of one another are important steps. However, ceramic tapes are also used for other applications, e.g., miniature ferrite memory cores (Sec. 4.8), and are of interest for other possible applications including structural composites, rapid prototyping (Sec. 7.4), and other possible applications such as fuel cells.

Several additional factors that indicate further potential for slip and tape casting should be noted. First, though not treated here, these processes, especially slip casting, are also important ceramic coating techniques for a variety of ceramic coatings for ceramic and metal substrates. Second, substantial work continues to expand the technical base of slip preparation (see [73,74]) and use, for example, regarding possible effects of using slips above room temperature via heating, including microwave heating of slips [75]. Third, slip and tape casting has significant potential for fabricating ceramic composites (particulate, whisker, and laminar) and functionally graded materials [76–78].

4.4.2 Electrophoretic Deposition (EPD)

Another important colloidal deposition process is electrophoretic deposition (EPD), which is most applicable to dielectric materials [79–82], but can be used some for deposition of electrically conducting particles. Electrophoretic deposition uses electrical charges that commonly form on the surface of colloidal powder particles in a liquid medium by applying an electric field between two electrodes with the colloidal medium between them. This typically uses electrical charges forming on the particle surfaces due to one of the following mechanisms:

1. Desorption of ions at the interface with the liquid, for example, on clay particles
2. Chemical reaction between the powder particle surface and the liquid, for example, of oxide powders and
3. Preferential adsorption of specific additives or impurity ions, for example, surfactants or polyelectrolytes

Thus, EPD uses an electrical field across a colloidal medium to drive the charged particles to the electrode where they will be neutralized and thus form a deposit of particles, a green body. This is analogous to electroplating except EPD uses an electric field to deposit charged powder particles in a fluid instead of a solution with ions to be deposited as atoms.

Electrophoretic deposition, like processes such as extrusion and slip casting, was originally discovered and developed with clay-containing bodies using water as the liquid, but has been applied to a variety of ceramics. Dielectric particles can be used in water, where the surface charge can be adjusted by changing the pH, as well as in many organic liquids, which can also be used with many oxide and nonoxide powders. Key factors increasing the rate of material deposition are increases in the density and size of colloidal particles, their charge, and the electrical field that can be applied between the electrodes. However, deposition is limited by too high a conductivity in the liquid, which can result in electrical breakdown, and more commonly by electrolysis in water-based systems and the release of O_2 and H_2, which can present limitations, for example, due to bubbles in the deposit, as well as possibilities, as discussed below.

There are some ways of avoiding or using effects of release of O_2 or H_2 when using water as the fluid medium as discussed below, as well as important possibilities of using such gas generation to form promising novel porous bodies (Sec. 7.3). However, a general way of avoiding such electrolysis limitations with aqueous suspensions is to use nonaqueous suspensions for EPD, for which there is a substantial literature [78]. EPD with nonaqueous suspensions requires higher, but manageable voltages—e.g., a few thousand volts—versus aqueous suspensions, where a few hundred to well less than 100 V are typical. In either case EPD is generally best for particles of ~ 1–20 μm and suspensions that are neither too stable nor too unstable, and often is aided by various additives, again for which there is a fair amount of documentation.

EPD, like slip casting, is basically a coating process that can be used simply for coating as well as for making various free-standing bodies; there are several similarities and differences between the two processes. Thus, while both use female molds, EPD often also uses male molds on which to deposit; and while both may be often done with binders, these may give too much adhesion to electrodes, especially male ones for EPD. However, EPD often needs little or no binder and low adhesion and burnable graphite (pencile lead) or sprayed electrodes can be used. Though actual EPD deposition rates slow at higher deposit thicknesses, they are theoretically linear in time as compared with a square root dependence of slip cast thickness with time. Thus, EPD generally has very reasonable deposition rates and times, rates of the order of a mm/min. Green densities from EPD may be lower than for slip casting, ~ 50% of theoretical density, but can be up to at least 60%.

Electrophoretic deposition is potentially ideal for forming both coatings on a variety of surface shapes as well as small, thin wall devices. Its advantages as a coating technique stem not only from the very low capital needed, (basically a modest power supply and plastic tanks) but also from the versatility of shape that can be coated. This versatility results from the fact that deposition is driven by the electrical field that is normal to the local section of the electrode, so the geometry of the deposit generally follows the global and local geometry of the electrode on which the deposit is made. Further, many parts, especially smaller ones, can be made simultaneously in one deposition cell. Electrophoretic deposition was developed as the fabrication process for β-alumina battery tubes [83], but greater potential is indicated by several other developments.

Work to use EPD to make fiber-optic ferrules, besides being another possible application, illustrates further potential for the process. Thus, Kerkar and Rice [84] used a male mold consisting of a noble metal coated on a much stiffer metal, (W) as the deposition electrode for EPD of zirconia ferrules. The purpose of the noble metal on the deposition electrode was to absorb the hydrogen produced by electrolysis of the water (and to be rechargable by thermally removing the hydrogen).

An important development in EPD and for its future potential was the demonstration by Aksay [79,85] of a machine (called the "Elephant") and technique of forming continuous tape or slabs from ~ 3 mm and up to ~ 15 mm thickness of clay containing bodies for tile and possibly dinnerware production, but potentially applicable to other bodies. This machine consisted of two large (1.5 m diameter) side-by-side hollow metal rolls that counter rotated at 5 r/hr, yielding tape or slab speeds of ~ 25 m/hr or ~ 40 cm/min. The rolls were separated at their closest approach by the thickness of the tape or slab to be produced and were the anodes on which the deposit was made on their outside surfaces. In the upper, curved V-shaped area between the two rolls is held a matching V-shape body that does not contact the rolls since it is the cathode, and the space between its curved surfaces is where the tapes begin to form on each roll. The lower portion of the V-shaped body is a pump supplied slip reservoir and distributor, and the upper portion is a collector and drain for excess slip to be recycled. The tapes formed on each roll over their length (30 cm in this case) are laminated together by the rolls and the moving takeup surface since this doubles output and also results in a symmetrical distribution of stress gradients in the two tapes, hence limiting their effects. The outer surfaces of the rolls are coated with Zn metal to absorb the oxygen given off there, converting it to ZnO, which must be removed periodically (e.g., once a year) or continuously by brushing, and the consumed Zn replenished. Annual consumption of Zn was 12 mm, which was ~ 0.5 kg of Zn per ton of tape produced, along with 22 kWh of electrical power per ton of tape to operate the machine. The machine has been produced and used commercially.

Another development of EPD to note is making bodies of laminated or graded structures in terms of composition and microstructure [86–88]. While any tape forming technique can be used for making such graded structures, EPD allows finer gradation steps, which can be in bodies of much more diverse configurations than can be made by conventional tape lamination. However, making tapes by EPD for lamination can also be advantageous for both the finer gradation feasible as well as speeding the process by making several tapes simultaneously, for example, in the same bath to compensate for moderate deposition rates. Also note that some experimental work has apparently shown potential of depositing more complex shapes and structure on an array of disposable (carbon) electrodes and by creating more complex electric fields within the deposition bath.

Thus, in summary, while there are constraints in using EPD for bodies with thicker cross sections, (due to deposition rates and electrolysis), there are a variety of capabilities that make it an attractive possibility for a variety of material and component applications. These capabilities include considerable shape, composition, and versatility, as well as some ways of reducing constraints on it; thus expanding its potential use and opportunities by taking advantage of its simplicity, modest cost, versatility, and potential for close control. Thus, EPD has considerable potential beyond its current use and investigation.

4.5 MISCELLANEOUS POWDER CONSOLIDATION TECHNOLOGIES

The first of a few less developed or used powder consolidation methods that are briefly noted is explosive compaction. Shaped explosive technology has been used to experimentally consolidate powder in a metal container to substantial density, but is seriously limited by not only safety and cost issues, but also by residual stress and cracking issues. Considerable use has been made of the ability to achieve high powder packing densities of powders of carefully designed, broad particle-size distributions via vibratory compaction, usually in tubes—e.g., for nuclear fuel rods and for electrical heating elements used both in home electric ranges and industrial heating elements. Tubes are typically swaged to high density, for example, to increase thermal conductivity, but in the case of nuclear fuel elements, to still allow escape of gases formed by nuclear fission. In this context it should also be noted that a very high percent of theoretical densities can be achieved by placing even very fine powders in steel tubes then cold rolling them fairly flat due to the very high local pressures between the rolls. Again, however, residual stress and cracking upon extraction, as well as cost issues, greatly restrict practical possible use.

Besides vibratory compaction and tube swaging, another less known compaction process of some established practical use is direct roll forming or compaction of powders without metal tubes. Such direct roll compaction has received considerable investigation for a variety of materials and applications, with some successful applications, and potential for considerably more. This basically consists of feeding powder between two counter rotating rolls that are usually horizontal with varying types of powder feed from the top and the compacted product, often tape or sheet exiting downward from the bottom of the rolls. However, specifics of the operation vary considerably with the material and application.

For ceramics, roll compaction has been mainly used to produce tapes, often thicker than typical tape casting (e.g., 0.5–1.5 mm and ~ $^1/_3$ m wide), such as for thicker electronic substrates, at substantial rates (e.g., several cm/sec [3,72]. It requires more rigid powder preparation, usually spray-dried (with binder) and screened between 80 and 325 mesh, and it is reported better for fixed high volume rather than small volume or experimental runs, and requires no drying step. Roll compaction is also used for pharmaceuticals, for example, to produce simple compacted shapes to be used for forming denser granules for final pill pressing [89], where an evacuated auger feeds de-aired powder between a fixed and a moveable rotor, which are counter-rotating at the same speed. Compaction pressures of ~ 140 MPa are indicated via use of plant air pressures on the rolls, with roll compaction taking only ~ $^1/_6$–$^1/_4$ the time for comparable die pressing. Other nonmetallic applications apparently include briquets (e.g., of charcoal) that are done at high volume and high rates.

Roll compaction has been more extensively investigated for consolidation of metal powders, where it has a long history and takes various manifestations as addressed in the comprehensive review by Dube [90]. Two of the most common manifestations are to produce tape or sheet from metal powder by either of two routes. These are by direct roll compaction of the powder without any binders or sintering, or by partial densification of powders with binders, that is, roll compaction of green tapes or sheets (followed by sintering and possibly cold or hot rolling). The latter compaction of green powder metallurgy sheets or tapes is directly analogous to ceramic roll compaction and thus a good source of information for such ceramic processing. However, even direct rolling of metal powders to high densities has relevance to roll compaction of ceramics since compaction of metals to ~ 91% of theoretical density is estimated to involve only about 12–15% bulk deformation. Metal results show that (1) particle morphology and surface roughness are important (e.g., while smooth spherical powders have good flow, they do not roll compact well); (2) larger rolls are better for higher densities and greater product thickness; and (3) there is a substantial decrease in density near the edges of roll compacted powders (also seen in ceramics) that must be addressed (removed or possibly reduced or eliminated by possible modifications of the roll compaction equipment). Note that while metal powder costs have often been a limitation (and binder costs for green sheet compaction a factor), there are existing applications of roll compaction of metals, and they are expected to expand. Further, some applications are for porous, not dense, metal sheets—e.g. for battery and fuel cell applications. There is some indication that roll compaction of strip with different surface versus interior composition is feasible.

Another possible way of forming ceramics is to have either a liquid precursor or a liquid slurry of ceramic powder where ingredients in the solvent can be polymerized, for example, catalytically or thermally, rigidizing the previously fluid system once poured in a mold, from which the rigid body can be removed. An example of this is sol-gel processing, where sols are poured in a mold then rigidized via polymerization (for alkoxide sols) or extraction of water (for colloidal sols). Thus, for example, Becher and coworkers [91] gelled alumina sols and sintered resultant bulk gel bodies (through various phase transformations) to bulk alumina bodies, approaching densities achieved in commercial sintering of alumina. However, the low alumina content of the sols and the resultant large shrinkages on gelling and on sintering were judged to make such routes at best limited technically as well as due to cost. Some industrial work was later conducted on use of sols as the binder for processing commercial alumina, but was abandoned.

Hench [92] more extensively investigated casting of SiO_2-based glasses from silica sols, developing additives to aid in drying. This technology apparently later became the basis of a commercial venture to produce low cost optical lenses that could be formed and sintered directly to final optical performance without any conventional lens machining and finishing. However, this approach was apparently unsuccessful due to difficulties in accurate control of the large shrinkages involved and was apparently replaced in this venture by use of low-melting glasses that could be molded by more conventional, but lower temperature, glass processing.

Significant interest later increased in rigidizing of liquid systems in a die or mold via use of organic binder gelling agents, given the name *gelcasting* [93]. This consisted of using acrylate monomers in organic solvents or acrylamides in water, with gelling via polymerization in either case. Promising results were demonstrated in casting various shapes from slurries of ~ 55 v/o alumina. Toxicity issues with these monomers led to development of other low toxicity systems [94]. Such liquid casting systems should be advantageous to preferred orientation of particles in electric or magnetic fields, with the latter recently being demonstrated [95]. PVA-based binders have also been reported for gelcasting use [96], and others are likely to be found, making this a possible alternative fabrication route, though more study and control of drying and firing shrinkages may be important. Such gelcasting processes, which entail gelling of organic materials, raise the question of whether observations of some pure polyacrylamide gels being greatly reduced in volume (e.g., by > 100-fold) by immersing them in acetone-water mixtures and applying small voltages (e.g., 2–2.5 V) across the gel [97], might be applicable to drying some gelcast bodies.

4. 6 BINDER SYSTEMS, DRYING, GREEN MACHINING, BINDER-BURNOUT, AND BISQUE FIRING/MACHINING

Binder technology is such an important aspect of ceramic fabrication that it deserves some added attention over and above that briefly given in the preceding sections on specific forming methods. This attention to consider some overall issues, needs, and similarities and differences in binder systems for different forming methods is still only a small fraction of this large and complex area. There has been much research attention in this area in recent years, but specific practical guidance, especially based on industrial practice is more limited since this is typically considered proprietary information. Some books on ceramic fabrication provide useful guidance on binders, [1,3,4,7]; some suggested literature articles are [98–106], and suppliers often provide useful information. A useful compilation of not only binders for various forming operations, but also general use of them and other organic additives in ceramic firming and processing is found in [106].

Overall binder needs are to meet enough of three criteria to be useable. The first and basic one is to provide either the "solid" deformation of the powder mass, or flowability of colloidal slurries or ceramic precursors, needed to form a consolidated powder compact suitable for the component, that is, near its shape and dimensions, allowing for shrinkage on firing. This deformation criteria generally significantly increases from least to most in the order of dry, isopressing or die pressing, extrusion, and injection molding, with significant increases in the extent of plastic deformation and, hence, total binder contents being required for the latter two fabrication processes. However, inadequate deformation or destruction of the relic structure of agglomerates, especially from spray-drying can be a problem. Injection molding has typically been with higher temperature, commonly thermoplastic based, binder systems, which have also been used for some extrusion, which is normally done with room temperature, water-based binder systems.

Binder contents for colloidal processing, e.g., slip, tape or pressure casting, or EPD, after drying are similar to, or less than, those for dry pressing. However, the amount of solvent needed for flow of such colloidal slurries is generally similar to or greater than that needed for extrusion, and binder plus solvent contents are similar to binder contents in injection molding. A critical difference, as discussed below, is removal of solvents versus other binder constituents and sinterability. The true green density, that is, the true volume fraction of the actual ceramic product in the green compact, is a key measure of the latter two issues, with green densities of at least 50–60% of theoretical being generally desired to necessary. Such densities are necessary for suitable firing, but can also be factors in limiting drying shrinkages and possible cracking from drying.

The second basic binder requirement is to provide sufficient strength for green body handling and subsequent stressing. The first challenge is removal of the part from the mold or die, for example, preventing cracking from springback or end-capping (in die pressing). However, beyond this, there can be a variety of demands, an important one being green machining, that is, machining with conventional tungsten carbide machine tools, as opposed to typical machining of fired ceramics which is much more expensive than green machining. Note that green machining parts, especially from isopressed bodies (logs) is an important method for making ceramic prototype parts (Table 4.1). Such machining may dictate the type of binder (e.g., acrylics) and their amount [102].

(In some cases, parts may be bisque fired beyond binder burnout with some sintering where they can still be machined with WC tooling due to sufficient strength from initial sintering.) Binder needs may also increase significantly for larger part sizes or extremes of shape due to stresses in handling such bodies and the limits in reducing such stresses by better handling methods. Finally, binder strengths can be a factor in limiting thermal stress cracking of green parts during heating for binder removal.

Actual binder removal is the third key factor in choosing and using binders, with part size, shape, and green microstructure being important factors along with acceptable firing atmospheres. Thus, binder removal is generally easier in oxidizing atmospheres since these allow oxidation of binder residues, which often occur since many binders do not completely break down to volatile and readily oxidized fragments as would be ideal. More challenging is removing binders from bodies that contain oxidizable constituents, such as bodies with metallization, for example, in cofired electronic ceramics, nonoxide ceramics, or composites with nonoxide constituents, where little or no oxidative character to the atmosphere can be tolerated. For such cases there is a general need and desire for binders that decompose by direct volatilization, pyrolysis, or both, and do so over a range of moderate temperatures. While individual binder constituents often burnout over a range of temperatures as desired, removal over a range of temperatures is often purposely enhanced by addition of constituents that extend the burnout temperature range. Thus, addition of limited binder constituents that burn out at lower temperature are often made to enhance burnout of the main binder components by the former's earlier burnout providing more ingress of the burnout atmosphere into the body and more egress of burnout products out of the body. It is also often feasible to control the oxidative character of the binder burnout atmosphere so it will oxidize the binder constituents, but not the most oxidizable body constituent, for example, by controlling water vapor content in firing ceramics with metalization.

Note that while some binder constituents may be added to aid binder burnout as noted above, there are some binder systems in which this is inherent. A large portion of these are binders with large solvent contents, in which case, solvent removal by drying thus provides easier burnout of remaining, more stable binder constituents. While this includes binders with organic solvents, those with water as the solvent are particularly advantageous. However, note that such drying as an early stage of binder burnout presents challenges of shrinkage cracking as does binder burnout and sintering. (Note that drying has received more recent research attention; see the review by [105].) There are, however, other, nonaqueous, binder systems with useful aspects of binder removal, of which binders based on polyethylene and mineral oil are a key example. This thermoplastic binder system phase separates on cooling from temperatures where it has suitable plasticity for forming bodies, with each phase being totally interconnected with itself. Thus, the mineral oil can be readily removed by evaporization or by solvent removal prior to, or separate from, subsequent pyrolysis of the polyethylene. Note that the polyethylene is both sufficiently strong to maintain component integrety as the sole remaining binder constituent, and that it is one of the cleanest burning binder in oxidizing or other atmospheres since it readily decomposes to volatile species at reasonable temperatures (e.g., being used for the initial development of large multilayer AlN electronic packages [107,108].

Besides the above key issue of actual binder removal during burnout, there is also the key requirement to do this without disruption of the body by generating pores or cracks. Such defects can arise due to either or both of two factors in binder burnout, which are

1. Excessive generation of gasses in the body with inadequate outgassing from the body
2. Differential ceramic-binder thermal expansions, during burnout

Thus, when local or global generation of burnout gasses in a body sufficiently exceeds the ability of the pore structure to allow gases to move out of the body cracks, larger pores, or both can develop. Such outgassing problems obviously increase with the rate of generation of gases which is both a function of the amount and type of binder and the heating rates. Thus, injection-molded bodies that commonly have more and higher temperature binder may require several days to a week of more for binder burnout (in a separate furnace from that for sintering to be cost-effective), while die-pressed bodies may have binder burnout during heating for sintering. Binder burnout problems also increase as the length and tortuosity of the pore channels for gas removal increase, which means increases with both the body dimensions, particularly the smallest ones, as well as with decreasing body particle sizes. Increasing body dimensions is also a key factor in problems from ceramic-binder expansion differences, which result from binder being mostly or completely removed from the outer portions of the body, with limited removal from the innermost portions of the body. Such outer portions are often weak and have the more modest thermal expansion of the ceramic, while the inner portions with substantial binder have higher thermal expansion dominated by the typically much higher binder (plastic) expansions. Thus, on continued heating the higher expanding interior puts the lower expanding, more friable exterior in tension which often cracks the exterior during binder burnout. Such cracks generally result in surface or near surface pores on sintering and can be a frequent problem in larger alumina wear tiles [109].

Finally, note a few other factors about binders. They are a measurable cost factor, some because of material costs, for example, where large contents are used, and also for the processing steps for both their use and removal. Mixing of the various binder-system constituents may require the addition of some constituents before others; e.g., surfactants may be denied powder surface sites needed to be effective if they are not added early, usually first. Additional additives are often needed to control side effects of other ingredients; for example, defoamers are commonly needed since foaming during milling of slurries can be a problem. Thus, interactions between various organic additives must also be considered [110]. Also, the presence of low levels of impurities or metallic constituents (e.g., from catalysts in preparing polymeric constituents) may have to be addressed. Some binders such as PVA can be measurably affected by humidity, and hence show seasonal and other variations [11,12]. The issue of deformation of spray-dried agglomerates and resultant defects in die pressing has attracted research [111–115]. While burnout is the most common method of removal, chemical removal is possible in some cases (e.g., with the oil-polyethylene binder), and removal by wicking out of the sample solvent extracted binder constituents is also feasible. The latter can be aided by burying components in powder that aids the wicking, but effects on binder removal time and net costs of the burying powder, its addition and removal must be considered. Also there may be effects of electric versus gas furnace heating on the amount and type of binder removal since gas combustion products may slow binder burnout at high furnace loadings. Controlling furnace and especially part temperatures during binder burnout may be a challenge due to exotherms from combustion of binder products in air firing.

4.7 SINTERING

Firing of ceramic bodies to uniformly sinter them to the desired dimensions and properties (and thus to a certain microstructural range) is commonly the last step in actually making a ceramic body, though other steps such as machining, metalization, coating, and inspection my be yet to come. As noted earlier, sintering may be after a separate binder burnout or bisque firing stage, or as a continuation of binder burnout. Successfully achieving this central role of sintering presents challenges of not only meeting the desired overall temperature-time schedule, which may vary for different components of the same body and with the atmosphere, temperature uniformity during the firing cycle, and handling of shrinkage and other deformation issues. All of these can vary with the size of the furnace and of the size and shape of the components being fired.

Consider first the furnace atmosphere and temperature. Common oxide firings are in air at temperatures to 1600–1700°C, which allows use of efficient tunnel kilns, as well as belt or periodic kilns of respectively decreasing cost effectiveness for volume production. Higher temperatures can be achieved with ZrO_2 resistive elements but at great reduction in furnace size and higher cost. Temperatures over the same range or substantially higher can readily be achieved in vacuum, inert, or reducing atmospheres, though furnace sizes tend to decrease and operation costs increase as temperature increases. Such furnaces, which as noted earlier present challenges for binder removal, are typically periodic, but belt furnaces to fire silicon nitride have been demonstrated. Firing of some nonoxide bodies such as Si_3N_4, SiC, or AlN requires limiting their vaporization, which can be accomplished using furnaces in a pressure vessel such as a HIP unit, or a vessel of less pressure capability (e.g., 100 atm or less), but both at higher costs. However, adequate atmosphere effects can often be obtained by firing parts in an inert or nonoxidizing (N2), atmosphere by burying the parts in a powder of similar composition as that of the bodies being fired. Thus, for firing Si_3N_4, Si_3N_4 powders with the same or similar sintering additives are used, but often with coarser particle sizes for the burying powder to limit it sintering to the parts, and potentially allowing its reuse since such burying powders are typically a measurable cost item. Some oxides also present vaporization problems, e.g., those such as PZT losing PbO. In such cases parts are often fired in compatible (e.g., MgO) crucibles with lids and some excess powder of the composition or the most volatile species such as PZT or the volatile species, PbO.

A critical factor for viable production is temperature uniformity on both a global and local scale in the furnace, particularly over the temperature range where measurable sintering occurs. Global uniformity is needed for adequately similar density, hence shrinkage, to be achieved in parts over most, preferably all, of the firing volume of the furnace, since successful firing of many specimens at once is commonly essential to the economical viability of products. Also of importance is the temperature uniformity locally, that is, on the scale of the parts being fired. Failure to have this will commonly result in differential sintering of some component areas, which commonly results in various combinations of not only variable microstructures, and hence variable local properties in a given component, to warping and distortion, as well as possible cracking of components. Kiln furniture, and use of trays or crucibles in which to fire parts, sometimes with burying them in powder, can be important to adequately uniform sintering. Such component uniformity often is also related to control of creep and of shrinkage of parts during firing as discussed below.

Creep of parts during sintering becomes an important firing problem as firing temperatures, part sizes, especially in one or two dimensions, and the extent of component creep increase, the latter typically being significantly exacerbated by use of most densification aids (e.g., in Al_2O_3 and Si_3N_4 with typical additives). To limit creep problems in firing, large components must be given substantial support to minimize deformation. However, note first that in some cases creep during sintering or after can sometimes be used to shape components (referred to as slump forming), but generally does not produce the highest performance components. Second, good support of larger components to minimize creep deformation during sintering may often exacerbate serious problems of constrained shrinkage.

Allowing free, unconstrained, shrinkage, which is essential to achieve the properties expected for the component and its processing, is often not a problem for small components. Thus the typical 15–20% linear shrinkage of green parts with maximum dimensions of ~ 1 cm means linear shrinkages of only 1.5–2 mm (actually usually half of these values since parts typically shrink about their center) against modest frictional forces from the small mass of the component. Often placing small components on some coarser grain or in a bed of burying grain of a compatible composition is enough, since for example the grain allows some movement by rolling, sliding, or otherwise shifting to accommodate the small shrinkages of small parts. However, even parts of modest size—dimensions of a few or more centimeters, with recesses, shoulders, or blind or through holes—may present shrinkage problems due to friction on the shoulder and higher frictional forces with greater mass, or constraints of grain particles in recesses or holes, especially as their dimensions increase.

Keeping constraints on sintering shrinkage small becomes an increasing critical problem for successful firing of larger components due to both the larger actual shrinkages and the greater component mass and resultant friction between the component and the surface supporting it. These problems and a clever engineering solution for some cases are illustrated by the firing of 90% alumina seal rings for a Tokomac fusion reactor that were pressure cast (Sec. 4.4.1) to ~ 54% of theoretical density with an OD and ID of ~ 122 and 114 cm and a thickness of ~ 2.5 cm. The radial firing shrinkage of ~ 20% thus represented radial contraction of the rings of ~ 12 cm! This large radial firing shrinkage was accommodated by placing the rings on a support structure made of a combination of two sets of flat, pie-shaped plates of SiC (similar to, but thicker than the pie-shaped pieces of Teflon sheet used to accommodate drying shrinkage); a large number of commercially produced ruby ball bearings (~ 2 mm dia.); and typical alumina grain used to accommodate sintering shrinkage. Larger pie-shaped SiC pieces were oriented radially as the base of the shrinkage accommodating support structure, with ~ 40 ruby balls placed on top of each SiC plate in this bottom array of SiC plates. Then the smaller pie-shaped SiC plates were placed on the ruby balls, and alumina grain was placed on top of the smaller SiC plates. The ruby balls were glued in place on each plate with model airplane glue, so the balls could not move during support assembly, placement of the seal ring on the support, and loading the system in the firing furnace, but the glue readily burned off during heating to allow the support to function as follows. Most of the radial shrinkage, hence also circumferential shrinkage, was accommodated by the ready radial motion of the top, smaller SiC plates over the larger, bottom SiC plates via the ruby balls between them, and the limited differential motion of the seal ring relative to each of the top SiC pieces was accommodated by the alumina grain between them and the seal ring.

The first of two important added aspects of sintering that are briefly noted here is alternative heating methods beyond those traditionally used for firing—i.e., gas firing, electrical resistance heating, or inductive heating. While microwave heating has received most attention, there are other newer heating methods that may have some future applicability, as discussed in Section 8.2.2. However, note that the advantage of rapid heating that alternate heating methods often offer is commonly limited by binder burnout, outgassing of powders, and thermal stresses. The second and larger topic is reaction sintering; that is, sintering constituents that not only are to densify, but also react during sintering to give the desired product composition—often a composite ceramic. Such reaction sintering, which is potentially important, is discussed in connection with the important broader topic of reaction processing in Section 6.5. However, note here that reactions typically involved commonly result in significant increases in constituent densities, which thus generates substantial new porosity in the reacting body. If such additional porosity is formed simultaneously with, rather than mostly after sintering of the starting constituents there may be greater limitations to achieving low porosity in the final body.

4.8 DISCUSSION AND SUMMARY

The substantial and growing diversity of powder-based fabrication options not only allow a broad range of components to be fabricated, but also often provides different fabrication options and thus the opportunity and need to make fabrication/processing choices. Some choices are often clearly based on component size, shape, or dimensional requirements. Thus, long straight cylinders or tubes with homogeneous character and a substantial range of cross-sectional shapes will often be made by extrusion. However, these might also be made by isopressing, slip, or pressure casting, and possibly also by EPD or even tape winding (for tubes), depending on availability of facilities and experience as well as the numbers to be produced. Change any of the above characteristics, and fabrication options change. Thus, short cylinders or tubes are likely to open some opportunities for die pressing and possibly injection molding (which often compete for many smaller components). Large diameters would begin to eliminate extrusion and favor at least some of the above options, and graded cross sectional character and varying cross-sectional dimensions, for example, for tubes with side openings or flared or closed focus on alternates to extrusion. The materials involved also can enter the evaluation—compositions with some mineral ingredients, especially clay, that enhance plastic forming can aid forming of more complex shapes, such as pipes with an additional entrance besides their two ends [1]. Besides availability of facilities and experience the many fabrication choices are driven basically by achieving the component character (configuration, dimensions, performance) and costs needed. These are interrelated, often in a complex fashion as discussed in Section 1.4; character variations increase costs to correct the deficiency or even more severely by rejection of components.

Briefly consider options for preparation of plates for electronic substrates, which are generally made by tape casting. They are also sometimes made by die pressing of thin plates, and can be made by extrusion, for example, some earlier development of AlN substrates (and subsequent development on multilayer AlN packages) used AlN tapes made by extrusion with a thermoplastic (polyethylene-mineral oil) binder. Both EPD and roll compaction may also be possibilities for some applications.

It is also important to consider variations in development, under competition, and with; for example, the important commercial development of automotive ceramic catalyst support honeycombs. At least two initial competitors pursued development via differing processes of making ceramic tapes (at least one via extrusion) and of forming them, for example, by calendering the tapes and rolling them into either individual monoliths or larger rolls from which individual monoliths could be cut. However, successful development was via extrusion, which required a great deal of development (done by one of the two above competitors), and remains the dominant method of manufacture (though apparently with changes over time, between ram and screw extrusion).

Next consider the first of two examples of fabrication changes over time. Ferrite memory (toroidal) cores have been an item of commerce for about 50 years or more, but their use and manufacture has changed over the years, mainly driven by minaturation. Thus, in the 1950s, cores with diameters of ~ 2 mm were produced by die pressing, but with diameters shrinking to ≤ 0.5 mm a switch was made to a high-speed tape casting process with cores then punched from the tape, which has continued with further reductions in size, to diameters of ~ 0.15 mm. Note that such minaturation is very common for many electrical, and especially electronic, components; for example, of ZrO_2 exhaust sensors in cars and many aspects of ceramic electronic packages.

The second case is the evolution of the manufacture of PZT rings for sonar transducers, where the rings are typically short sections of tubes ~ 1–2-cm height, < 1 cm wall thickness, and ~ 4–10-cm diameter (Fig. 4.4). Earlier use was made of slip casting or die pressing, but plant area and drying issues (of time) of the former and pressing defects of the latter led to production by iso-pressing tubes that were then gang sliced to length after sintering.

Briefly consider now the size capabilities of the various fabrication methods. These are determined by limitations of both the fabrication process and sintering, with the two often interacting—injection molding is limited by both adequate die fill and binder burnout-sintering, both of which are significantly dependent on the binder behavior. As noted earlier, sizes are generally not limited to specific dimensional limits, but instead are limited by either increasing costs to achieve the desired larger parts, or by reductions of part capabilities due to sacrifices in microstructure and hence performance of larger components, or commonly some combination of both, all of which are generally impacted by the material, the fabrication process and its parameters, and specifics of the component size and configuration. Thus, for example, clever engineering was noted for solving shrinkage challenges for both the drying and firing of large pressure cast alumina rings for a Tokomak reactor. However, this was at the expense of increased costs, and the success of these methods would most likely be substantially reduced for other similar geometries; for example, shrinkages of solid discs of the same thickness and diameter would have been much more difficult to accommodate due to the much greater contact area between the part and its support and the greater part mass both greatly increasing frictional resistance to the shrinkages of the parts. As summarized in Table 4.1, large parts of substantial mass—a meter or more in one dimension and tens of centimeters in lateral dimensions—can be made by, in approximate order of decreasing size capability, die pressing, isopressing, slip and some pressure casting, and extrusion. Longer lengths with more moderate lateral dimensions can often be made by extrusion, isopressing, some casting methods, EPD, and possibly roll compaction.

The above are guidelines for selection of fabrication based on component geometry. Where more than one process is suitable from the geometry and related cost perspective, selection then is made on a performance basis. However, this is more complex since the property capabilities of the processing methods of this chapter are often competitive, and vary with a variety of property, material, and processing factors; nevertheless there are some guidelines. The promise of pressure-cast bodies over those from injection molding noted earlier is one important example. Further, effects of the amount, location and character of residual porosity and of grain size and orientation on component performance, noted in Table 4.1, though often complex, can often be important guides. Thus, for example, axial porosity in extruded bodies are detrimental to axial electrical breakdown, but more benign for axial stressing. However, competition between fabrication methods of this chapter and those of Chapter 6 will often favor some of those of Chapter 6, provided they meet geometry and cost requirements.

Thus, in summary, there is a diversity of fabrication methods and processes from which to choose. Choices are impacted by materials and microstructures to be fabricated, as well as availability of facilities and experience, but particularly by component character (shape, size, and properties) and costs, with the two being interrelated. Also note (1) additives often play an important role in various fabrication (e.g., in colloidal technology; [73,74]) and in densification (Chap. 5), and (2) there are also other important fabrication methods (Chap. 6) that, though often not given as much attention, have significant potential to compete with typical powder sintering methods of this chapter.

REFERENCES

1. F. Singer, F.F. Singer. Industrial Ceramics. London: Chapman and Hall, 1984.
2. W. Ryan. Properties of Ceramic Raw Materials. New York: Pergamon Press, 1968.
3. J.S. Reed. Introduction to the Principles of Ceramic Processing. New York: John Wiley & Sons, 1988.
4. D.W. Richerson. Modern Ceramic Engineering. New York: Marcel Dekker, Inc., 1992.
5. P.E.D. Morgan. Chemical processing for ceramics (and polymers). In: H. Palmour III, R.F. Davis, T.M. Hare, eds. Processing of Crystalline Ceramics, Materials Science Research, Vol. 11. New York: Plenum Press, 1978, 67–77.
6. R.W. Rice. Ceramic processing: an overview. J. AiChE, 36(4):481–510, 1990.
7. M.N. Rahaman. Ceramic Processing and Sintering. New York: Marcel Dekker, Inc., 1995.
8. C.R. Veale. Fine Powders Preparation, Properties and Uses. London: Applied Science Publishers LTD, 1972.
9). D. Segal. Chemical Synthesis of Advanced Ceramic Materials. Cambridge: Cambridge Univ. Press, 1989.
10. R.A. DiMilia, J.S. Reed. Effect of Humidity on the Pressing Characteristics of Spray-Dried Alumina. In: J.A. Mangles, G.L. Messing, eds. Forming of Ceramics. Adv. in Ceramics. Vol. 9. Westerville, OH: Am. Cer. Soc. 1984, pp. 38–46.
11. N. Shinohara, M. Okumiya, T. Hotta, K. Nakahira, M. Naito, K. Uematsu. Effect of seasons on density, strength of alumina. Am. Cer. Soc. Bul. 78(2):81–84, 1999.
12. D.W. Whitman, D.I. Cumbers, X.K. Wu. Humidity sensitivity of dry-press binders. Am. Cer. Soc. Bul. 74(8):76–79, 1995.
13. A.B. Van Groenou. Compaction of ceramic powders. Pwd. Tech. 28:221–228, 1981.
14. R.A. Thompson. Mechanics of powder pressing: I. Model for powder densification. Am. Cer. Soc. Bul. 60(2):237–243, 1981.
15. J.J. Lannutti, T.A. Deis, C.M. Kong, D.H. Phillips. Density gradient evolution during dry pressing. Am. Cer. Soc. Bul. 76(1):53–58, 1997.
16. T.A. Deis, J.J. Lannutti. X-ray computed tomography for evaluation of density gradient formation during the compaction of spray-dried granules. J. Am. Cer. Soc. 81(5):1237–1247, 1998.
17. B.K. Molnar, R.W. Rice. Strength anisotropy in lead zirconate titanate transducer rings. Am. Cer. Soc. Bul. 52 (6):505–509, 1973.
18. H.A.M. Al-Jewaree, H.W. Chandler. Air entrapment: a source of laminations during pressing. Br. Ceram. Trans. J. 89:207–210, 1990.
19. R.A. Thompson. Mechanics of powder pressing: II. Finite-element analysis of end-capping in pressed green powders. Am. Cer. Soc. Bul. 60(2):244–247, 1981.
20. R.W. Rice. Processing Induced Sources of Mechanical Failure in Ceramics. In: H.

Palmour III, R.F. Davis, T.M. Hare eds. Processing of Crystalline Ceramics. New York: Plenum Press, 1978, pp. 303–319.
21. R.W. Rice. Comparison of physical property-porosity behavior with minimum solid area models. J. Mat. Sci. 31:1509–1528, 1996.
22. R.W. Rice. Porosity of Ceramics. New York: Marcel Dekker, Inc., 1998.
23. R.E. Fryxell, B.A. Chandler. Creep, strength, expansion, and elastic moduli of sintered BeO as a function of grain size, porosity, and grain orientation. J. Am. Cer. Soc. 47(6):283–291, 1964.
24. D.B. Quinn, R.E. Bedford, F.L. Kennard. Dry-bag isostatic pressing and contour grinding of technical ceramics. In: J.A. Mangles, G.L. Messing, eds. Forming of Ceramics. Adv. in Ceramics. Vol. 9. Westerville, OH: Am. Cer. Soc. 1984, pp. 4–15.
25. G.A. Fryburg, F.B. Markar. Improved fabrication of Al_2O_3 tubing for sodium vapor lamps. In: A. Mangles, G.L. Messing, eds. Forming of Ceramics. Adv. in Ceramics. Vol. 9. J. Westerville, OH: Am. Cer. Soc. 1984, pp. 32–37.
26. B.J. McEntire. Tooling design for wet-bag isostatic pressing. In: J.A. Mangles, G.L. Messing, eds. Forming of Ceramics. Adv. in Ceramics. Vol. 9. Westerville, OH: Am. Cer. Soc., 1984, pp. 16–31.
27. T.M. Wehrenberg, E.R. Begley, R.F. Patrick. Isostatic pressing large refractory blocks. Am. Cer. Soc. Bul. 47(7):642–645, 1968.
28. G.D. Kelly. Effects of Hydrostatic Forming. Am. Cer. Soc. Bul. 40(6):378–382, 1961.
29. T. Nishimura, K. Jinbo, Y. Matsuo, S. Kimura. Forming of ceramic powders by cyclic-CIP-effect of bias pressure. J. Cer. Soc. Jpn. Intl. Ed. 98:133–, 1990.
30. O. Abe, S. Kanzaki, H. Tabata. Structural homogeneity of CIP formed green compacts (2)-influence of applying pressure. J. Cer. Soc. Jpn. Intl. Ed. 97:30–35, 1989.
31. O. Abe, S. Kanzaki, H. Tabata. Structural homogeneity of CIP formed green compacts, (Part 3)-influence of applying pressure on mechanical properties of sintered silicon carbide. J. Cer. Soc. Jpn, Intl. Ed. 97:423–428, 1989.
32. Y. Matsuo, T. Nishimura, K. Yasuda, S. Kimura. Development of cyclic-CIP and its application to powder forming. Yogo-Kyokai-Shi 95:82–85, 1987.
33. T. Nishimura, K. Kubo, K. Jinbo, Y. Matsuo, S. Kimura. Forming of ceramic powders by cyclic-CIP-effect of frequency. J. Cer. Soc. Jpn., Intl. Ed. 98:1380–, 1990.
34. G.A. Ackley, J.C. Reed. Body parameters affecting extrusion. In: J.A. Mangles, G.L. Messing, eds. Forming of Ceramics. Adv. in Ceramics. Vol. 9. Westerville, OH: Am. Cer. Soc., 1984, pp. 193–200.
35. B. C. Mutsuddy. Hot Ceramic Extrusion. In: J.A. Mangles, G.L. Messing, eds. Forming of Ceramics. Adv. in Ceramics. Vol. 9. Westerville, OH: Am. Cer. Soc., 1984, pp. 212–219.
36. D. Kovar, B.H. King, R.W. Trace, J.W. Halloran, Fibrous monolithic ceramics. J. Am. Cer. Soc. 80(10):2471–2487, 1997.
37. C.V. Hoy, A. Barda, M. Griffith, and J.W. Halloran. Microfabrication of ceramics by co-extrusion. J. Am. Cer. Soc. 81(1):152–158, 1998.
38. I.M. Lachman, R.D. Bradley, R.M. Lewis. Thermal expansion of extruded cordierite ceramics. Am. Cer. Soc. Bul. 60(2):202–205, 1981.
39. R.W. Rice. Mechanical Properties of Ceramics and Composites, Grain and Particle Effects. New York: Marcel Dekker, Inc., 2000, p. 95.

40. A.Y. Chen, J.D. Crawley. Extrusion of α-Al_2O_3-boehmite mixtures. J. Am. Cer. Soc. 75(3):575–579, 1992.
41. C.S. Kumar, U.S. Hareesh, A.D. Damodaran, K.G.K. Warrier. Monohydroxy aluminium oxide (boehmite, AlOOH) as a reactive binder for extrusion of alumina ceramics. J. Eur. Cer. Soc. 17:1167–1172, 1997.
42. S. Blackburn, T.A. Lawson. Mullite-alumina composites by extrusion. J. Am. Cer. Soc. 75(4):953–957, 1992.
43. D. Muscat, M.D. Pugh, R.A.L. Drew, H. Pickup, D. Steele. Microstructure of an extruded β-silicon nitride whisker-reinforced silicon nitride composite. J. Am. Cer. Soc. 75(10):2713–2718, 1992.
44. J.A. Mangels, W. Trela. Ceramic components by injection molding. In: J.A. Mangles, G.L. Messing, eds. Forming of Ceramics. Adv. in Ceramics. Vol. 9. Westerville, OH:Am. Cer. Soc., 1984, pp. 220–233.
45. M.J. Edirisinghe, J.R.G. Evans. Review: fabrication of engineering ceramics by injection molding. I. Materials selection. Int. J. High Tech. Cers. 2:1–31, 1986.
46. M.J. Edirisinghe, J.R.G. Evans. Review: fabrication of engineering ceramics by injection molding. I. Materials selection. Int. J. High Tech. Cers. 2:249–278, 1986.
47. M.J. Edirisinghe, J.R.G. Evans. Properties of ceramic injection molding formulations Part 1. Melt technology. J. Mat. Sci. 22:269–277, 1987.
48. M.J. Edirisinghe, J.R.G. Evans. Properties of Ceramic Injection Molding Formulations. Part II. Integrity of Moldings. J. Mat. Sci. 22:2267–2273, 1987.
49. B.C. Mutsuddy. Study of ceramic injection molding parameters. Adv, Cer. Mats. 2(3A):213–218, 1987.
50. R.M. German, K.F. Hens, S.-T.P. Lin. Key issues in powder injection molding. Am Cer. Soc. Bul. 70(8):1294–1302, 1991.
51. J.G. Zhang, M.J. Edirisinghe, J.R.G. Evans. A catalog of ceramic injection molding defects and their causes. Ind. Cers. 9(2):72–82, 1989.
52. J.R.G. Evans, M.J. Edirisinghe. Interfacial factors affecting the incidence of defects in ceramic moldings. J. Mat. Sci. 26:2081–2088, 1991.
53. G. Bandyopadhyay, K.W. French. Injection-molded ceramics: critical aspects of the binder removal process and component fabrication. J. Eur, Cer. Soc. 11:23–34, 1993.
54. K. Uematsu, S. Ohsaka, N. Shinohara, M. Okumiya. Grain-oriented microstructure of alumina ceramics through the injection molding process. J. Am. Cer. Soc. 80(5):1313–1315, 1997.
55. J.A. Mangels. Low-pressure injection molding. Am. Cer, Soc. Bul. 73(5):37–41, 1994.
56. A.J. Fanelli, R.D. Silvers, W.S. Frei, J.V. Burlew, G.B. Marsh. New aqueous injection molding process for ceramic powders. J. Am. Cer. Soc. 72(10):1833–1836, 1989.
57. T. Zhang, J.R.G. Evans. The properties of a ceramic injection molding suspension based on a preceramic polymer. J. Eur. Cer. Soc. 7:405–412, 1991.
58. I. Tsao, S.C. Danforth. Injection moldable ceramic-ceramic composites: compounding behavior, whisker degradation, and orientation. Am Cer. Soc. Bul. 72(2):55–64, 1993.

59. R.D. Nelson, Jr. Dispersing Powders in Liquids. New York: Elsevier, 1988.
60. R.R. Rowlands. A review of the slip casting process. Am. Cer. Soc. Bul. 4591):16–19, 1966.
61. H. Taguchi, Y. Takahashi, H. Miyamoto. Slip casting of partially stabilized zirconia. Am. Cer. Soc. Bul. 64(2):325, 1985.
62. Z. Zhang, L. Hu, M. Fang. Slip casting nanometer-sized powders. Am. Cer. Soc. Bul. 75(12):71–74, 1996.
63. M. Okuyama, G.J. Garvey, T.A. Ring, J.S. Haggerty. Dispersion of silicon carbide powders in nonaqueous solvents. J. Am. Cer. Soc. 72(10):1918–1924, 1989.
64. W. Huisman, T. Graule, L.J. Gauckler. Centrifugal slip casting of zirconia (TZP). J. Eur, Cer. Soc. 13:33–39, 1994.
65. G.A. Steinlage, R.K. Roeder, K.P. Trumble, K.J. Bowman. Centrifugal slip casting of components. Am Cer. Soc. Bul. 75(5):92–94, 1996.
66. A. Salomoni, I. Stamenkovic. Pressure casting offers possibilities for technical ceramics. Am. Cer. Soc. Bul. 79(11):49–53, 2000.
67. K.-H. Kim, S.-J. Cho, D.-J. Kim, K.-J. Yoon. Centrifugal casting of large alumina tubes. Am. Cer. Soc. Bul. 79(11):54–56, 2000.
68. B.A. Freed, H. Dusseldorp, J. Klieb, B. Woods, S. Oda, Sr. Microwave drying automates slip casting process and enhances overall economy. Ind. Heating 32–34, 1990.
69. O. Lyckfeldt, E. Liden, M. Presson, R. Carlsson, P. Apell. Progress in the fabrication of Si_3N_4 turbine rotors by pressure slip casting. J. Eur. Cer. Soc. 14:383–495, 1994.
70. P.H. Rieth, J.S. Reed, A.W. Naumann. Fabrication and flexural strength of ultrafine-grained yttria-stabalized ziriconia. Am. Cer. Soc. Bul. 55(8):717–, 1976.
71. V. Pujari, D.M. Tracey, M.R. Foley, N.I. Paille, P.J. Pelletier, L.C. Sales, C.A. Wilkens, R.L. Yeckley. Development of improved processing and evaluation methods for high reliability structural ceramics for advanced heat engine applications. Phase I. Norton Co. Advanced Ceramics final report for U.S. Dept. Energy contract DE-AC05-84OR21400, 1993.
72. T.P. Hyatt. Electronics: Tape casting, roll compaction. Am. Cer. Soc. Bul. 74(10):56–59, 1995.
73. R. Moreno. The role of slip additives in tape technology. Part I. Solvents and dispersants. Am. Cer. Soc. Bul. 71(10):1521–1531, 1992.
74. R. Moreno. The role of slip additives in tape technology. Part II. Binders and plasticizers. Am. Cer. Soc. Bul. 71(11):1647–1657, 1992.
75. J.G.P. Binner. The effect of rising temperature using microwaves on the slip casting of alumina-based ceramics. Br. Cer. Trans. J. 91:48–51, 1992.
76. M.J. Hoffmann, A. Nagel, P. Greil, G. Petzow. Slip casting of SiC-whisker-reinforced Si_3N_4. J. Am. Cer. Soc. 72(5):765–769, 1989.
77. H. Takebe, K. Morinaga. Fabrication and mechanical properties of lamellar Al_2O_3 ceramics. J. Cer. Soc. Jpn. 96:1122–1128, 1988.
78. L. Shaw, R. Abbaschian. Fabrication of SiC-whisker-reinforced $MoSi_2$ composite by tape casting. J. Am. Cer. Soc. 78(11):3129–3132, 1995.
79. S.N. Heavens. Electrophoretic deposition as a processing route for ceramics. In: J.G.P. Binner, ed. Advanced Ceramic Processing and Technology. Vol. 1. Park Ridge, NJ: Noyes Pub., 1990, pp. 255–283.

80. R. Moreno, B. Ferrari. Advanced ceramics via EPD of aqueous slurries. Am. Cer. Soc. Bul. 79(1):44–48, 2000.
81. J.B. Birks. Electrophoretic deposition of insulating materials. In: J.B. Birks, J.H. Schulman, eds. Progress in Dielectrics. Vol. 1. NY: John Wiley & Sons Inc., 1959, 273–312.
82. L.M. Apininskaya, V.V. Furman. Electrophoretic deposition of coatings (a literature review). Pwd. Met. 122(2):61–64, 1973.
83. M.A. Andrews, A.H. Collins, D.C. Cornish, J. Dracass. The forming of ceramic bodies by electrophoretic deposition., Fabrication Science 2. Proc. Brit. Cer. Soc. No. 12, 1969, pp. 211–229.
84. A.V. Kerkar, R.W. Rice, R.M. Spotnitz. Manufacture of Optical Ferrules by Electrophoretic Deposition. U.S. Patent 5,194,129, 1993.
85. I. Aksay. Processing and equipment for the production of materials by electrophoresis ELEPHANT. Interceram. 27(1):3334, 1978.
86. M. Nagai, K. Yamashita, T. Umegaki, Y. Takuma. Electrophoretic deposition of ferroelectric barium titanate thick films and their dielectric properties. J. Am. Cer. Soc. 76(1):253–55, 1993.
87. P.S. Nicholson, P. Sarkar, S. Datta. Producing ceramic laminate composites by EPD. Am. Cer. Soc. Bul. 75(11):48–51, 1996.
88. P.S. Nicholson, P. Sarkar, X. Haung. Electrophoretic deposition and its use to synthesize ZrO_2/Al_2O_3 micro-laminate ceramic/ceramic composites, J. Mat. Sci. 28:6274–6278, 1993.
89. V. Shulman, A.E. Hodel. Roll compaction system cuts changeover time and increases production by up to 75%. Chem Proc. 40–42, 1987.
90. R.K. Dube. Metal strip via roll compaction and related powder metallurgy routes. Intl. Mat. Revs. 35(5):253–291, 1990.
91. P.F. Becher, J.H. Sommers, B.A. Bender, B.A. MacFarland. Ceramics sintered directly from sol-gels. In: H. Palmour, III, R.F. Davis, T.M. Hare, eds. Processing of Crystalline Ceramics. New York: Plenum Pub. Corp, 1978, 79–85.
92. L.L. Hench. Use of drying control chemical additives (DCCAs) in controlling sol-gel processing. In: L.L. Hench, D.R. Ulrich, eds. Science of Ceramic Chemical Processing. New York: John Wiley & Sons, 1986, 52–64.
93. O.O. Omatete, M.A. Janney, R.A. Strehlow. Gelcasting—A new ceramic forming process. Am. Cer. soc. Bul. 70(10):1641–1649, 1991.
94. M.A. Janney, O.O. Omatete, C.A. Walls, S.D. Nunn. R.J. Ogle, G. Westmoreland. Development of low-toxicity gelcasting systems. J. Am. Cer. Soc. 81(3):581–591, 1998.
95. M.H. Zimmerman, K.T. Faber, E.J. Fuller, Jr. Forming textured microstructures via the gelcasting technique. J. Am. Cer. Soc. 80(10):2725–2729, 1997.
96. S.L. Morissette, J.A. Lewis. Chemorheology of aqueous-based alumina-poly(vinyl alcohol) gelcasting suspensions. J. Am. Cer. Soc. 82(3):521–528, 1999.
97. T. Tanaka, I. Nishio, S.-T. Sun, S. Ueno-Nishio. Collapse of gels in an electric field. Science 218:467–469, 1982.
98. M. Harley. Polymers for ceramic and metal powder binding applications. Chem. Ind. 51–55, 1995.

99. N.R. Gurak, P.L. Josty, R.J. Thompson. Properties and uses of synthetic emulsion polymers as binders in advanced ceramics processing. Am. Cer. Soc. Bul. 66(10):1495–1497, 1987.
100. S.L. Bassner, E.H. Klingenberg. Using poly(vinyl alcohol) as a binder. Am. Cer. Soc. Bul. 77(6):71–75, 1998.
101. X.K. Wu, D.W. Whitman, W.L. Kaufell, W.C. Finch, D.I. Cumbers. Acrylic binders for dry pressing ceramics. Am. Cer. Soc. Bul. 76(1):49–52, 1997.
102. X.K. Wu, W.J. McAnany. Acrylic binders for green machining. Am. Cer. Soc. Bul. 74(5):61–64, 1995.
103. A.S. Barnes, J.S. Reed, E.M. Anderson. Using cobinders to improve manufacturability. Am. Cer. Soc. Bul. 76(7):77–82, 1997.
104. H.M. Shaw, M.J. Edirisinghe. Removal of binder from ceramic bodies fabricated using plastic forming methods. Am. Cer. Soc. Bul. 72(9):94–99, 1993.
105. G.W. Scherer. Theory of drying. J. Am. Cer. Soc. 73(1):3–14, 1990.
106. T. Morse. Handbook of Organic Additives for Use in Ceramic Body Formulation. Butte, Montana: Montana Energy and MHD Research and Development Inst., 1979.
107. J.H. Enloe, J.W. Lau, C.B. Lundsager, R.W. Rice. Hot Pressing Dense Ceramic Sheets for Electronic Substrates and for Multilayer Electronic Substrates. U.S. Patent 4,920,640, 1990.
108. J.H. Enloe, J.W. Lau, R.W. Rice. Electronic Package Comprising Aluminum Nitride and Aluminum Nitride- Borosilicate Glass Composite. U.S. Patent 5,017,434, 1991.
109. R.W. Rice. Failure analysis of ceramics. In: J. Varner, G. Quinn, eds. Fractography of Glasses and Ceramics IV. Westerville, OH: Am. Cer. Soc., 2001.
110. B.R. Sundlof, C.R. Perry, W.M. Carty, E.H. Klingenberg, L.A. Schultz. Additive interactions in ceramic processing. Am. Cer. Soc. Bul. 79(10):67–72, 2000.
111. W.J. Walker, Jr., J.S. Reed, S.K. Verma. Influence of granule character on strength and Weibull modulus of sintered alumina, J. Am. Cer. Soc. 82(10):50–56, 1999.
112. H. Takahashi, N. Shinohara, K. Uematsu, T. JunIchiro. Influence of granulae character and composition on the mechanical properties of sintered silicon nitride. J. Am. Cer. Soc. 79(4):843–848, 1996.
113. J. Zheng, J.S. Reed. The different roles of forming and sintering on densification of powder compacts. Am Cer. Soc. Bul. 71(9):1410–1416, 1992.
114. J.Y. Wong, S.E. Laurich-McIntyre, A.K. Khaund, R.C. Bradt. Strengths of green and fired spherical aluminosilicate aggregates. J. Am. Cer. Soc. 70(10):785–791,1987.
115. M. Takahashi, S. Suzuki. Deformation of spherical granules under uniaxial loading. Am. Cer. Soc. Bul. 64(9):1257–1261, 1985.

5

Use of Additives to Aid Densification

5.1 INTRODUCTION

Chemical species, other than those constituting the body composition sought, may be added purposely or occur as inadvertent or unavoidable constituents, which may affect densification, as well as resultant properties, positively or negatively. The focus here is on constituents that are intentionally added since their presence or the nature of that presence give some advantage to the resultant body; or more commonly, give some advantageous trade-off in properties, fabricability, costs, or combinations of these. There are also some cases where inadvertent species, not at first identified, are subsequently identified and studied by intentional addition.

Treatment of the densification and property effects of such species is challenging because of the scope, extent, complexity, and incomplete understanding of the topic. The challenges arise from various factors, such as the diversity of additives and impurities (that is, inadvertent species and their frequent variability), their interactions with one another, and the nature and parameters of the processing and raw materials; for example, particle size and its distribution. However, such challenges are an integral part of the engineering tasks to make useful bodies via trade-offs in materials, fabrication/processing, properties, and costs to meet a specified function. While, the availability of finer, purer, more agglomerate-free powders has substantially increased to reduce the need for processing additives, there is still extensive need for and use of them. Some of this is historical (continued use of older processes), but much is still needed due to performance and/or cost trade-offs.

A few overall comments on densification mechanisms are in order, since, while often not clearly identified for a given additive, material, and process, general trends are at least partly defined. There are three basic densification mechanisms, namely: (1) liquid-phase sintering; (2) enhancing solid diffusion or favorably changing the balance of surface, grain boundary, and bulk diffusion; and (3) controlling grain growth, primarily suppressing the breakaway of grain boundaries from pores, so they can be eliminated while still at grain boundaries. Possibly the most widely used and most effective mechanism is liquid-phase sintering. Though there are only three mechanisms, complexities arise since the details of the mechanism(s) for any specific application are generally not defined, may not be only a single mechanism, and may change for different materials, other additives or impurities, and temperature ranges (e.g., due to different materials or particle sizes). Thus, a small amount of particle solubility in an additive may be effective for fine particles, but not for larger particles. Further, while solubility generally increases with increasing temperature, which may often enhance the liquid-phase mechanism, this may not always be so, due to probable increases in reactivity and vaporization of the liquid. Additionally, limited amounts of liquid phase may be more effective in pressure sintering, for example, hot pressing, than in pressureless sintering, and enhanced diffusion may be more important in pressureless sintering and less important in pressure sintering. Finally, note that control of grain growth in pressureless sintering generally implies controlling grain growth at high temperatures, which still entails increased grain sizes, and is often important where materials with sufficient optical transparency are needed, such as for lamp envelopes.

Though widely neglected in most treatments of ceramics, the use of additives and some effects of impurities are extensively addressed in summary form in this chapter and Chapter 3 because of the importance of additives. The focus is on reported effects rather than mechanisms since the latter are often ill defined and inadequate for practical guidance; mechanisms are noted where there are reasonable indications that they provide some guidance. Treatment is in the order of additives for single oxide, then mixed-oxide bodies followed by nonoxide bodies, then composites with first-oxide additives then nonoxide, mixed, or other additives addressed for each of the material classes. Some of the property effects are noted in presenting the additive use, but some broader observations on additive effects are summarized in a subsequent section.

5.2 ADDITIVES FOR DENSIFICATION OF ALUMINUM OXIDE

From both a historical and economical standpoint, liquid-phase sintering of oxides, especially Al_2O_3 via use of SiO_2-based glass phases (e.g., to produce commercial bodies of. nominally, 85%, 90%, 94%, 96%, and 99% alumina), is most important. While Al_2O_3 can be sintered with SiO_2 (and other additives to form a glass phase at lower temperature) [1], industrial practice is to use natural minerals, mainly clays and talcs, as the sources of the desired in situ formed glasses. Although these glass-phase sources can present some issues, they are generally advantageous due to very low cost, ready glass formation (and without mullite formation, which can inhibit densification), and on balance have fewer health issues, for example, the absence of SiO_2, especially quartz (J. Rubin, personal communication, 1999). Another source of much lower temperature (e.g., 450–800°C) "densification" (actually bonding not sintering) of Al_2O_3, is phosphate bonding using H_3PO_4 to form $AlPO_4$, which often forms at least some glassy phase [3].

The most extensively used additive for sintering Al_2O_3, beyond use of silica-based glassy phases, is MgO (e.g., 0.1–0.3 w/o), which was originally developed to produce envelopes for high-pressure sodium vapor lamps. Such applications require near-zero porosity, and thus sintering at very high temperatures, which, even with grain growth control, results in grain sizes a few to several times greater than in normal sintering, and thus generally less strength and related properties than with much conventional processing. (Note that the grain growth with MgO additions probably reflects, at least in part, loss of MgO by vaporization.) The discovery of MgO as an additive resulted from study of the sinterability of various lots of fine, high-purity Al_2O_3 powder to give bodies approaching transparency for lamp envelope application indicated better results as limited contents of MgO increased and SiO_2 decreased [4,5]. Subsequent studies of purposely doped (via nitrate additions), high-purity α-Al_2O_3 powder (Linde A, particle size 0.3 μm) sintered for 3 hr at 1900°C (in H_2) corroborated that inhibition of grain growth, that is, of grain boundary mobility, was a major factor in achieving dense, pore-free Al_2O_3 [6]. Thus, while firing without additives yielded 2.3% porosity, mostly within the large, 90 μm, grains, firing with MgO additions yielded 99.9% of theoretical density with only isolated pores, mostly at or near the boundaries of grains averaging 12 μm. Tests of similar separate additions of CaO, SrO, BaO, Al_2O_3, and ZrO_2, all with lower liquefying temperatures than with MgO, gave densities and grain sizes similar to those obtained without additives [7]. Tests of combined addition of 0.1 w/o each of MgO and Y_2O_3 gave similar results as with MgO alone, thus showing that some combined additives were successful (but other tests show combination with SiO_2, as well as by itself [4], were unsuccessful). Hwang and coworkers [1] also showed that limited additions of SiO_2 (from TEOS) on forming mullite or Al_2O_3-SiO_2 liquid inhibited both densification and grain growth. More recent tests with purer Al_2O_3 powders—for example, by Bae and Baik [8]—showed more details of the interactions of additives and impurities. Thus, in sintering of > 99.99% pure Al_2O_3 powder in sapphire tubes for 1 hr at 1900°C in argon with controlled additions of MgO, SiO_2, or CaO, they showed that sintered densities increased to a plateau of 96% of theoretical values at 300–500 ppm MgO; SiO_2 additions gave a maximum density of nearly 94% at a level of 100 ppm; then decreased to 90% or more at 500 ppm, while CaO additions may have resulted in a maximum of 92% of theoretical density with 25 ppm CaO, and then clearly decreased to only 88% at 300 ppm CaO. They also showed that codoping with MgO and CaO modestly but progressively reduced densities achieved, and that while grain size increased modestly from 20+ to nearly 26 μm as MgO doping levels increased from 0 to 200 ppm, CaO gave bimodal grain sizes or exaggerated grain growth, which required at least an equal level of MgO addition to suppress such CaO induced growth.

Other subsequent studies have corroborated the benefits and effects of MgO and alternative additions; e.g., Warman and. Budworth [9] confirmed that NiO additions also work and showed that $MgAl_2O_4$ did as well, ZnO, CoO, and SnO_2 additions also could work, but CaO and Cr_2O_3 did not. They also confirmed that theoretical density could be obtained by sintering in pure O_2 or H_2 atmospheres as well as vacuum, as had other investigators, since O_2 and H_2 could be diffused out of the pores [10]. However, they also showed that there were practical size limits, above certain specimen sizes, green densities, and heating rates, and O_2 and H_2 atmosphere pressures had to be reduced to avoid bloating similar to that with other gases when the effective outward distances for diffusion of O_2 or H_2 from the specimen interior had been exceeded. More recent discussions of effects of MgO on sintering Al_2O_3 are found from a conference on Al_2O_3 and MgO [11–13].

Consider effects of additives on sintering of Al_2O_3 at lower temperatures where the goal is not necessarily full density, but limited grain size with limited porosity. This is commonly sought for a favorable balance of properties, especially mechanical ones, from their general increase with decreasing residual porosity as sintering time and especially temperatures increase, versus general decreases with increasing grain size from increased firing to reduce porosity. Use of Cr_2O_3, MgF_2, or $AlPO_4$ have been recommended for industrial use for this purpose [14]. MgF_2 should typically form MgO, and $AlPO_4$ will result in phosphate bonding (discussed above).

Use of additives in hot pressing (and hot isostatic hot pressing HIPing) has the potential of compounding the enhanced densification and reduced grain growth of both additives and pressure sintering; though, as will be shown below, additive effects may be different between pressureless sintering and hot pressing or other pressure sintering. However, use of MgO additions works well in hot pressing, as shown in earlier work, e.g., of Gazza et al. [15]. They showed that hot pressing of a high-purity, fine γ-Al_2O_3 in vacuum gave higher densities than hot pressing in air, but that additions of 1.25 m/o of MgO or 1.35 m/o CoO powders aided densification, especially in vacuum hot pressing. However, more significant was the inhibition of grain growth with the additives. Grain sizes without additives were 1–3 μm with isolated grains of 5–10 μm at 1400°C and 20–25 μm at 1600°C, but 0.5–1 μm and 1 μm at 1400°C and 1600°C, respectively, with MgO. The CoO was as effective at 1400°C, but was less so at 1600°C; e.g., giving isolated grains of ~ 5 μm at 1600°C, which may reflect effects of some probable reduction of CoO in the reducing environment due to the vacuum and, especially, the use of graphite dies and resultant CO formation. Harmer and Brook [16] studied the kinetics of hot pressing of Al_2O_3 with MgO additions in the solid solution range, interpreting their results in terms of a diffusional creep process. More recently Bateman and coworkers [17], showed that Al_2O_3 with ~ 0.25 v/o silicate-based impurities could be vacuum hot pressed to full density at 1600°C, but that MgO was still of value in suppressing formation of elongated grains (e.g., aspect ratios of > 3) associated with the silicate content.

Turning to other additives, again focusing on pressureless sintering, the use of TiO_2 additions (0.2%, i.e., within its limited range of solubility in Al_2O_3) to densify Al_2O_3 (at 1900°C) was noted in Ryshkewitch's original book [18] and has received considerable study, but has little or no industrial use (apparently in part due to its frequent limitation of strength and presumably related mechanical properties due to formation of Al_2TiO_5). TiO_2 addition by itself or in combination with either MnO or CuO have both been studied. Various papers [1,19–22] focusing more on initial, intermediate, or both stages of sintering, for example, at lower temperatures of 1100–1600°C, generally attribute enhanced sintering to enhanced diffusion. Defect models for Ti in alumina have been presented [23]. Direct comparison of TiO_2 and MgO additions have been reported by Harmer and coworkers [21], who noted both increased final density for short firing at 1850°C, and Ikegami and coworkers [24], who noted final densities on sintering with TiO_2 at 1600°C were less that for undoped alumina, which was itself less than for MgO addition. Watanabe and Nakayama [25] noted that sintering with TiO_2 (or Cr_2O_3) additions in a reducing atmosphere at 1500–1600°C resulted in reduced densities attributed to gas evolution from additive reduction. Bettinelli and coworkers [26] also noted differences with TiO_2 additions under air and hydrogen firing. These differences again indicate differing effects due to additive type, amount, temperature range, heating rate, and atmosphere (e.g., TiO_2 is fairly readily reduced to Ti_2O_3, which is highly soluble in Al_2O_3, but has no valence difference); similarly, note that reduction of ZrO_2, e.g., as in hot pressing in graphite dies, can increase the stability of the cubic phase. More serious effects of densification atmosphere are shown by Wakamatsu and coworkers' [27] study showing that up to 2 w/o V solution doping of high purity Al_2O_3 sintered at 1650°C in air formed an $AlVO_4$ grain boundary phase that depressed densification, grain growth, and strength; while firing in a reducing atmosphere, which resulted in most V in solid solution, had little effect on densification, grain structure, or mechanical properties.

Note that there is some industrial use of TiO_2 additives to produce 90–92% "black" sintered alumina to opacify the resultant body for microelectronic packaging applications. This is done with combined additions of MnO and Fe_2O_3 or of Cr_2O_3 and MoO_3 additions with firing in a reducing (H_2 containing) atmosphere to produce the desired black color. This opacification is desired to avoid possible photoelectric effects of semiconductors in the package from occurring and alterating the electronic behavior of the package by preventing stray optical radiation from reaching and stimulating the semiconductors. Such black aluminas also avoid cosmetically undesired metal markings that can be left on normal white alumina from rubbing contact with metallized areas. Strength limitations due to Al_2TiO_5 formation are apparently not as severe as for other applications, or are reduced by the use of the other additives limiting the extent of the titanate phase formation, its grain growth, or combinations of these. There is however some interest in reducing or removing TiO_2 due to its high dielectric constant.

Use of Y_2O_3, which was noted earlier and is noted below, has also been studied [28], and commercial high purity, for example, 99% alumina bodies are commonly sintered with small Y_2O_3 additions [2]. An important benefit of Y_2O_3 additions is their effectiveness at very low levels and temperatures. Thus, for example, Delaunay and coworkers [29] reported that very small additions, 0.002 a/o, of very fine (10 nm) Y_2O_3 were very effective in enhancing sintering at 1300°C versus much less benefit from 1 a/o of 100 nm Y_2O_3.

Study of ZrO_2 toughened alumina (ZTA) composites revealed inhibition of the growth of the fine starting grains of either phase by the other. The resultant finer ZrO_2 particle, and especially the finer alumina matrix grain sizes are important factors in increased strengths of such composites, and, in fact, resulted in considerable strength increase at low levels of ZrO_2 additions, where transformation or microcracking toughening is not significant. Lange and Hirlinger [30] were apparently the first to explicitly show this, but they reported that at least 2.5 m/o ZrO_2 was needed to control exaggerated grain growth (and more with less uniform mixing). Hori and coworkers [31,32] corroborated such strength benefits from grain growth control, even for lower levels of fine ZrO_2 additions, despite some lowering of toughness. Thus, they showed that the Al_2O_3 grain size dropped from 4+ to ~ 2.5 μm as the ZrO_2 content went from 0 to 0.1 w/o (with a ZrO_2 particle size of 0.2 μm or larger) and decreased less rapidly as the ZrO_2 content was further increased, reaching grain size (G) ~ 1.5 μm at 5 w/o ZrO_2 (with its particle size of ~ 0.3 μm). Composite strengths increased rapidly at lower additions, then more slowly; that is, starting from ~ 310 MPa and reaching ~ 570 MPa as the matrix grain size decreased, consistent with behavior of pure Al_2O_3. The strength increases occurred despite the indentation fracture (IF) toughness decreasing from ~ 4.8 MPa•m$^{1/2}$ with no ZrO_2 to a minimum of ~ 3.9 MPa•m$^{1/2}$ at ~ 1.1 w/o ZrO_2, then rising to ~ 4.6 MPa•m$^{1/2}$ at 5 w/o ZrO_2. Such grain growth benefits are obtained in many other composites where fine particles of a second phase with limited solubility or reaction with the matrix are used. This includes not only other composites with ZrO_2, e.g., in MgO, as shown by Nisida and coworkers' study of composites with up to 10 w/o of either fine tetragonal or cubic ZrO_2 particles [33], and 3–20 v/o of Al_2O_3 in a cubic ZrO_2 matrix by Lange and Hirlinger [34]—but also many other matrix-particle composites, such that significant fractions of composite strength increases are commonly due to such grain growth limitation.

Other oxide additives investigated include single oxides Cr_2O_3 [35], MnO [36], as well as Cr_2O_3 and MoO_3 [37], the latter two again in air and hydrogen firing, respectively, at 1400°C. Nagaoka and coworkers [38] showed that addition of 0.5–3 w/o of CaO (from $CaCO_3$) to a low sodium, submicron, commercial α-Al_2O_3 sintered at 1650°C in air for 4 hr gave ~ 1.5% porosity for 0, 0.5, and 1% CaO, increasing to 2.7 and 7.1% porosity, respectively, at 2 and 3 w/o CaO. The density decreases correlated with large increases in the amount of Ca aluminate formed, but while Young's modulus also decreased, flexural strength increased from 386 to 585 MPa at 0 and 2.0 w/o CaO. Additions of 1000 ppm of either Y_2O_3 or La_2O_3 to a high-purity, submicrom, commercial alumina sintered in air at 1350°C both showed substantial inhibition of grain growth as density increased (to ~ 99% of theoretical), with La_2O_3 limiting growth more than Y_2O_3 [39]. FeO has also been investigated, showing that while it was not an effective sintering aid by itself, it can aid the sintering of MgO-doped alumina [40]. Note also the combination of MgO and Y_2O_3 additions by Rossi and Burke [6] and more recent more detailed studies of Sato and Carry [41] on combinations of MgO and Y_2O_3 at the 500 or 1500 ppm level to a submicron commercial α-Al_2O_3 sintered in air at 1700°C. The latter study showed that Y_2O_3 segregation to grain boundaries delayed densification and increased the apparent activation energy, then increased densification as Y_2O_3 segregation approached saturation, then decreased again as Y_2O_3 precipitation occurs. Also note variations in low levels and ratios of CaO to SiO_2 can have significant effects on the microstructure of the resultant dense alumina; for example, giving many platelet-shaped grains [41] and Tomaszewski's evaluation of effects of Cr_2O_3 additions in conjunction with the glass phase in an otherwise 96% alumina body [35].

The above use of mixed- or compound-oxide additions has increased. Successful use of $MgAl_2O_4$ in limited trials of sintering Al_2O_3 to transparency by Warman and Budworth [8] is one indication of success. For sintering at lower temperatures, Wroblewaska [42] reported that addition of 2 m/o of Mg, Ca, Sr, or Ba titanates lowered alumina firing temperatures to 1500°C, with the latter additive forming aluminates. He subsequently reported some, but less benefit of somewhat smaller additions of $MgNb_2O_6$, which was found to decompose and produce $AlNbO_4$ [43]. Wanqui and coworkers [44] showed that separate additions of 0.15–1 w/o of Ta_2O_5 or MgO to high-purity Al_2O_3, Ta_2O_5 was more effective in densification by sintering, while MgO was more effective in grain growth control (e.g., indicating more solid solution of the Ta_2O_5), suggesting possible advantageous combination of the two additives. Some benefit of combinations with more Ta_2O_5 than MgO was reported for sintering at 1550–1800°C. On the other hand, Xue and Chen [45] more recently reported that high-purity Al_2O_3 powders with combined additions of 0.9 m/o CuO + 0.9 m/o TiO_2 + 0.1 m/o B_2O_3 + 0.1 m/o MgO allowed > 99% of theoretical density to be obtained in 1 hr at 1070°C.

Turning to other, mostly nonoxide, additions, considerable data exists for composites as noted earlier, as well as from earlier studies not focused on composites, with only modest levels of addition or both. Such additions typically require firing in nonoxidizing atmospheres, and thus have some flexibility and cost limitations relative to air firing, but not relative to vacuum sintering. Substantial strength benefits were shown via addition of fine Mo [46,47] or W [48] particles in sintered Al_2O_3, as well as with substantial AlON addition [49]. The strength increases were due to finer resulting Al_2O_3 grain size, e.g., diminishing at higher additions (to ~ 16%) of Mo [46]. Many of the above additives for grain growth control are also usable in hot pressing with similar, though possibly, reduced benefits (due to typically less grain growth in hot pressing), provided that the additive is not seriously reactive with typically used graphite dies. An example of this is hot pressing Al_2O_3 with Si_3N_4 additions [50].

The lower temperatures and applied pressures of hot pressing can also allow use of liquid-phase additives, whose liquid phase forms at lower temperatures, and may not be as effective at higher temperatures. Thus, Rice [51] showed that 2w/o LiF additions allowed fine grain α- or γ-Al_2O_3 (e.g., respectively, Linde A or B) to be hot pressed to ≥98% of theoretical density at ≤1100°C (Fig. 5.1), which is at least 200–300°C less than for hot pressing these (and similar powders) without additives. Resultant bodies had useful, but reduced, strengths; for example, 250–400 MPa at 22°C, which are one-third to one-half that at similar densities without the LiF addition (much of which remained at the grain boundaries [52]). However, about half of this strength limitation was probably due to the formation of some exaggerated grains (~ 10 μm in a matrix of grains ≤ 1 μm), which should be suppressible with addition of MgO which is compatible with use of LiF. That the effect of LiF is via a liquid phase is consistent with its melting temperature (870°C) and expected dissolution/reaction of Al_2O_3 with LiF (and related fluorides, e.g., LiF and MgF_2 are the most reactive fluorides and hence more effective as fluxes in welding Al metal) [53], and the

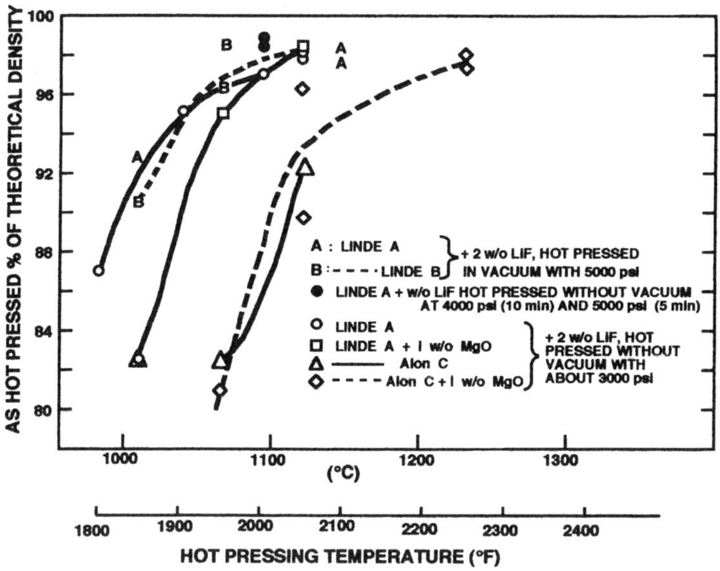

FIGURE 5.1 Densification of Al_2O_3 powders with LiF additions versus hot pressing temperature (with 21–28 MPa pressure). Note similar low temperature densification of fine α- (Linde A) and γ- (Linde B) powders, but higher temperatures required for a fumed alumina (Alon C). (From Ref. 51.)

inhibition of sintering of Al_2O_3 with related fluorides of Ca and Al, especially the latter [54]. Note the higher temperatures required for fine, fumed Al_2O_3 powder derived from gaseous oxidation of $AlCl_3$ vapor, which was attributed to effects of residual Cl [51].

5.3 OTHER OXIDES

BeO, while often densified by either sintering or hot pressing without additives, has often been densified by either process with 0.5–2 w/o MgO. Thus, Carniglia and Hove [55] reported benefits of 1 w/o MgO in hot pressing BeO, but noted possible effects of anion impurities, especially fluorine, as possible factors in the benefits of MgO. The possibility of F impurities aiding densification was supported by Adams and Stuart [56] who showed that exposure of high surface area (~ 280 m^2/gm) powder to HF, then hot pressing at 1140°C, resulted in enhanced densification at higher levels of adsorbed F (while adsorption of sulfate ions via powder exposure to H_2SO_4 inhibited densification). Duderstadt and White [57], reported that pure UOX (sulfate derived) powder sintered to only 79 and 95% of theoretical density by itself at 1500°C in H_2 for respectively 1 and 100 hr, while addition of 0.5 w/o MgO gave respective values of 93 and 98% (and respective values for 3 w/o ZrO_2 addition by themselves via respectively as a sol or as a powder were 87 and 98% and 81 and 98%). Hill and coworkers [58] also reported enhanced densification of BeO hot pressed with 2 w/o MgO by itself or with 8 w/o SiC added for grain growth control. Later Greenspan [59] showed substantial advantage in hot pressing BeO by addition of 3–20 (commonly 7) w/o Al_2O_3; for example, near theoretical density with Al_2O_3 at 1350°C versus at ≥ 1600°C without Al_2O_3 addition. Besides nearly full density, use of Al_2O_3 gave fine grain size (1–2 μm) and higher strengths in hot pressing at ~ 1350°C. MgO typically gives larger grains, but how much of this is intrinsic, that is, at comparable densities, or due to lower porosity levels is not clear. Simultaneous addition of fine SiC particles and MgO has been successful, as other combinations might be. Commercially, most sintered BeO is produced from UOX BeO with 0.5% magnesium trisilicate.

Brown [60] showed that additions of 1–5 m/o SrO inhibited sintering of CaO, while Petersen and Cutler [61] showed that firing in an atmosphere with H_2O progressively enhanced the initial rate of densification with increasing H_2O pressure above 0.01 atm. Rice [62] showed that > 99% of theoretical density (i.e., some degree of transparency) could be achieved by hot pressing CaO without additives at 1200°C, but that addition of 2 w/o LiF or NaF gave similar or higher density at 1000°C (i.e., 200°C reduction in temperature needed) with LiF additions judged to be somewhat superior. Analysis showed substantial residues from LiF remained at grain boundaries, with Li being more removable than F with heat treatment (similar tests with NaF were not given) [52]. However, while heat treatment at higher temperature to remove most or all of the residues of the additive appeared to give strengths higher than for bodies tested as hot pressed, grain size increased (e.g., from 10–20 μm to ~ 200 μm) as a result of heat treatment, giving lower strengths that could be obtained by hot pressing at lower temperatures [63]. Rice [63] reported that liquid-phase densification was probably occurring, but that densification decreased at and near the melting point of LiF (870°C). He also noted that the melting and wetting of LiF could enhance release of residual gas species that often limit important for densification. Later Deruto and coworkers [64] showed the LiBr-$CaCO_3$ eutectic enhancing decomposition of the $CaCO_3$, supporting this in view of similarities expected between LiF and LiBr (e.g., as indicated with MgO densification below). Gupta and coworkers [65] subsequently reported significantly improved optical transparency of CaO vacuum hot pressed with small (e.g., 0.2–0.6 w/o) addition of CaF_2.

Baumard and coworkers [66] noted a surprising lack of investigation of additives to aid sintering of CeO_2 and investigated effects of oxide additions of Nb_2O_5 or TiO_2 on sintering in air at 1200–1480°C. They showed that while sintering without additives gave ~ 97% of theoretical density at 1480°C (with ~ 30μm grains), higher densities were achieved at 1300–1400°C, with these increasing with Nb_2O_5 addition levels over the range studied (0.1–3 w/o) to 99.5–99.9% of theoretical density. The same additions of TiO_2 showed similar, but more spectacular effects—e.g., 99.4% of theoretical density with 3 w/o addition at only 1200°C (with ~ 1 μm grains)—with substantial increase in densification occurring before formation of either $CeTi_2O_6$ or a liquid phase. In a subsequent broader study of CeO_2 Chen and Chen [67] investigated a series of oxide additions at 0.1 or 1% levels in order of increasing charge/size of Mg^{2+}, Ca^{2+}, Sr^{2+}, Sc^{3+}, Yb^{3+}, Y^{3+}, Gd^{3+}, La^{3+}, Ti^{4+}, Zr^{4+}, and Nb^{5+} in sintering at 1270 or 1420°C. They concluded the higher addition level suppressed grain growth via solute drag, otherwise severely undersized dopants (Mg, Sc, Ti, and Nb) enhanced grain growth, but was also charge dependent. Ovcrall 0.1% Mg and 1.0% Sc increased grain growth the most while 1.0% Y decreased it the most. Yoshida and coworkers [68] recently reported that samaria-doped ceria, of interest as a solid electrolyte for solid-oxide fuel cells, requires sintering at ≥1600°C, but that 1% addition of gallia allowed comparable microstructures and properties could be obtained at ~ 150°C lower temperatures. Commercially, CeO_2 is sintered with CaO addition, but selection and effects of additives are often compromised by the presence of sodium picked up in the commercial production of CeO_2 powder by carbonate precipitation.

Cr_2O_3, which is limited in its sinterabiliy due to volatization (probably as CrO_3) at higher temperatures, was shown by Ownby and Jungquist [69] to be sinterable to essentially theoretical density with 0.1 w/o MgO at 1600°C with 2×10^{-12} atmosphere O_2 pressure, and reduced grain sizes from those in pure Cr_2O_3 alone (attributed to formation of Mg-Cr spinel grain boundary particles). Use of 1% MgO resulted in densities between those with 0.1% MgO and none. Roy and coworkers [70] reviewed some data on sintering aids for Cr_2O_3 and corroborated and extended the above results; for example, besides the similarity of benefits of MgO in both Al_2O_3 and Cr_2O_3, there is also a similar bcnefit to use of TiO_2 additions (and also marked effects on electrical properties), but with greater benefits in Cr_2O_3 as shown by Callister and coworkers [71] and Nagal and Ohbayashi [72] (e.g., ~92% dense with 1 w/o TiO_2 at 1200–1300°C with proper control of the O_2 partial pressure). Commercially, large quantities of Cr_2O_3 blocks are produced by sintering with TiO_2 additions for use as liners in tanks for commercial melting of corrosive lead- and boron-containing glasses.

Eu_2O_3, of interest for a fast neutron absorber in reactors, presents serious processing challenges due to its propensity to spontaneously microcrack as a result of its substantial thermal expansion anisotropy due to its noncubic (monoclinic) structure. At grain sizes > 12 μm from sintering temperatures of 1800–1900°C microcracking occured [73]. Malarkey and Hunter [73] showed that while lower levels of Ta_2O_5 addition resulted in a maximum in grain size of 38–55 μm at 1.0 a/o Ta, increasing addition over the range of 0.2–0.6 a/o Ta resulted in progressively finer grain size, e.g., of 6–10 μm at 0.6 a/o Ta, and thus no microcracks.

Fe_2O_3 was reported by Wright [74] to show enhanced sintering with 0.5 to 2% additions of CaO (as $CaCO_3$), with acceleration of sintering on forming a liquid phase. Fe_2O_3 and Fe_3O_4 are more sensitive to effects of atmosphere on sintering than many binary oxides, from some lower temperature study by Hayes [75] and effects on grain size of dense sintered Fe_3O_4 by Yan [76]. Chlorine in the firing atmosphere can enhance grain growth [77].

Consider next densification of MgO with oxide additives, which has received attention in part due to its important use for refractories, as has Al_2O_3. Ryshkewitch [18] cites FeO as one of the most effective and commonly used mineralizers for MgO (e.g., giving 90–92% dense MgO at 1500°C), but variations can occur due to stoichiometry changes with variations of the firing atmosphere. Kriek and coworkers [78] studied effects of CaO, Fe_2O_3, TiO_2, Al_2O_3, and SiO_2 (mostly 1–10% additions in firing dead-burned MgO between 1250 and 1400°C. CaO was the least beneficial (actually inhibiting sintering) and TiO_2 the best, but with an optimum concentration of 1–2% or more, and the optimum level tending to decrease as the temperature increased. Similar trends for an optimum additive level were also found for Al_2O_3 and SiO_2. Brown [79] reported that 0.01% vanadium oxide (from ammonium vanadate) aided densification on sintering in the 1200–1450°C range, but with benefits apparently diminishing above 1250°C where there was a liquid phase. Budnikov and coworkers [80] reported that addition of 0–0.3 m/o HfO_2 to fine active MgO powder significantly enhanced densification, generally with modest increases in grain growth. Densification increased as the level of addition increased until 99–100% density was achieved at 0.1% addition and 1400°C firing, and thus no further benefit with greater addition. These results were extended and compared with effects of 0.1–0.5 a/o additions of Fe^{3+}, Zr^{4+}, Sc^{3+}, or Ni^{2+} (whose radii are similar to that of Mg^{2+} as well as Hf^{4+}), showing that Hf addition was superior. Among the latter four, Zr and Sc were much less effective at lower levels of addition, but were superior at higher levels [81]. The utility of TiO_2 additions (e.g., 0.2w/o) was corroborated in more recent studies of sintering magnesite for refractories at 1600°C by Chaudhuri and coworkers [82], with its enhanced sintering being attributed to a combination of enhanced diffusion and liquid-phase sintering (from liquids formed with impurities and TiO_2). Finally, note that phosphate bonding of MgO is of interest, but the use of H_3PO_4, as with some other refractories, results in too vigorous a reaction; use of $Mg(H^2PO_4)_2 \bullet 2H_2O$ can give substantial densification as shown by Itatani and coworkers [83]. They showed that while < 0.5 m/o addition retarded sintering, it was promoted by ≥1 m/o addition, especially as sintering temperatures increased—3 m/o addition showed porosity ~ 2% lower than the 35% without addition at 1200°C, but a greater than threefold porosity reduction (from 22 to 5%) at 1400°C.

Hamano and coworkers [84–87] conducted a series of investigations of effects of additives on MgO sintering. In earlier work, they showed that an optimum level of addition of B_2O_3 was 0.2 m/o at lower temperatures and 0.5 m/o at 1500°C, but was complicated by formation of $3MgO \bullet B_2O_3$, which first increased, then decreased, as temperature increased [84]. They later showed that 2m/o of Sm_2O_3, ZnO, or NiO aided densification at 1400°C, with Sm_2O_3 and ZnO being particularly effective in earlier stage sintering, and all, especially Sm_2O_3, limiting grain growth [85]. Subsequently, they showed that ZrO_2 was also beneficial over the 1000–1500°C sintering range investigated, with addition via an oxychloride [86] precursor giving somewhat higher densities than via a sulfate precursor [87] added to $Mg(OH)_2$ before calcining. Sintered densities passed through modest maxima (mainly for the oxychloride precursor at intermediate sintering temperatures) or essentially plateaued at 1.5 and 1.5–3 w/o, respectively. Note some of the benefits of the ZrO_2 additions were from benefits on green density, and hence via powder effects, with more powder effects also shown in earlier work [88].

Turning to nonoxide additions to densify MgO, LiF has received extensive attention and use, following apparently independent discovery by Rice ([89–92], with initial aid of V. Edlin) and by Carnall and coworkers [93], with most attention to its benefits in hot pressing, resulting in moderate to high transparency. This discovery was in part based on the work of Atlas [94] showing that small (e.g., 0.5%) additions of LiF, LiCl, LiBr, or LiI, gave 3–10% porosity on sintering at 1400°C (and 13 and 18% porosity for Li_2SO_4 and Li_2CO_3 additions, respectively) versus 38% porosity without additives. However, Atlas's results were for sintering and LiF was only the third best (i.e., the next to the worst) of the four halide salts (leaving ~ 9% porosity), while hot pressing yields transparency over a range (e.g., 0.5–2 w/o) addition of LiF (which avoids deliquescence of the other Li salts) at lower temperatures [90,92]. (Atlas also found addition of NaF had no benefit or slightly retarded, sintering contrary to results below, which also show other effects of processing.) Thus, Rice [90,92] showed that submicron MgO powder without LiF cold pressed at 20–200 MPa to, respectively, 30–50% density sintered to, respectively, 60–90% of theoretical density, while powders with 2 w/o LiF sintered to 94–97% density at 1200°C and hot pressing with 0.5–2 w/o LiF at ~ 1000°C yields transparent bodies. The very rapid densification in the vicinity of melting of LiF (Fig. 5.2) is vividly shown in operating a manual hydraulic pump to maintain hot pressing pressure. During the nearly vertical trend of ram travel versus temperature, one cannot pump fast enough to maintain the initial pressing pressure until most of the densification has occurred, that is, until final densification. Rice also showed that heat treatment with very slow heating rates can remove most or all of the residue from the LiF addition with limited or no reduction in transparency [91, 92].

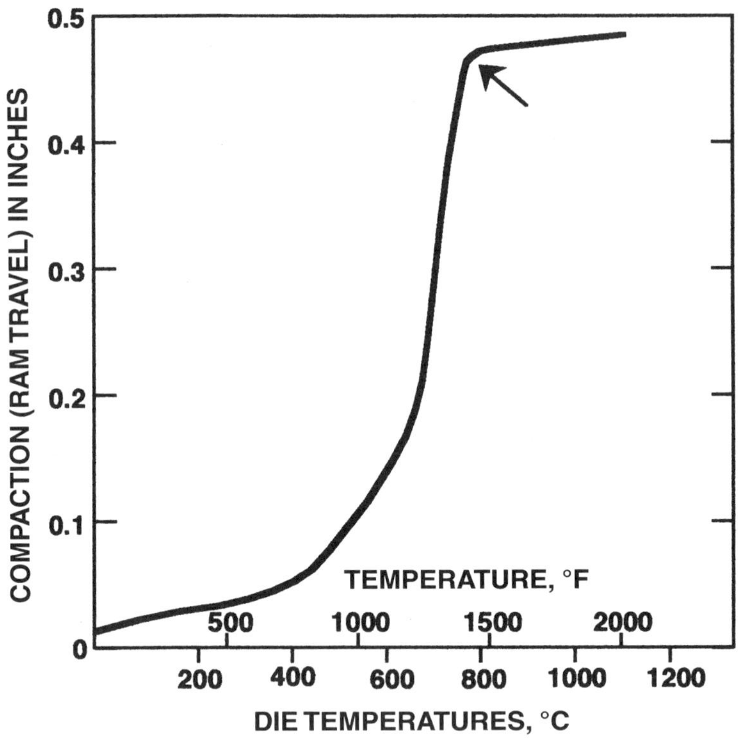

FIGURE 5.2 Plot of ram travel (hence densification) versus die temperature in hot pressing MgO+ 2 w/o LiF in graphite dies. While the exact relation between die and hot-pressed body temperature is uncertain, it is clear that rapid densification occurs near, but probably somewhat below, the melting point of LiF, then ceases fairly abruptly (arrow) where final densification begins. (Rapid densification somewhat below the melting pont of LiF may reflect its interaction with surface hydroxide and carbonate species.) (From Ref. 92.)

Miles and coworkers [95] further developed the process, showing that improved mixing and use of vacuum hot pressing improved MgO transparency, its homogeneity, or both, with vacuum being more important as size increases. Smethurst and Budworth [96] also used LiF additions, noting the value of fine particle size—below ~ 0.5 μm—and many others have used or studied the process ([97–101]; for example, corroborating that densification is accelerating well below the melting point of LiF [97,100].

Razinger and Fryer [102] investigated mechanisms of densificaion of MgO by hot pressing MgO with LiF, Li_2O, or MgF_2 additives, concluding that an important aspect of the effect of LiF on densification was via control of the microstructure. Shimbo and coworkers [103] later reported that grain growth of MgO treated with either AlF_3 or especially MgF_2 was either decreased or increased depending on concentration of the additive and the temperature.

The process has been extended to other sources of fluorine or of other related constituents. Thus, Rice [104] showed that other than requiring about 100°C higher temperature, 2 w/o NaF worked nearly as well as LiF in hot pressing and tends to give somewhat finer grain sizes, though often with some larger grains—that is, a bimodal distribution. Banerjee and Budworth [105] showed that incomplete sintering trials with 0.25 w/o NaF grain boundaries did not separate from the intergranular pores due to less grain growth than with LiF, and thus bodies could be sintered to transparency by first presintering in pure O_2 at 600°C (i.e., below the liquefying temperature), then long sintering at 1600°C. Use of vacuum and presintering at 1000°C (above the liquefying temperature) resulted in only 97% density.

Watanabe and coworkers [88] showed that additions of 1, 4, 5, or 8 w/o MgF_2, 4 w/o gave the best results and that while use of fine particles (<2 μm) accelerated sintering (for 1 hr) above 1000°C to the limits of testing, 1400°C, coarser particles (10–20 μm), though making some reduction in green density, accelerated sintering more above ~ 1050°C. The greater benefit of larger MgF_2 particles was attributed to their slower decomposition. They also showed that firing MgO compacts without particulate additions of MgF_2, but in a closed platinum crucible with MgF_2 powder not in contact with the MgO resulted in accelerated sintering nearly as high as that with the coarser MgF_2 particle addition, from 1000–1300°C (the latter temperature being where the decomposition of MgF_2 was believed to be complete). However, benefits of MgF_2 additions were found to occur for only short firing times (≤ 1 hr); higher densities were obtained without MgF_2 additions with longer sintering at 1400°C. Ikegami and coworkers [106] showed that as little as 0.02 w/o F^- introduced via MgF_2 or HF addition to MgO (with particle sizes < 0.15 μm) then calcined, at 900°C, noticeably increased densification so transparent MgO could be obtained by vacuum sintering at 1600°C, with most densification complete by ~ 1250°C where most outgassing was completed. Similar additions of Cl^- were also beneficial, but much less so than F^- additions, though combined F^- and Cl^- additions were quite effective. Ikegami and coworkers [107] later extended these results showing that $Mg(OH)_2$ treated with HF solution, then calcined at 900°C showed accelerated shrinkage from ~ 800 to 1200°C, with grain growth increasing at ~1100°C, then accelerating significantly between 1200 and 1250°C, while F loss begin rapidly at ~ 1200 and was nearly complete by ~ 1600°C. This is generally consistent with more extensive studies of removal of additive residues by Johnson and coworkers [52], who also showed that more F than Li or Na remained after hot pressing. Leipold and Kapadia [108] showed that addition of S^{-2}, Cl^-, F^-, or OH^-

generally at levels of 0.2–0.9 a/o, except 1.8–2.2 a/o OH, (added, respectively, as elemental sulfur, $MgCl_2$, MgF_2, and $Mg(OH)_2$) all inhibited densification of MgO, with the extent of inhibition increasing in the order listed, in hot pressing at 850–1200°C. Hana [109] reported that treatments of MgO powder to add ~ 0.15 w/o NH_4F gave translucency on hot pressing at ≥1500°C and that HF treatment of the MgO powder had similar effects.

The above studies of MgO with LiF or other related additives show that vaporization aids distribution of the additive (as well as its loss) and that, while there may be some activated sintering before a liquid forms, the latter is a critical step; for example, it allows relative particle motion for consolidation in hot pressing and liquid-phase sintering (possibly pressure enhanced in hot pressing). LiF and NaF have been shown to be good lubricants for MgO, with LiF best at ~ 900–1100°C and NaF at 1300—1500°C [145], consistent with LiF giving more densification at lower temperatures and NaF probably being better at higher temperatures. Vacuum atmosphere can be beneficial or necessary, especially for larger bodies, and may aid removal of additive residues, but may also be detrimental, possibly due to premature removal of additive species. Liquid-phase formation may be extended by interaction with impurities and use of pressure, e.g., in hot pressing, as well as reduced by oxidation or compound formation. Though often neglected, and generally secondary, other factors besides the amount of additive are important, such as the distribution of the additive phase, and its particle size, and that of the MgO, which reflect not only starting parameters, but also changes in various stages of the process—for example, on green density. Again, direct characterization shows residual fluorine from fluoride additions in as-hot pressed MgO [52].

One study shows CoO additions to NiO aid its sintering [110]. Varela and coworkers [111] showed that 2 m/o CuO addition aided sintering of SnO_2, which is typically limited by serious vaporization. Dry air was better than dry argon since the latter inhibited densification due to reduction of CuO and vaporization of SnO_2 over the 900–1250°C range investigated. Drozd, Degtyareva, and coworkers [112,113] also reported benefits of CuO in sintering SnO_2 at 1550°C as well as of V_2O_5, Bi_2O_3, CoO, MnO_2, ZnO, or Sb_2O_3. Commercially SnO_2 is sintered with ZnO additions, for example, for large electrodes for electrical melting of glasses.

ThO_2, being the most refractory oxide, is clearly a candidate for using sintering aids. CaO is an additive commonly used for its sintering, as noted by Ryshkewitch [18] and indirectly by Singer and Singer [14] in noting use of CaF_2, $CaBr_2$, or $CaCl_2$ (as well as SrF_2, NH_4I, V_2O_5,SnO_2, Bi_2O_3, or ZrO_2). Curtis and Johnson [114] were an early user of CaO (or CaF_2) additions, investigating 0.5–3% additions and showing that 0.5–1% additions to powder with particles < 1 μm gave 97% dense ThO_2 at 1800°C. Later Jorgensen and Schmidt [115] showed that 2 m/o addition was optimum for sintering to theoretical density (i.e., transparency) by the second phase inhibiting grain growth (giving grain sizes of 5–10 μm). Stuart and coworkers [116] reported that adsorbed F on ThO_2 increased the rate of densification by hot pressing at temperatures as low as 700°C, but that adsorption of SiO_2 with or without F inhibited densification. Halbfinger and Kolodney [117] showed that sintering in air at 1370°C with 2 w/o NiO, Co_2O_3, ZnO, or Cu_2O all gave < 6% porosity, while additions of Cr_2O_3, SiO_2, TiO_2, MgO, Ta_2O_5, ZrO_2, Al_2O_3, CdO, BaO, and SrO all gave > 20% porosity. NiO was particularly effective—its effects plateaued above ~ 0.3 w/o—and produced even greater benefits when combined with limited amounts of Y_2O_3, which by itself gave limited increase of sintering. This is in contrast to results of Greskovitch and coworkers [118], who achieved transparency with 5–15 m/o Y_2O_3 by sintering at 2380°C in H_2, with 5 m/o being preferred since it gave the smallest grain size (~ 100 μm). The differences between this and the work of Halbfinger and Kolodney reflect changes due to differing amounts of additive, and especially atmosphere and temperature, though temperatures as low as 1850°C were used [119]. (Note that yttria-doped thoria was commercially produced for a while, and that there were reports that ZrO_2 additions could also give transparent ThO_2.)

Chen and Wu [120] reviewed some effects of additives on sintering of TiO_2, and also experimentally examined effects of Nb_2O_5 and CaO, since such additives are of interest for some electrical applications where differences in conduction between the grain boundary region and the grain interior are critical, that is, grain boundary concentration is important. They showed that addition of 0.25 m/o Nb_2O_5 alone to rutile powder modestly inhibited sintering over the temperature range investigated (1100–1350°C), while additions of 0.1 or 1 m/o CaO with the Nb_2O_5 gave higher densities at 1150 and 1250°C, the latter giving the maximum density of ~ 4 gm/cc for all rutile-based bodies studied. In contrast, anatase powder gave densities increasing from ~ 4.15 to 4.2 gm/cc as sintering temperatures increased from 1100 to 1350°C. All additions fell somewhat below this trend at 1050°C, and the 0.25 m/o Nb_2O_5 + 0.1 1 m/o CaO combination gave somewhat lower density at 1350°C, with all additions giving comparable fired densities as TiO_2 alone at 1150 and 1250°C. Higher densities (~ 98% of theoretical) were reached using anatase powder, despite its giving 5–10% lower green densities, versus use of denser rutile powder giving densities of only ~ 90% of theoretical. This was attributed to the onset of abnormal grain growth commencing in rutile-derived bodies before densification was completed, while in anatase-derived bodies such growth did not commence until densification was complete, despite their conversion to the rutile structure during sintering. Work at Bell Labs showed that very small amounts of Cr_2O_3 added to TiO_2 allowed it to be hot pressed to theoretical density, though reduced some, i.e., being black in color. However, full oxygen stoichiometry could be achieved throughout the body by subsequently annealing in an oxidizing atmosphere, while bodies hot pressed without the small Cr_2O_3 addition could not be returned to full oxygen stoichiometry over much of the interior, which thus remained black.

The one study known for UO_2 is that of Radford and Pope [121] showing that sintering in NH_3 (versus the normally used wet H_2) or treatment of powders with various ammonia salts increased sintered densities modestly, but consistently—e.g., from 97 to 98% of theoretical density, along with increased grain size.

The first pressureles sintering of Y_2O_3 to full density was with the use of 10 m/o ThO_2, reversing the roles of matrix and additive for densifying ThO_2. Jorgensen and Anderson [122] reported this and that ThO_2 both enhanced diffusion and retarded grain growth due to segregation at grain boundaries. Greskovich and Woods [123] reported further observations and process development with ThO_2 additions. Commercialization of this material was apparently pursued unsuccessfully, possibly due to the minor radioactivity of ThO_2, which today would be a larger impediment. Rhodes [124,125] later reviewed sintering Y_2O_3 to full density, including use of Al_2O_3 or MgO, and also demonstrated a novel process via addition of 8.1–12.1 m/o La_2O_3 to be feasible because of a two- rather than a single-phase field existing at higher temperatures. Thus, by sintering at 2170°C where a second phase exists, full density could be achieved, but then obtain complete solid solution (to remove the second phase and thus give good optical transmission) by annealing at 1920–2020°C. While such phase relations are not typical, this may provide a model for some other bodies. Subsequently, Katayama and coworkers [126] reported that additions of 1–10 m/o CaO to Y_2O_3 gave increased densification on sintering at 1500–1700°C, with optimum results at 1600°C with 1 m/o CaO. Lefever and Matsko [127] reported that addition of LiF to fine Y_2O_3 yielded transparent bodies by press forging at 950°C with pressures of 70–94 MPa for two days.

Volatization of ZnO makes it a good candidate for use of sintering aids, and some have been demonstrated. Moriyoshi and coworkers [128] reported that addition of ~ 1 a/o phosphate via treatment of ZnO powder in H_3PO_4 yielded quite translucent bodies on sintering at 1200°C, which may have involved some liquid phase, and was accompanied by considerable grain growth. Of four additions to ZnO, 0.5 w/o Sb_2O_3 retarded densification substantially on sintering at 1180°C for 2 hr, while Li_2O accelerated densification on heating to 1180°C but gave slightly lower density after holding there 2 hr than pure ZnO [129]. On the other hand, K_2O addition inhibited sintering on heating some, but gave slightly higher final density than pure ZnO, and TiO_2 addition inhibited densification on heating somewhat less, while giving the highest density achieved after sintering 2 hr at 1180°C. Recently Luo [130] reported substantial sintering of ZnO with Bi_2O_3 additions of > 0.23 m/o, with ~ 0.58 m/o apparently being optimum. Increased densification occurred with increasing temperature beginning at > 600°C and accelerating at ~ 700°C to the eutectic temperature (740°C), with the shrinkage increasing inversely with heating rate; that is, much greater at 4°C/hr versus 0.5°C/min. This occurrence of sintering before a liquid forms and enhancement with lower heating rates indicates a solid-state transport mechanism.

An important application of ZnO is for varistors, which are highly nonlinear resistors that have very high resistance at normal voltages, but low resistance at excessive voltages. This highly nonlinear resistive behavior has important applications, especially for shunting voltage surges; for example, from lightening strikes, to protect electronic and electrical systems from such surges. The development of ZnO varistors is a good example of both the multiple and critical roles that additives often play, as well as the frequent and important role of that serendipity often plays in such developments. Such varistors (as well as apparently some other oxidc varistor materials) depend on the amount and nature of grain boundary phase that results from the use of oxide additives used [130]. Such varistors commonly contain several additives, e.g., 97m/o ZnO with Sb_2O_3, Bi_2O_3, CoO, MnO, and Cr_2O_3, to tailor properties. Originally, ZnO was manufactured as a linear resistor using an electrode metallization that had Bi_2O_3 and other glass-forming constituents that allowed low-temperature firing at several hundred degrees centigrade. However, a firing accident at Matsushita resulting in gross overfiring of the electrode material was found to have produced substantial nonlinear resistance behavior. This behavior was found to be due mainly to interdiffusion of Bi_2O_3 from the electrode composition into the ZnO forming a grain boundary phase later found to be the basis of the resistance nonlinearity. Note that the Bi_2O_3 dopant also plays an important role in the densification (as noted above) as well as the electrical behavior. (Note that in contrast to the role of grain boundary phases ZnO [and probably other oxide] varistors has no bearing on the other important commercial varistor material, namely SiC, where the resistivity nonlinearity results directly from the contacts between the SiC particles—that is, resistive nonlinearity occurs in a bed of loose, contacting SiC particles. The use of the glassy bond phase of SiC varistors is to provide physical integrity of the varistor and consistency of the SiC-SiC particle contacts and their freedom from moisture and other possible contaminants.)

Shakelford and coworkers [131] showed that addition of SiO_2 to (or the presence of SiO_2 as an impurity in) ZrO_2 with CaO stabilizer enhanced sintering shrinkage as the SiO_2 content increased steadily over the range studied from 0 to 2 w/o, except for a more rapid increase at ~ 1% addition. Net densities approximately plateaued from 1–2% addition due to green density decreases as SiO_2 addition increased and were highest for powder naturally containing ~ 1 w/o SiO_2. Benefits were greatest for partial (3.5 w/o CaO) than full (> 5 w/o CaO) ZrO_2), and were similar for MgO stabilization. Wu and Yu [132] noted earlier literature showing benefits of small additions such as of Al_2O_3, Bi_2O_3, Fe_2O_3, SiO_2, or TiO_2, and investigated the addition of Al_2O_3 or SiO_2 to ZrO_2-MgO. Mu and coworkers [11] have shown a definite maximum of densification rate of ZrO_2 with CaO stabilization at

12 m/o CaO. Chiou and coworkers [133] reported optimum densification of ZrO_2+ 15 m/o CaO with tailored silica-based glasses as 1350°C. Lu and Bow [134] reported that ZrO_2+ 3 m/o Y_2O_3 with additions of up to 1 m/o MgO enhanced sintering at 1350°C due to reaction of the MgO with CaO and SiO_2 impurities, but also were accompanied by grain and local-phase changes; for example, bimodal grain sizes. Also note that Scott and Reed [135] reported that residual Cl in oxychloride derived ZrO_2+ Y_2O_3 powders inhibited initial and intermediate stages of sintering and sinterability was increased by laundering the powders to remove residual Cl and by deagglomeration. An example of commercial practice is the fairly recent announcement of sintering ZrO_2 with 3 m/o Y_2O_3 and 0.25 w/o Al_2O_3 to produce ultra-fine-grain fiber-optic ferrules that do not have problems on exposure to water.

5.4 MIXED OXIDES

While mixtures of oxides such as in the stabilization of ZrO_2 discussed above could come under this heading, the focus is primarily on ternary compounds formed by reaction between two binary-oxide compounds, e.g., MgO and Al_2O_3 to form $MgAl_2O_4$. While such mixed-oxide compounds offer more opportunities to enhance densification by varying stoichiometry than binary oxides, additives are frequently used in addition to or instead of such compositional variations. This section provides a fairly extensive summary of additive densification of important members of ternary oxide families of aluminates and silicates (structural materials), and ferrites, niobates, and titanates (magnetic and electronic materials), which are treated in the order listed.

5.4.1 Aluminates

Consider first magnesium aluminate, i.e., magnesia alumina spinel, $MgAl_2O_4$, which is of interest for both structural and especially IR window applications. One of the earlier successes in obtaining translucent to transparent $MgAl_2O_4$ was work of Rhodes and coworkers [136]; they used various, mostly fine $MgAl_2O_4$ powders or mixtures of MgO and Al_2O_3. Dense bodies were obtained by hot pressing without additives at 1300–1400°C with 70–100 MPa pressure with or without vacuum, or by press forging at 1650–1850°C with 35 MPa pressure, but the most transparent bodies were made by press forging with 0.8% SiO_2+ 2% Li_2O (at 1750°C and giving grain sizes of ~ 60 μm).

Rhodes and coworkers also reported high densities obtained with 1% LiF additions hot pressed at 1200–1300°C, which had finer grains (~ 0.2 μm) and lower room temperature strengths than bodies of similar grain size made without additives, but higher strengths than bodies with large grains. After this, McDonough and Rice [137], as well as investigators at Coors [138,139], also obtained substantial densification with LiF additions, with the latter group focusing on higher temperature densifications for high transparency with larger grains (and some reduction in mechanical properties due to this and residual boundary phase from the additive). Lange and Clarke [140] showed the presence of a grain boundary film in commercial spinel densified with LiF and interpreted the nature of this film and its boundary coverage to indicate a solidified melt.

De With and coworkers [141,142] reported vacuum sintering translucent $Y_3Al_5O_{12}$ at 1700–1800°C with small additions of MgO or SiO_2 (respectively, ~ 500 and 2000 ppm), giving grain sizes of ~ 3 and ~ 10 μm, respectively.

Though effects on densification were not reported, Hashiba and coworkers [143,144] reported effects of various fluorides on enhancing formation of $ZnAl_2O_4$ from ZnO and Al_2O_3 (and cite similar investigation of $MgAl_2O_4$ formation. LiF was found most effective for forming $ZnAl_2O_4$.

5.4.2 Silicates

Turning to silicates, consider mullite ($3Al_2O_3{\bullet}2SiO_2$), which is often difficult to sinter unless much of the densification is achieved before extensive mullite formation or unless quite fine mullite particles are used, so densification aids have been sought. Besides adjusting the stoichiometry, some use of oxide additives has been investigated. Thus, Baudin and Moya [145] reported that addition of TiO_2 below the solubility limit (~ 2.9 ⊕ 0.2 w/o at 1600°C) significantly aided sintering, while additions above the solubility limit formed Al_2TiO_5 and inhibited sintering. While Mitamura and coworkers [146] investigated additions of Y_2O_3, Er_2O_3, Sm_2O_3, La_2O_3, and CeO_2, noting benefits of 0.5 w/o Y_2O_3 or CeO_2, Mori and coworkers [147] reported benefits of Y_2O_3 additions, as did Hwang and Fang [148,149]. Hwang and Fang showed benefits of higher levels of Y_2O_3 additions (2–10 w/o) and evidence for a liquid phase. Prereacted fine mullite powder has apparently been hot pressed to full density using AlF_3 addition.

Modest additions of some oxides have been reported to aid sintering of zircon, $ZrSiO_4$, typically in the 1450–1550°C. Thus, Hayashi and coworkers [150] reported that TiO_2 additions resulted in formation of $ZrTiO_4$ on firing in O_2, increased early stage densification on firing in H_2, and dense bodies on sintering in N_2 (with some reduction of the TiO_2). Later De Andres and coworkers [151] reported achieving > 98% density with TiO_2 additions, attributed to formation of a liquid phase. Kim [152] reported increased sintering with addition of MgO, MgO + CaO, or Al_2O_3 + CaO, which apparently reacted with SiO_2 from zircon dissociation, which did not occur with pure zircon.

5.4.3 Ferrites

Consider next magnetic materials, starting with hard ferrites. Arendt [153] investigated use of several liquid-phase sintering aids for $BaFe_{12}O_{19}$ and $SrFe_{12}O_{19}$, reporting that several additives in the systems M-A-S and M-A-B-S, where M=

Pb, Ba, or Sr oxides (with the Ba or Sr choice corresponding to the Ba or Sr ferrite being sintered), and A=Al_2O_3, S= SiO_2, and B= B_2O_3, were effective for densification and magnetic properties—the latter especially for sintering at ≤ 1200°C where grain growth was limited. Lucchini et al [154] showed that low-melting silicate glasses (glazes) were effective for sintering dense hexaferrite bodies at 1150°C (e.g., 150–200°C lower than without additions) with limited grain growth. Fang and coworkers [155] subsequently reported that small additions (0.3 w/o) SiO_2 in sintering barium ferrite effectively inhibited grain growth, even within the solubility range of SiO_2 in the ferrite. Beseniar and coworkers [156] reported that additions of ZrO_2 can enhance sintering of Sr ferrites via a liquid phase and give optimum mechanical and magnetic properties at ~ 0.3 w/o.

Turning to soft ferrites, an earlier investigation of a Li ferrite showed that increased addition of NiO resulted in limited decreases in density for firing at 1150°C in O_2 (and with packing powder) from 99% dense at zero addition to 98% dense at 20 m/o NiO and discontinuous grain growth [157]. Additions of $NiFe_2O_4$ decreased density a bit more, but suppressed discontinuous grain growth. Both additives had mixed, but often different, effects on properties. Peshev and Pecheva [158] reported that modest additions of Bi_2O_3 yielded high density Li ferrites with good properties at 900–1000°C due to liquid-phase forming. Kim and Im [159] reported that additions of Nb_2O_5 or V_2O_5 to a lithium ferrite substantially enhanced densification, especially with V_2O_5, attributed to both additives increasing O diffusion and decreased Li loss and possible formation of a liquid phase with V_2O_5. More recently, Rezlescu and Rezlescu [160] investigated effects of CaO, Na_2O, Sb_2O, and ZrO_2 additions (0.3 w/o) to a Li ferrite all resulted in enhanced sintering over undoped bodies, especially on firing at ≥1050°C and achieving > 97% density at 1140°C, 98% with CaO with additives versus 94% without additives. All additives limited grain growth and had varying effects on properties.

Bando and coworkers [161] reported that simultaneous additions of up to a few m/o of CaO and SiO_2 to a Mn-Zn ferrite resulted in increasing discontinuous grain growth as the addition increased over the 1100–1300°C temperature range studied, and was associated with liquid-phase sintering. Later, however, Toolenaar [162], while showing exaggerated grain growth with SiO_2 additions, claimed it was due to grain boundary chemistry effects, not due to liquid-phase formation. Jain and coworkers [163] showed that addition of 0.5 w/o of MoO_3 to a Mn-Zn ferrite aided densification, giving ~ 94% of theoretical density at 1065°C with controlled grain growth (and the removal of the MoO_3 due to volatilization). More recently, Drofenik [164] showed that sintering of a Mn-Zn ferrite with Bi_2O_3 additions at ~ 1350°C, while slightly increasing density from 0.02 to 0.2 w/o, first decreasing density slightly to 1 w/o, and more seriously at higher levels, had more complex effects on grain growth and magnetic permeability. Thus, at low additive levels (~ 0.03 w/o) a high permeability and normal grain structure were obtained, while higher additive levels (> 0.05 w/o) gave good permeability with anomalous grain growth. Higher additions (> 0.08w/o) again gave a homogeneous microstructure along with intragranular porosity and lower permeability. They noted that addition of the Bi_2O_3 increased FeO concentration at grain boundaries and decreased that of Ca and Si.

Based on reports of improved performance of Co-Ni, Zn-ferrite cores with BaO additions, Drofenik and coworkers [165] showed that small additions (< 0.5 w/o) limited grain growth due to forming a Ba-Co hexaferrite layer on the grain boundaries. Mahloojchi and Sale [166] showed that additions of CuO to MgO-based soft ferrite improved sintering.

5.4.4 Electrical Ceramics

Turning to electrical and electronic ceramics, Shimada and coworkers [167] reported that sintering of $LiNbO_3$ was significantly accelerated by modest additions of CdO (e.g., 3 w/o). This was attributed to both a large increase in oxygen diffusion as well as a reaction at 850–900°C, with the resulting second phase apparently hindering exaggerated grain growth that occurs at 1050–1100°C in pure $LiNbO_3$. Ye and coworkers [168] reported that ~ 5 w/o addition of LiF+ MgF_2 allowed densification of $LiTaO_3$ to ~ 97% of theoretical density on sintering in air at 900°C, especially in using prereacted $LiTaO_3$ rather than simultaneous reaction and sintering. The resultant bodies showed increasing Curie temperatures with increasing additive content, but piezoelectric and pyroelectric coefficients were ~ 30 and 20% of single-crystal values, respectively.

Consider now various titanates for electrical applications starting with $BaTiO_3$. A variety of additions to aid densification were evaluated by Swilam and Gadalla [169] and discussed in terms of cations with higher valence substituting for Ba^{2+} and cations of lower valence replacing Ti^{4+}, with sintering being controlled by Ba diffusion. They showed additions of ZnO, CdO, $PbTiO_3$ and CuO in amounts above the solubility limits at 1200°C and > 3% NiO at 1300°C caused rapid densification due to formation of a reactive liquid phase, while additions of MgO, NiO, $CaTiO_3$ and Bi_2O_3 in amounts above the solubility limits at 1200°C increased densification due to grain boundary precipitation. Howing and McCutcheon [170] discussed firing of PTCR compositions with Pd, and also Ca in some cases, via a liquid phase due to 2m/o SiO_2 + 0.5 m/o TiO_2 additions. Burn [171] reported sintering of $BaTiO_3$ modified with Ba and Sr zirconates by firing with liquid phase (of CdO, Bi_2O_3, B_2O_3) at 1100°C. Srakar and Sharma [172] reported sintering of $BaTiO_3$ with B_2O_3 or PbB_2O_4 glasses at 800–900°C (instead of 1400°C for $BaTiO_3$ alone) significantly increased dielectric breakdown, but with equal or greater reductions in dielectric constant. An important application of sintered $BaTiO_3$ bodies is as the dielectric in multilayer capacitors where compatibility with the sintering requirements of the electrode material is critical. Thus, switches from more refractory and expensive Pd- or Pt-based electrodes to lower cost and lower sintering Ag-based electrodes required further lowering of the sintering of the titanate dielectric, which is done commercially by combinations of ZnO, CdO, PbO, GeO_2, SiO_2, B_2O_3, Bi_2O_3, and WO_3.

Significant benefits of densifying $BaTiO_3$ with LiF additions were apparently first reported by Walker and coworkers [173–175]. They showed that rapid densification via hot pressing began at lower temperatures for the following additives: LiF(600°C), LiF-NaF-KF eutectic (650°C), NH_4F (800–900°C), and NaF, LiCl, MgF_2, or BaF_2 (1000–1200°C), versus ~ 1200°C for $BaTiO_3$ alone. Focusing on LiF as the most promising, they also showed 90–95% or more of theoretical density could be achieved by sintering in air with 1–2 w/o LiF addition, versus ~ 1300°C without it (Fig. 5.3). Densities could often be first increased a few percent then decreased by a similar amount with subsequent annealing of hot-pressed (and some sintered) bodies, but annealing generally exacerbated grain growth beyond normal large grain sizes (~ 50 μm) commonly resulting from densification. Useful restriction of such grain growth was obtained by also adding 2 w/o MgO, which also resulted in increased strengths over those for pure $BaTiO_3$ of the same grain size. They concluded that the primary mechanism was liquid-phase densification, but a distinct (translucent) yellow color indicated some solid solution effect.

Several investigators further studied and developed densification of $BaTiO_3$ with LiF. Haussonne and coworkers [176] corroborated significant benefits of densification with LiF: that an important factor was transient liquid-phase densification, and Li is removed faster than F. They also showed that the process was more complicated in the details of its progress since it involved some phase transformations, could vary some with the amount of LiF and with modest changes in the Ti/Ba ratio, and involved some substitution of Li for Ti and F for O. They observed that $BaLiF_3$ should be a desirable additive, and subsequently showed this to be an advantageous additive [187]. Potin and coworkers [178] confirmed significant lowering of the sintering temperature (e.g., to 930°C and that some excess of Ba was a factor. They also showed the presence of a liquid phase below the melting point of LiF, that the Curie temperature was reduced from 120 to 10°C, and questioned the mechanism of forming and effects of Li_2TiO_3 as suggested by others. Lin and Wu [179] further confirmed the formation of a liquid phase as well as its wetting of $BaTiO_3$, and reported the formation of $LiTiO_2$ when using LiF, but not when using $BaLiF_3$, further supporting its use. Guha and Anderson [180] also corroborated significant enhancement of sintering $BaTiO_3$ with a few percent of LiF and reported that LiF first reacts with $BaTiO_3$ to form a cubic solid solution and Li_2TiO_3 then forms a liquid phase at 740°C, which leads to high density (e.g., at 900°C) followed by considerable volatile loss.

Ueda [181] reported a substantial study of the effects of 11 binary oxide

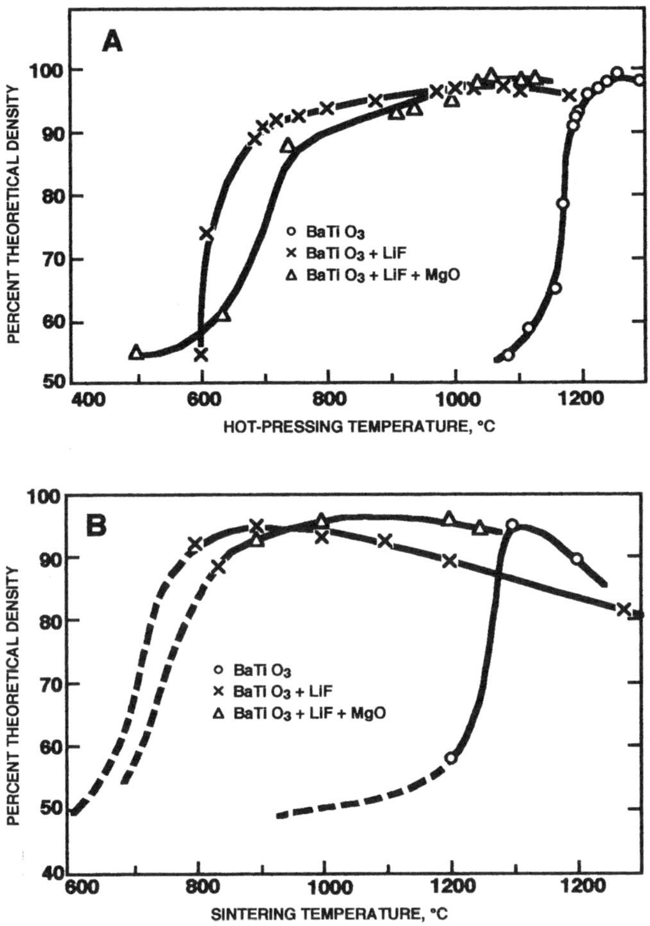

FIGURE 5.3 Densification of $BaTiO_3$ by itself, with 2 w/o LiF, or with 1 w/o LiF + 2 w/o MgO during (A) hot pressing and (B) sintering in air. (From Ref. 173. Published with permission of the Am. Cer. Soc.)

additives on the densification and properties of $PbTiO_3$, noting that Li_2CO_3, NiO, Fe_2O_3, and MnO_2 gave densities of 97–99% of theoretical density (with fine grain sizes of 0.2–1.8 μm, which is important to limit microcracking) on firing at 1180–1200°C for 1 hr in platinum crucibles in air, while other additives gave only 75–95%. Wittmer and Buchanan [182] showed that 0.1 to 6 w/o V_2O_5 addition allowed dense lead zirconate titanate (PZT) to be achieved at 960°C in 15 min (thus eliminating the need for PbO atmosphere control) versus 4 hr at

1280°C with no additive. The optimum range of addition of 0.25 to 1 w/o gave comparable properties as obtained without the additive. The enhanced densification was attributed to enhanced surface activity and liquid-phase sintering. However, the commercial sintering of PZT and PLZT remains using a PbO-rich atmosphere provided by excess PbO.

Chen and coworkers [183] showed that 4 w/o V_2O_5 addition aided densification of $SrTiO_3$ at 1250°C but gave fine-grain bodies with large pores and TiO_2 precipitation. Addition of excess SrO reacted with the precipitated TiO_2. Cheng and coworkers [184] reported sintering $SrTiO_3$ at 960°C for 2–3 hr with addition of Li_2CO_3 3 w/o, due to formation of a liquid phase. Laurent and coworkers [185] investigated use of LiF, Li_2CO_3, or $LiNO_3$ to densify doped $SrTiO_3$ while maintaining desired property levels, reporting best results with $LiNO_3 + Bi_2O_3$.

Clarke and Hirschfeld [186] have shown that sintering of slip cast $(Ca_{0.6},Mg_{0.4})Zr_4(PO_4)_6$ powders was significantly aided by addition of ZnO, giving > 99% theoretical density with 1.5–2% addition with temperatures of 1170–1200°C for short times (e.g., 0.5 hr). The mechanism was interpreted to be liquid-phase sintering, based in part separate observations of partial melting of the phosphate. However, rapid coarsening of the microstructure and excess grain boundary phase occurred with marked decreases in room temperature flexure strength (< 30 MPa) and changes in thermal expansion—for example, negative expansion at 1000°C with substantial hysteresis. A fine microstructure with normal low expansion (~ 2 ppm/°C) and reasonable strengths of ~ 100–110 MPa or greater were achieved by limiting exposure to the liquid phase by limiting the amount of ZnO additive, the firing temperature, and time at temperature.

5.5 NONOXIDES

Consider nonoxides, for which additive development and use is primarily for binary borides, carbides, and nitrides, which are covered in the order listed. Within each of these families, individual compounds are generally taken in alphabetical order, except where some modification of this order is advantageous because of related technical or editorial factors. The focus is on additives whose key function is to aid densification whether they aid or diminish some other aspects of performance other than effects due to reduced porosity; that is, densification benefits in making a composite bodies are not addressed here. Many of these materials for which additives have been used to aid densification do not absolutely require such additives, but they have been used because they offer advantages, most commonly easier densification. Though improved powders, especially finer, more uniform ones, and in some cases purer ones (where impurities are detrimental rather than of some advantage) have reduced some of the driving force for use of additives, but practicality still is important in their use. Further, there are more nonoxides of significant importance which generally cannot be sintered, or sinter poorly, without additives, namely diamond, SiC, AlN, BN (hexagonal and cubic), B_4C, and Si_3N_4.

Turning first to borides, LaB_6 was sintered by Shlyuko and coworkers [187] at 2000–2200°C (buried in LaB_6 powder) using Y_2O_3. While the Y_2O_3 addition enhanced sintering, it also reacted to form yttrium boride with the release of gaseous B_2O_3, which above 4 w/o Y_2O_3 was progressively more detrimental giving optimal results at 2 w/o Y_2O_3. Additions of Al_2O_3 substantially inhibited sintering. Hot pressing of NbB_2 without additives at 2100°C by Murata and Miccioli [188] yielded large grains (~ 20 μm) and a few percent of inter- and intragranular pores as well as considerable (mostly grain-scale, i.e., micro-) cracking. Screening studies with 5 w/o additives of borides, carbides, or nitrides of Ti, Zr, or W hot pressed at 2100°C showed most were ineffective in improving the NbB_2, with WC reacting to leave WB at grain boundaries and ZrN segregating at grain boundaries. However, the Ti additives gave benefits in the order TiN > TiC > TiB_2, with all eliminating cracking and most or all intragranular porosity as well as reducing grain size, i.e., ~ 4, 9, and 14 μm, respectively, with total residual porosity scaling with grain size. Optimum results were obtained with ~ 9 w/o TiN which gave a minimum in grain size of ~ 2 μm. Low and McPherson [189] reported reaction sintering of SiB_6 with either ZrO_2 or $ZrSiO_4$ in air at 1300–1500°C to fabricate ZrB_2-based bodies.

TiB_2, which is one of the most important refractory borides, has received considerable investigation and densification development with mainly Fe, or especially Ni, or some ceramic additions [190,191]. Champagne and Dallaire [192] briefly reviewed use of metal additions and reported that TiB_2 powder made using ferrotitanium (FeTi) powder with excess Fe allowed HIPing to high density at 1300°C. Shim and coworkers [193] showed that TiB_2 sintered with Fe additions increased to a maximum of 89% of theoretical density with 0.4 w/o at 1800°C, decreasing with higher addition levels or sintering temperature (due to vaporization). They also showed that Ni additions gave nearly the same maximum density at 1800°C, with higher temperatures giving poorer results due both to vaporization and microcracking (due to larger grain size). Ferber and coworkers [194] have shown that strengths and toughness of pure TiB_2 hot pressed to ≥98% of theoretical density at 1800–2000°C and bodies > 99% of theoretical density at 1425°C with 1.4 or 7.9 w/o Ni correlated in an inverse fashion with their grain sizes, due in part to increased microcracking at larger grain sizes, with the lower Ni addition giving the best results due to its smallest grain size, ~ 4 μm. Einarsrud and coworkers [195] have shown that 1.5 w/o additions of Fe or Ni were effective in sintering in argon or vacuum to > 94% of theoretical density at 1500–1700°C (with no significant benefits of higher additive contents) but larger grains. They reported similar results with similar NiB additions. Some limited work has been conducted on use of intermetallic additives, for example, NiAl [196] in hot pressing at 1450°C giving limited grain growth but also an intergranular phase.

Beside the NiB addition noted above, a variety of ceramic additions have been investigated for TiB_2. Watanabe [197] showed 12 w/o Ni•P additions yielded high (and a maximum) density and strength on hot pressing in vacuum at 1300°C. Watanabe and Kound [198] investigated effects of nine refractory boride additions of 0–10 w/o on density, grain size, strengths, and hardness of vacuum hot pressed TiB_2+ 1 w/o CoB, showing diborides of Hf, Nb, Cr, Mo, Ta, V, and Co gave < 1% porosity minima at 3 to 7 w/o, while additions of MnB_2 and Mo_2B_5 reached minimum porosities of ~ 1 and 0.5%, respectively. TiB_2+ 1 w/o CoB gave a minimum porosity of zero at 1800°C, while 5 w/o Ta_2B_2 gave this value at 1700°C and 5 w/o of W_2B_5 gave a minimum porosity of ~1 % at 1600°C. Strengths of 900–1100 MPa were achieved with several additions, in part due to significant grain growth limitations, especially with 5 w/o TaB_2 and W_2B_5. Matsushita and coworkers [199] reported substantial improvements in sintered densities at 1900°C with ~ 5 w/o Cr_2O_3. Torizuka and coworkers [200] reported vacuum sintering TiB_2 + 2.5 w/o SiC giving 96% of theoretical density and 99% with HIPing, which was attributed to a liquid phase formed via reaction with the ~ 1.5 w/o oxygen content in the TiB_2 powder.

Thévenot [201] listed several oxides such as silicate glasses, Al_2O_3, Fe_2O_3, and MgO (added directly as Al or fluorides), as well as a variety of metals, excess carbon, and refractory ceramics or combinations of these as additives for sintering or hot pressing of B_4C, with TiB_2, CrB_2, W_2B_5, and Be_2C, or SiC. SiC was added directly or as a polycarbosilane, and some additives were combined with carbon (e.g., Al or Si). He noted that while many improved densification, most resulted in larger grains and lower strengths. Most additives also left a few percent porosity. More recently Lee and Kim [202] briefly reviewed additive sintering of B_4C as well as reporting further study of densification by sintering in argon at 2150°C with Al_2O_3 additions. A maximum of 96% of theoretical density was achieved with 3 w/o addition (attributed to forming a liquid phase), while exaggerated grain growth was observed with > 4 w/o addition. Kanno and coworkers [203] showed that additions of TiB_2, AlF_3, and especially Al significantly enhanced densification, giving 95% sintered density at 2200°C in argon, but Mg or MgF_2 and especially SiC were not effective. Sigl [204] reported that addition of 13–16 w/o TiC allowed B_4C to be sintered to > ~ 98% of theoretical density at 2150–2000°C due to formation of C and TiB_2, both of which aid densification.

SiC is the most extensively studied carbide for additive effects on densification because of its desirable properties and it by itself is very resistant to sintering, for example, as shown by Nadeau's observations on limited self-bonding with hot pressing with 20–50 kbars until temperatures of over 1500°C with Al_2O_3 additions [205]. The first breakthrough was successful hot pressing of either α- or β- SiC powder with addition of ~ 1m/o Al (added as Al_2O_3 in making SiC powder from Si and C) by Alliegro and coworkers [206]. Use of the Al addition via hot pressing was commercialized, but has continued to be studied some, for example, by Misra [207]. More recently a superior commercially produced hot pressed SiC has been produced that uses only a small amount of AlN to aid densification.

Additions of Li, Ca, Cr, and Fe also aided densification, and SiC bodies densified with modest addition levels had good strengths, even at elevated temperature [207]. Subsequently, hot pressing with Al_2O_3 was further demonstrated (at 1950°C), as well as successful pressureless sintering, by Lange [208] and Mulla and Krstic [209], respectively. Extension to sintering with additions of Al_2O_3+ Y_2O_3 [210] followed. Better results were obtained with additions via sols rather than powders (with temperatures as low as 1800°C) [211], and with other additions (e.g., SiO_2 [212]. Liquid-phase densification is generally believed to occur. Subsequent work has also included combined additions with at least one of the additions being a rare earth oxide to yield elongated grains as has become common in processing Si_3N_4. Thus, for example, Chen [213] used 6 m/o additions of Al_2O_3+ Gd_2O_3 to pressureless sinter SiC to ~ 98% of theoretical density, with best results at the eutectic mixture (Gd_2O_3/Al_2O_3 molar ratio of ~ 0.3) and 1950°C, which also increased room temperature toughness, attributed to increased crack deflection.

The other major development with SiC was the demonstration of its pressureless sintering using B+C additions in various forms (including as B_4C) and amounts, initially by Prochazka and coworkers at temperatures ~ 2100°C [214–216]. Further study indicated a liquid-phase sintering mechanism [217] and that the amount of B needed is ~ 1 w/o when there is enough carbon to remove the silica coating on the SiC particles [218]. Further developments have included other combinations of additives, for example, Al, B, and C [219–222], other sources of more uniform distribution of B [223], as well as of carbon by itself [224]. Hot pressing of SiC with substantial AlN additions indicates some advantage of AlN in densifying SiC [225], which is consistent with small amounts of only AlN being used to densify SiC by hot pressing (e.g., for ballistic armor and high-temperature semiconductor processing).

Most other carbides, other than WC, are normally made without additions, but some use of additions, especially metals, has been investigated. Thus, Klimenko and coworkers [226] investigated use of 5–30% Ni binder with chromium carbide for hot pressing which is greatly facilitated by formation of a liquid phase at ~ 1200°C. Small (0.1–0.2%) additions of P to the Ni further lowered the liquefying and densification temperatures by an additional 100–200°C, that is, to ~ 1000°C. Cermets of TiC and various metals, such as Fe and Ni, have been made [227–230], as well as of metal—e.g., nickel-based alloys or Ni with additions of Mo or Nb [230]—but such bodies are generally not as good as cermets based on WC. Eyre and Bartlett [231] noted that small [e.g. 0.1-0.2] additions to both UC and U; PuC promote sintering and can yield consistently higher densities, and further studied details of the effects of Ni additions on UC. Mathers and Rice [232] conducted screening evaluations of additions of metals (Al, B, Cr, Co, Cu, Fe, Mo, Si, Ti, Zr) and of silicides of Fe, Mo, and Nb individually by themselves, or with added C for signs of wetting and densification at 1900°C with negative results.

As noted above cermets based on WC, especially with Co, have been widely investigated and are in wide production for cutting tools and wear applications, with various compositions from a few to about 20% metal are commonly produced, mainly by sintering [227,233], e.g., in the 1350–1450°C range. The presence of a liquid phase in WC-Co compositions, which has been directly observed [234], is generally accepted, e.g., at ~ 1280°C, but can be lowered by additives such as B and Ni [235]. Other sintering aids/bonding agents, especially Ni by itself or in combination with other metals have been investigated [235-238], but use of Co still predominates despite safety precautions required with its use in sintering.

Carbon bodies, for example, of graphite or diamond, also do not sinter. While graphite and other nondiamond carbon powders are commonly adequately densified by use of carbon producing polymers or CVD, there has long been a need for sintering aids for high-pressure hot pressing of diamond powders (which themselves are made via high-pressure conversion from graphite using additives such as Ni, Sect. 3.2). Initially small amounts (e.g., 1 m/o) of B, Si, or Be [239] or additions of some boride, carbide, nitride, or oxide powders were demonstrated [240], but much attention and use has been focused on Co (e.g., ~ 10%) additions [241-245], similar to WC practice. However, additives in addition to Co have been investigated, for example, graphite [245] or WC, the latter to suppress excessive grain growth that can occur [246]. Reaction sintering, that is, conversion of graphite to diamond via hot pressing in one step using a graphite precursor with added diamond powder with Ni as both a promoter of graphite to diamond conversion and as a sintering aid/binder [247] has also been demonstrated. While earlier temperatures and pressures of 1800–1900°C at 6–7 GPa were used, better materials (finer powders) and technology have lowered densification temperatures to 1400–1700°C. (Graphite apparently dissolves in liquid Ni and then precipitates as diamond.)

Turning to refractory nitrides, AlN, BN, and Si_3N_4, which are all important and undergo little or no sintering without additives, making additives to aid densification important for them. Earlier investigations of sintering AlN focused on use of metal additions (e.g., of Ni, Co, or Fe [248,249], but attention shifted first to oxide (and later to some oxide producing) additives due substantially to Komeya and Inoue [250,251], who successfully hot-pressed AlN with 5–10 w/o Y_2O_3 additions at ~ 1800°C, which was motivated by possible uses for structural applications, (e.g., in engines). Subsequently, increased recognition of the role of oxygen surface species [252], and use of other or combined additives occurred

[253–258], with much of the motivation being high thermal conductivity for electronic applications, such as substrates, and processing more by sintering than hot pressing. While use of either CaO or Y_2O_3 (or them combined) is established for both research and some commercial production, further work continued on other oxide additives, mixtures or ternary compounds of them, use of small additions of carbon, and addition of desired cations with other anions such as C, and especially F to further lower firing temperatures to ≤ 1600°C [259–261]. The high level of activity is reflected in the extensive and often complex literature, especially the patent literature in this field, but the broad consensus is that liquid-phase sintering is generally an important, if not key, factor in densification with additives. Some of the above results as well as several additive tests are covered in a brief paper by Schwetz and coworkers [261]. Further work continues on additive mixtures; for example, 0.5 w/o TiO_2 + 1.5 w/o Y_2O_3 + 0.4 w/o CaO of Nakahata and coworkers [262], and of known beneficial cations with other anions, such as Neuman and coworkers [263] using CaF_2, CaC_2, CaH_2, Ca_3P_2, and CaS, with all of these except the latter converting to CaO and being beneficial. Commercially produced AlN for the electronics industry is sintered with Y_2O_3 and Al_2O_3 (the latter from the oxide layer on AlN powder particles), with quality AlN being commercially produced by hot pressing with as low as $^1/_4$% Y_2O_3, meeting specifications calling for 98.75% AlN components.

Consider next hexagonal BN, that is, the analog of graphite, which is also not very sinterable, especially without the common oxygen/water surface contamination. BN also lacks the readily available polymeric precursors for processing analogous to graphite and often requires substantially lower porosities than are obtained by preceramic polymer impregnation and pyrolysis. Thus, considerable attention has been focused on densification with additives, mainly by hot pressing, which has been in commercial production for a number of years. The focus is on use, and often extension, of the B_2O_3 on the powder surface with additional addition of B_2O_3, and small (e.g., 0.2–0.3 %) additions of oxides such as Al_2O_3, phosphates, alkaline earth oxides, especially CaO, or SiO_2 with formation of a borate liquid phase an important factor [264,265]. Analysis of commercial samples shows 2–9% residual B_2O_3 and 2–7% residual fine porosity, both mainly intergranular, with the porosity and residual B_2O_3 amounts generally being inversely related. Higher purity BN grades with less residual B_2O_3 are produced by post densification annealing at high temperatures in vacuum to reduce B_2O_3 contents via volatilization. More recent studies of Hubáek and coworkers [266] showed that use of small additions of metallic Cu increased hot-pressed densities modestly (from ~ 90 to ~ 94% of theoretical density) but decreased flexural strengths from ~ 46 to ~ 30 MPa (with ~ half of this decrease being recovered by using up to 10 w/o Cu). However, very small Cu additions had a pronounced effect on grain structure, especially a high degree of preferred orientation, which rapidly decreased with increased Cu additions.

Fabrication of cubic BN bodies from cubic BN powder (which like diamond powder is made via use of catalysts, such as Mg-B-N or Ca-B-N compounds, see Sec. 3.2) is greatly assisted by use of additives as are diamond compacts. However, as with forming the cubic phase, densification of cubic BN emphasizes use of different additives than for diamond. Thus, while some ceramic binders have been used for diamond, the emphasis has been on metallic additives (binders), especially Co, while the focus of additives for desifying cubic (or wurtzite) BN powder compacts have been ceramics such as AlN [267,268], AlB_2 [267], TiC [269–270], TiN [268,269], TiC-TiN solutions [270,271], and TiB or TiB_2 [271].

Silicon nitride can be pressure sintered to high density and reasonable properties without additives, for example, by hot pressing with high pressures (3 GPa) at 1900°C [272] or HIPing at 1800°C with 0.17 GPa pressure [273]. While such densification may be aided by surface oxide contents, it nonetheless, reflects a significant need for densification aids, which has received substantial attention, possibly more than for any other ceramic. However, until densification aids were discovered, the focus in Si_3N_4 processing was via reaction sintering (or bonding—RSSN or RBSN, respectively) by in situ nitridation of Si-powder compacts, which is also aided by some additives—especially iron oxide, also a common impurity in Si powders (see Sec. 3.2—as well as inhibited by other compounds, such as Al_2O_3 [274]. (The focus on reaction processing of Si_3N_4 was due in part to the discovery that such processing could be carried out with the fortuitous aspect that despite the substantial [~23%] volume expansion of Si on conversion to Si_3N_4, compacts can be reacted to as low as ~ 20% porosity and reasonable properties with dimensional tolerances of ~ 0.5% between the green and reacted bodies, as well as done with reasonable costs.)

A significant step in densification of Si_3N_4 to low to zero porosities was the study of Deeley and coworkers [275] on effects of various additions (e.g., commonly of 4–10%) on the hot pressing of Si_3N_4 powder at 1800–1850°C. They showed that Al_2O_3, BeO, and MgO (as well as Mg_3N_2) additions each gave < 1% porosity, while additions of ZrO_2, CaO, Fe_2O_3, ZnO, Cr_2O_3, (as well as of B, $MoSi_2$, TiN, Cr_2N) gave porosities increasing from ~ 14–40% in the order listed, and generally with lower porosity as the amount of additive increased over the range evaluated. The focus in their further investigation was on MgO since it gave somewhat lower porosity, the second highest strength (> 100% higher than the Al_2O_3 additions), while BeO presents a health hazard and Mg_3N_2, though giving the highest strength by ~ 20%, is hydroscopic (and also probably reacts to form MgO and $MgSiN_2$). This focus on MgO additions was also the case in much of the subsequent work of others, becoming the basis of the initial commercialization of hot-pressed Si_3N_4, which is apparently still in production. The discovery that poor high temperature strengths of hot pressed Si_3N_4 was due to Ca impurities in the Si_3N_4 powder also added interest to use of MgO for densification since with powders with low or no Ca impurities, good high temperature strengths were obtained. (Note that Fe impurities also common in some Si_3N_4 powders can be beneficial, e.g., for nitriding Si, as noted earlier and in Sec. 3.2.) Subsequent expanding efforts lead to other hot pressing additives to give near-zero porosity and broader recognition of the benefit of starting with high α-Si_3N_4 content and its conversion to β-Si_3N_4.

The next step was broadening the range of useful densification aids for hot pressing, initially to use of Y_2O_3, apparently discovered independently by Gazza [276] and Tsuge and coworkers [277], which also gave better properties, such as increased high temperature strengths (and an oxidation problem as noted below). Subsequent developments included use of Zr-based additives [278], mainly oxides, and more extension into use of rare earth and related oxide additives, such as CeO_2 [274]. Thus, Rice and McDonough [278], reported that ZrO_2 without stabilizer or with Y_2O_3 stabilization, as well as with $ZrSiO_4$ (or ZrN or ZrC) gave near theoretical density (especially with $ZrSiO_4$ or ZrC) and good strengths, with some Si_3N_4 powders, and somewhat poorer results with other powders (which was the probable reason why Deeley and coworkers' [274] results with ZrO_2 additions were intermediate in their survey). Others, for example, Dutta and Buzek [279], also reported success with ZrO_2-Y_2O_3 additions. Various investigators also reported excellent densification with CeO_2 additions (e.g., Smith and Quackenbush [280], which was offered commercially but then withdrawn due to serious, often catastrophic, effects of oxidation on strengths and integrity of such bodies after intermediate temperature (at ~ 1000°C) oxidation [281]. This and similar oxidation problems with some Y_2O_3 densified bodies [282] are a severe reminder of the need for comprehensive characterization of new materials, not just room and high temperature tests.

Further development of densifying Si_3N_4 with additives was along several avenues, a major one being increasing attention and success in pressureless sintering with the same or similar additions used in hot pressing (or HIPing). Other important and often interrelated developments were more use of combined additions, more attention to interaction of additives with impurities, and the use of this to crystallize glassy grain boundary phases to improve high temperature properties, and use of additives (with Y_2O_3 + other additions, such as Yb_2O_3) to enhance development of elongated β-Si_3N_4 grain structures (often also aided by seeding with fine β-Si_3N_4 particles) [283]. Examples of mixed additives are physical mixtures, for example, Y_2O_3 + Al_2O_3 [282], and chemical mixtures, for example, $YAlO_3$ [284] and celsian ($BaAl_2Si_2O_8$) [285]. The former also includes benefits of MgO+ Fe_2O_3 (the latter also stimulating Si nitriding) [286]. For more details on part of the development of additives for Si_3N_4, readers are referred to Popper's review [287].

Additives for Si_3N_4 have also included some nonoxide additives, in combination with oxide additives or by themselves. The former includes MgO + CaF_2

[288] or MgF_2 [289], which further lower densification temperatures (e.g., to 1400°C), and of $(Y,La)_2O_3$+ AlN [290]. Limited investigations of nonoxide additives by themselves show some failures [232] and some successes [291–293], but many, if not all, of the latter reflect conversions to beneficial oxide additions, such as nitrides of Mg, Ca, and Sr.

Some general results can be seen from the substantial investigations of densifying Si_3N_4. Oxide additives, which are almost exclusively used singly or in combinations, usually at levels of a few percent, generally function via a liquid phase that lowers densification temperatures by 200–400°C [273]. The liquid phase becomes a solid grain boundary phase that is commonly glassy, but can be crystallized by composition control (including impurities) and heat treatment. The liquid/grain boundary phase observations are supported by phase data, direct TEM identification and effects on properties. The latter include enhanced intergranular fracture and the occurrence of moisture driven slow crack growth, but generally with, often significantly, improved strength and toughness at room temperature [294–296], but decreased strengths and enhanced slow crack growth (from grain boundary sliding) at higher temperatures as compared with Si_3N_4 made without additives (by CVD or high pressure densification). These property changes are corroborated by the one case of successful densification with a nonoxide additive, $SiBeN_2$, which shows no grain boundary phase and none of their effects [291,297]. Note that other trials of a variety of nonoxide additives by Mathers and Rice [232] were unsuccessful, as have been most other attempts like those of Deeley and coworkers above [275]. Exceptions have been where nonoxide additives either are believed to react to form oxides (e.g., Mg_3N_2, ZrC or ZrN), or react with oxygen to form a solid solution with Si_3N_4 [291,297]. The latter may or may not be related to the use of B, C, B_4C, or combinations with SiC, but in either case raise the question of why nonoxide densification aids are not feasible or not found. Commercially, several densification routes are followed that include reaction sintering of Si with added Y_2O_3 + Al_2O_3 for subsequent sintering of the resultant Si_3N_4 to near theoretical density and sintering of Si_3N_4 powder with additives such as Y_2O_3 + Al_2O_3 + TiO_2. Commercial hot pressing still includes either MgO or Y_2O_3 additives.

TiN has been densified by high-pressure hot pressing at 1800°C and 5 GPa pressure [298] and at 2100–2200°C and 14 MPa pressure [299], but can be aided by various additives, which can also aid pressureless sintering. Thus, like TiC, Ni additions (5 w/o) have been used with reasonable success, especially in sintering with some other additives such as (V,Ta)C, as well as some oxygen [300]. Hot pressing with 5 or 10 w/o additives of Al_2O_3, Y_2O_3, or MgO, or BN, AlN, or Si_3N_4, or SiC or B_4C at 1950°C with 14 MPa pressure was investigated with 10 w/o of Y_2O_3, Al_2O_3, or B_4C giving 96–98% of theoretical density, with B_4C being the most promising based on both density and property evaluations [301]. Kamiya and Nakano [302] report that 5 w/o Al benefits hot pressing of coarse

(75 μm) TiN powder at 1400°C in a N_2 atmosphere and fine (1.4 μm) powder in a vacuum (the latter apparently required to accommodate gas evolved from the Al-TiN interfacial reaction).

Turning to $MoSi_2$ bodies, SiO_2, which is a common impurity, has also been added (for example, as up to 15% polysilicic acid, then sintering in argon at 1550–1650°C giving 6% minimum porosity [303]. This gives finer grain sizes and less strength decrease at higher temperatures than the use of ~ 8% clay in forming heating elements. Suzuki and coworkers [304] showed that addition of ~ 5 m/o of Y_2O_3 gave maximum densification in hot pressing at 1600°C and good high temperature strengths, especially due to the additive and SiO_2 forming a refractory silicate. However, they also found that while ~ 1 m/o addition of Sc_2O_3 gave somewhat lower density it gave higher mechanical properties at ~ 22°C and comparable strengths at higher temperatures as the Y_2O_3 additions.

While ZnS is readily hot pressed to transparency at modest temperatures (e.g., 800°C), Uematsu and coworkers [305] report that small (0.01–1%) additions of Bi_2S_3, Al_2S_3, or Li_2S, though not significantly affecting densification, limited grain sizes, especially the Bi_2S_3.

5.6 CERAMIC COMPOSITES

Ceramic composites present particular challenges to densification, especially by pressureless sintering, since the presence of substantial second phase typically seriously inhibits sintering approximately in proportion to the volume fraction of dispersed phase. The difficulty of sintering composites also increases on progressing from particulate to platelet or whisker to fiber, especially continuous, and particularly multidirectional fiber composites, as well as the volume fraction of nonoxide constituents increases. Thus, as discussed further in Section 6.2, most ceramic composites are made by pressure sintering, mainly by hot pressing, which is often aided by use of additives, especially for particulate composites.

As noted above, ceramic particulate composites are commonly processed, mainly by hot pressing, with additives. Again, additives are often used to improve densification since two- or multiple-phase bodies often are more difficult to densify, especially when at least one phase is a refractory nonoxide phase with more limited densification, particularly at temperatures where oxides are typically densified. However, densification with additives also often has other benefits, such as retention of finer particle and matrix grain size to increase properties often higher at finer grain or particle sizes. While maintenance of finer particle and matrix grain sizes is typically an important consequence of composites, especially particulate composites (and to some extent also platelet and, especially, whisker composites), some additional microstructural control is often very beneficial. Composites present additional challenges to the use of additives since additive compatibility with at least two phases are required. Data are primarily available for particulate composites since whisker, platelet, and fiber composites provide progressively increasing challenges to sintering, and thus are densified under pressure, generally requiring less densification aid. Also, additives may result in undesirable effects in such composites. However, on the positive side some composite constituents are densification aids for the other phase, for example, Al_2O_3 for some composites with nonoxides, such as SiC. Further, an additive for one phase may also often be effective for another phase—for example, Y_2O_3 for both SiC and Si_3N_4 and with Al_2O_3.

Little or no data exists on additive effects in densification of composites consisting of oxide particles dispersed in an oxide matrix, in part, due to generally easier sintering of oxide materials. (In the case of YTZP sintered with 0.25w/o Al_2O_3 for optical ferrules [Sec. 5.3], the Al_2O_3 apparently is mainly to maintain fine grain size.) Thus, the focus is on composites with nonoxide phases starting with those with nonoxide particles dispersed in an oxide matrix, followed by such particles in nonoxide matrices—in both cases in alphabetical order of the matrix material. Again, oxide additives are considered first, followed by nonoxide additives.

Thus, first consider Al_2O_3-TiB_2 composites, densification of which are briefly reviewed by Stadlbauer and coworkers [306], who also corroborated benefits of small addition (1% MgO)—for example, giving 98% dense Al_2O_3 without TiB_2 at 1800°C, and 96 and 90% density with 5 and 40 w/o TiB_2, respectively. Cutler and coworkers [307] showed that addition of 3.7 w/o TiH_2 to Al_2O_3 + ~ 30w/o TiC allowed sintering to ~ 94% of theoretical density at 1860–1890°C due apparently to a transient liquid Ti-based phase, and thus allowed cladless HIPing to full density at ~ 1600°C, that is somewhat better than by hot pressing at 1700°C. Chae and coworkers [308] showed that Al_2O_3 + 30 w/o TiC could be sintered to a maximum of ~ 97% of theoretical density at 1700°C at an optimum Y_2O_3 addition of 0.35 w/o. They subsequently reported that composites with 50 w/o TiC could be sintered to ~ 99% of theoretical density at 1750°C with higher (3.5 w/o) Y_2O_3 [309]. Sintering of both compositions was attributed to a liquid phase that crystallizes to YAG on cooling. Again, composites with an alumina matrix can be "densified" via phosphate bonding, as discussed by Karpinos and coworkers [310]. Finally, composites of an alumina matrix with up to 30 w/o Ti-C-N, densified with the aid of 0-5 w/o Ni by itself or with other metals, were sintered to 90–100% of theoretical density at 1750°C by Ekström [311].

TiB_2 matrix composites with ~ 19 w/o 2YTZP were reported by Torizuka and coworkers [312)] to be pressureless vacuum sintered to 96.7% of theoretical density at 1700°C with 2.5–5 w/o SiC, and only 63.7% dense without the SiC. The SiC addition was also effective in limiting growth of the TiB_2 grains and of the TZP particles. Zhang and coworkers [313] reported that TiB_2 + SiC composites could be reaction hot pressed to 99% or above theoretical density at 2000°C with a few percent Ni additions, but that 2 w/o addition gave maximum hardness and flexure strength (but a minimum, though substantial, toughness). Hayashi and coworkers [314] also reported that additions of Ni (with C) were used as densification aids for pressureless sintering of TiB_2-B_4C composites at 1600°C. Dai and coworkers [315] reported densifying nominally 58 w/o TiB_2 + 40 w/o Ti(CN) with 2 w/o Ni at 1850°C.

Turning to SiC matrix composites, Cho and coworkers [316] reported hot pressing of composites with 0–70 w/o of TiB_2 or TiC at 1850°C via liquid-phase densification from (7 w/o) Al_2O_3 plus + (3 w/o) Y_2O_3 additions (plus SiO_2 on the SiC surface). Cho and coworkers [317] extended this work by annealing composites with TiB_2 at 1950°C to obtain exaggerated growth of α-SiC from the β-SiC. Lin and Iseki [318] showed that SiC with 0–40 v/o of TiC could be hot pressed to 97.1–99.7% of theoretical densities at 1800°C with 5 w/o Al-B-C addition (without which densities were only 58–66% of theoretical densities), and Endo and coworkers [319] used B_4C and C as densification aids in hot pressing SiC with 0–100% TiC.

SiC-AlN bodies have been hot pressed to near theoretical density between 1950–2100°C [225] and have been similarly sintered using 2 w/o Y_2O_3 addition by Lee and Wei [320] (while the same level of CaO or Al_2O_3 additions gave only ~ 65% of theoretical density). Pan and coworkers [321] reported similar hot pressing of such bodies with 0.5 w/o Y_2O_3. In some contrast to these SiC-AlN results, Mazdiyasni and coworkers [322] reported that hot pressing of AlN+ 0–30 BN at 2000°C gave 92–98% of theoretical densities, with best densification with 15 CaH_2 addition, intermediate densification without any additive, and the lower levels of densification with 5% Y_2O_3 addition. Also note that Karpinos and coworkers [323] reported phosphate bonding of AlN and Si_3N_4 (by themselves and with) Al_2O_3 via H_3PO_4 reaction and heating to 1250°C.

Mathers and Rice [232] tried reactive hot pressing of Al and Si with Si_3N_4 to produce Si_3N_4 with AlN and $MoSi_2$ at 1820°C, which resulted in reaction, but with essentially no densification, yielding only ~ 60% of theoretical density. Hot pressing tests at 1925°C resulted in substantial Si_3N_4 decomposition. Si_3N_4 with 0–50 v/o TiC was hot pressed by Mah and coworkers [324] to near theoretical density at 1750°C, using 5.5 w/o CeO_2 as a hot-pressing aid. Mazdiyasni and Ruh [325] also hot pressed Si_3N_4 with 0–50 BN and 6% CeO_2 at 1750°C. Si_3N_4 matrices with 9–33 w/o AlN were sintered at 1900°C under 1 MPa N_2 with addition of La_2O_3 by Zhung and coworkers [326]. While no additive gave only ~ 68% of theoretical density with 25% AlN, addition of La_2O_3 first increased densities rapidly, e.g., to 94% at 0.5 w/o, then more slowly to 97–98% of theoretical density at 1–2 w/o La_2O_3, then plateauing or slightly decreasing at the maximum used (7 w/o) (but with a strength maximum at 5 w/o and of toughness at 2–5 w/o).

Investigators fabricating composites of Si_3N_4 with SiC particles have frequently done so using either Al_2O_3 or Y_2O_3 (or related compound) additions, or combinations of these. Thus, Tanaka and coworkers [327] showed that sintering with 10.5 w/o Al_2O_3 + 4.5 w/o Y_2O_3 gave ~ 97% theoretical density at 0% SiC, then decreasing slowly with ~ 20 SiC then more rapidly as SiC content further increased. Kim and coworkers [328] used 6w/o Y_2O_3 + 2 w/o Al_2O_3 to sinter bodies with 20 w/o SiC particles at 1750°C to 95% of theoretical density. Cheong and coworkers [329] used 6 w/o Y_2O_3 + 2 w/o Al_2O_3, or 4 or 8% Y_2O_3 in hot pressing at 1800°C, finding best results with 4% Y_2O_3 in densifying their nanocomposites with 20 v/o SiC. Park and coworkers [330] instead used 8–16 w/o Yb_2O_3 additions to hot press nanocomposites with 20 v/o SiC, finding best results with 14% Yb_2O_3. Similar to the above, composites of reaction processed powders of Si_3N_4 + 35–40% TiN were sintered by Hillinger and Hlavacek [331] with a gas pressure of 2.5 MPa between 1710 and 1740°C using 6 w/o Y_2O_3 + 4 w/o Al_2O_3.

Petrovic and coworkers [332] hot pressed Si_3N_4 + 0–50 v/o $MoSi_2$ using 1 w/o MgO at 1750°C to obtain 94–97% of theoretical density. Zhang and coworkers [333] hot pressed $MoSi_2$ with 0.1 to 10 w/o Al metal (to react with the oxygen on the $MoSi_2$ particles and form Al_2O_3) by hot pressing at 1600°C, with the highest density being obtained at 5 w/o Al addition. Ting [334] hot pressed composites of $MoSi_2$ + 20 v/o SiC with and without 500 ppm B_2O_3 at 1600 and 1750°C respectively, giving 97.4 and 98.1% of theoretical density, respectively, with the B_2O_3 giving finer grain size and glassy areas and higher strength at 22°C.

In some composites, the added dispersed phase may also aid densification. Thus, Zakhariev and Radev [335] reported that addition of 10–30 w/o of WC to B_4C resulted in enhanced densification. Shobu and coworkers [336] reported that substantial, e.g., 20 w/o addition of Mo_2B_5 to $MoSi_2$, sintered to full density at 1500°C.

5.7 DISCUSSION AND CONCLUSIONS

Three aspects of using densification aids need further discussion, namely mechanisms, effects, and further opportunities. Consider first mechanisms of actual densification, where the interest is not in detailed mechanisms, but in those mechanistic aspects that indicate practical engineering guidance in selection and development of densification via additives. While it is clear that much remains to be understood, an overall separation into mechanisms of probable or known liquid-phase densification (which can entail liquid-phase sintering as well as interparticle sliding, especially in pressure densification such as hot pressing) and other non–liquid-based mechanisms, which generally entail enhanced diffusion. The latter, though not as effective as the former, can be useful by themselves, and often may accompany liquid-phase mechanisms, e.g., before the liquid forms. Mechanisms operative in the solid state typically require some solid solution, usually of ions of different valance than their counterparts in the material to be densified, while liquid-phase mechanisms clearly require a second, grain boundary, phase. While the melting point of the additive is a key factor in the temperature of liquid-phase formation, effects of grain boundary impurities often play an important role in this as well as another key aspect, namely, wetting of the grains of the material to be densified by the liquid phase. However, other key aspects about which information is incomplete are the extent of solubility of the material to be densified by the liquid phase and precipitation of the former from the latter.

Note that while metals are generally more easily sintered than many ceramics, and hence rely less on densification aids, nonetheless some metallic densification aids are used in sintering some metals, especially for refractory metals. The common use of a few to several percent of Ni in sintering W powder is a clear example of this. More directly pertinent to some ceramic processing is significant enhancement of sintering of Si shown by Greskovich [337] in adding small (e.g., 0.4 w/o) additions of B, which also limited grain growth (while Sn retarded densification and enhanced grain growth).

Two further considerations are central to selecting and using additives as densification aids. The first is the presence and effects of impurities, including those on powder particle surfaces. As noted earlier impurities often play an important role in densification with additives, frequently being a route to discovering additives, or an important reason for materials specifically added working as well as they do. It is always better to monitor impurity levels that effect additive effectiveness and adjust additive amounts accordingly. However, this is often not done, which is a factor in varying densification. Second is additive effects on properties. Additives often have dual effects, one on densification and one on resultant properties. The latter generally involve intrinsic compositional and microstructural effects. Besides the obvious impact of reduced porosity is the frequent control of grain growth, which is also often important. While many second phases at grain boundaries are effective at controlling grain size, there can be substantial variation with temperature, impurities, and the microstructural relations between the body and additive phases. Since there are frequently performance trade-offs due to effects of additives (discussed below), this is an important factor in additive selection.

Thus, while basic additive data such as solubility, ionic sizes, melting, and reaction between additive and material to be densified are important in aiding additive selection, much still needs to be aided by actual densification results. While this is generally true for a single compound to be densified, it is particularly true for densification of composites. It is believed that the review of this chapter is a help in this selection. It is also believed that some selection insight can be gained from consideration of other additive uses in Chapter 3, especially flux growth of crystals, V-L-S growth of whiskers, and stimulants or inhibitors to grain growth. Also note that either or both the atmosphere provided in the densification furnace as well as that generated in the compact being densified and their interactions can also be important.

Additives effects on properties is an important factor needing some further comment since almost invariably their use involves some trade-off between improvements due to reduced porosity versus reductions due to other effects of the additive. Additives that form a solid solution with the material to be densified generally have limited effects on most mechanical properties, but may have some to substantial effects on electrical, thermal, magnetic, and especially some electromagnetic properties (such as color). On the other hand, grain boundary phases, which often have approximately a rule of mixtures effects on sonic, elastic, and magnetic properties, can have little, to substantial effects on thermal and electrical conductivity, and especially on dielectric breakdown and varistor behavior. A key limitation of liquid-phase densification is high-temperature strength and deformation due to residual boundary phase near and especially above the melting point of the remaining grain boundary phase. However, note that solid phases, especially at grain boundaries, can present serious problems if their structure and size result in reduced strength; for example, due to possible microcracking as densifying Al_2O_3 with TiO_2 due to formation of Al_2TiO_5 [27,338] or of using V_2O_3 additives. The latter is a good example of complications of other variables since the phase formed is dependent on the processing environment, that is, lower strength in Al_2O_3 + V_2O_3 sintered in air due to formation of an $AlVO_4$ grain boundary phase despite some inhibition of grain growth, but a solid solution of the V without significant strength degradation on firing in a reducing atmosphere. Residual fluoride phases, for example, from use of LiF, can lower strengths and toughnesses at room temperature as well as change electrical properties, cause bloating or blistering from volatilization at higher temperatures, especially at higher heating rates and larger sizes, and greater reductions of high temperature strengths [91,294,339] and creep [340]. Residues of additives can also have other positive or negative effects on behavior, such as oxidation, corrosion, or other environmental effects and electrical properties [341].

Some of the above, as well as other effects may arise from other effects of additives. Thus, for example, some additives may increase green densities achieved, as reported by Udalova and coworkers [88,342] for LiF additions to some oxide powders. Densified microstructures may also be modified by the use of additives, for example, effects of small Cu additions to BN [266]. Phase transformation may also be altered by the use of densification aids, as reported in Si_3N_4 (see also Sec.3.3) [343].

Three opportunities can be noted. The first is that while much yet needs to be established, there is an increasing database and increased understanding, the latter in part due to the availability of purer materials, and thus less confusion due to impurity effects. Second, analytical tools have greatly increased in capability and availability to aid in understanding and development. Third, both the increased database and the frequent commonality of additives across various compounds, allow reasonable selection of additives for densifying composites as demonstrated above. Thus, for example, note the common use of Y_2O_3 (or other rare earth oxides) for several oxides and nonoxides and, to a lesser extent, for use of metals such as Fe, Ni, and especially Co. However, a basic question that remains is Why are there so few nonoxide additives, besides metals, even for nonoxide materials to be densified, such as SiC and Si_3N_4? While penetration of oxide particle coatings is probably a factor, why B, Al, Si, or B_4C are the only established additives for SiC and $BeSiN_2$, respectively, and the only one for Si_3N_4 is still a basic question. This is important since these additives do not result in an oxide (if any) grain boundary phase, and thus do not cause slow crack growth, e.g., due to H_2O at low temperatures, or enhanced grain boundary sliding at high temperatures in contrast to oxide additives resulting in such limitations. However, such nonoxide additives also result in normal toughness and strength, rather than increases in these often obtained with oxide additives, thus again being a reminder of further needs for understanding of effects of additives on densification and property trade-offs.

Finally, a few words of caution, primarily about possible significant size effects that may occur in processing larger bodies. Most additive development, as well as much additive usage, is done with bodies that are small—a few millimeters—in at least one dimension, which allows for removal of much of the additive or its residues. This is often important, e.g. especially for high thermal conductivity AlN. However, for preparation of larger bodies significant less such removal may occur, especially in the interior. Also in some cases some oxidation of the additive may be important, for example, with nonoxide additives for AlN, which may occur if there is sufficient material on the particle surfaces for this or one body dimension is small enough to allow potential oxygen diffusion from outside of the body. Where surface oxidation is limited, there is incomplete internal oxidation, especially in larger bodies. Such size effects are also often limitations on removal of gases from the interior of the body, even if the gases are soluble in the ceramic, as shown with MgO.

REFERENCES

1. C.-S. Hwang, C.-Ti Lin, H.H. Lu. Effect of SiO_2 and TiO_2 on the solubility and microstructure of alumina ceramics. J. Cer. Soc. Jap. Intl. Ed. 101:632, 1993.
2. Reference deleted.
3. S.J. Lukasiewicz, J.S. Reed. Phase development on reacting phosphoric acid with various bayer-process aluminas. Am Cer. Soc. Bul. 66(7):1134–1138, 1987.
4. R.L. Coble. Sintering crystalline solids. II. Experimental test of diffusion models in powder compacts. J. Appl. Phy. 32(5):793–799, 1961.

5. R.L. Coble. Sintering alumina: effect of atmospheres. J. Am. Cer. Soc. 45(3):123–127, 1962.
6. G. Rossi, J.E. Burke. Influence of additives on the microstructure of sintered Al_2O_3. J. Am. Cer. Soc. 56(12):654–659, 1973.
7. I.B. Cutler, C. Bradshaw, C.J. Christensen, E.P. Hyatt. Sintering of alumina at temperatures of 1400°C and below. J. Am. Cer. Soc. 40(4):134, 1957.
8. So Ik Bae, S. Baik. Critical concentration of MgO for the prevention of abnormal grain growth in alumina. J. Am. Cer. Soc. 77(10):2499–2504, 1994.
9. M.O. Warman, D.W. Budworth. Criteria for the selection of additives to enable the sintering of alumina to proceed to theoretical density. Trans. Brit. Cer. Soc. 66(6):253–264, 1967.
10. M.O. Warman, D.W. Budworth. Effects of residual gas on the sintering of alumina to theoretical density in vacuum. Trans. Brit. Cer. Soc. 66(6):265–271, 1967.
11. S. Mu, E. Gilbart, R.J. Brook. Solid-state sintering: the attainment of high density. In: W. D. Kingery, ed. Structure and Properties of MgO and Al_2O_3 Ceramics, Advs. Cer., 10, Westerville, OH: Am. Cer. Soc., 1984, pp. 574–582.
12. M.P. Harmer. Use of solid-solution additives in ceramic processing. In: W.D. Kingery, ed. Structure and Property of MgO and $A1_20_3$ Ceramics, Advs. Cer. 10, Westerville, OH: Am. Cer. Soc., 1984, pp. 679–696.
13. A. Cohen, C.P. Van der Merwe, A.I. Kingon. Effect of MgO dopent dispersing method on density and microstructure of alumina ceramics. In: W.D. Kingery, ed. Structure and Property of MgO and $A1_20_3$ Ceramics, Advs. Cer. 10, Westerville, OH: Am. Cer. Soc., 1984, pp. 780–790.
14. F. Singer, S.S. Singer. Industrial Ceramics. New York: Chapman and Hall, 1984, pp. 505,1139.
15. G.E. Gazza, J.R. Barfield, D.L. Preas. Reactive hot-pressing of alumina with additives. Am. Cer. Soc. Bul. 48(6):606–610, 1969.
16. M.P. Harmer, R.J. Brook. The effect of MgO additions on the kinetics of hot pressing in Al_2O_3. J. Mat. Sci. 15:3017–3024, 1980.
17. C.A. Bateman, S.J. Bennison, M.P. Harmer. Mechanism for the role of magnesia in the sintering of alumina containing small amounts of a liquid phase. J. Am. Cer. Soc. 72(7):1241–1244, 1989.
18. E. Ryshkewitch, D.W. Richerson. Oxide Ceramics. Orlando, FL: Academic Press, Inc. 1985.
19. R.D. Bagley, I.B. Cutler, D.L. Johnson. Effect of TiO_2 on initial sintering of Al_2O_3. J. Am .Cer. Soc. 53(3):136–141, 1970.
20. R.J. Brook. Effect of TiO_2 on the initial sintering of Al_2O_3. J. Am. Cer. Soc. 55(2):114–115, 1972.
21. M. Harmer, E.W. Roberts, R.J. Brook. Rapid sintering of pure and doped α-Al_2O_3. Trans. & J. Brit. Cer. Soc. 78(1):1979.
22. C.-S. Hwang, Z.-e Nakagawa, T.-H. Sung, Y. Ohya, K. Hamano. Effect of TiO_2 on the initial sintering of Al_2O_3. Report of Res. Lab. of Eng. Materials, Tokyo Inst. Tech., No. 12, 95-102, 1987.
23. F.A. Kröger. Defect models for sintering and densification of Al_2O_3: Ti and Al_2O_3: Zr. J. Am. Cer. Soc. 67(6):390–392, 1984.

24. T. Ikegami, K. Kotani, K. Eguchi. Some roles of MgO and TiO_2 in densification of a sinterable alumina. J. Am. Cer. Soc. 70(12):885–890, 1987.
25. T. Watanabe, C. Nakayama. Sintering of alumina green sheet with TiO_2 and Cr_2O_3 in reducing atmosphere. J. Cer. Soc. Jap. Intl. Ed. 101:544, 1993.
26. A. Bettinelli, J. Guille, J.C. Bernier. Densification of alumina at 1400°C. Cer. Intl. 14:31–34, 1988.
27. M. Wakamatsu, S. Ishida, N. Takeuchi, T. Hattori. Effect of firing atmosphere on sintered and mechanical properties of vanadium-doped alumina. J. Am. Cer. Soc. 74(6):1308–1311, 1991.
28. M.A. Gülgün, V. Putlayev, M. Rühle. Effects of yttrium doping α-alumina: I, microstructure and microchemistry. J. Am. Cer. Soc. 82(7):1849–1856, 1999.
29. D. Delaunay, A.M. Huntz, P. Lacombe. The influence of yttrium on the sintering of Al_2O_3. J. Less. Com. Met. 70:115–117, 1980.
30. F.F. Lange, M.M. Hirlinger. Hinderence of grain growth in Al_2O_3 by ZrO_2 inclusions. J. Am. Cer. Soc. 67(3):164–168, 1984.
31. S. Hori, R. Kurita, M. Yoshimura, S. Somiya. Suppressed grain growth in final-stage sintering of Al_2O_3 with dispersed ZrO_2 particles. J. Mat. Sci. Let. 4, 1067–1070, 1985.
32. S. Hori, R. Kurita, M. Yoshimura, S. Somiya. Influence of small ZrO_2 additions on the microstructure and mechanical properties of Al_2O_3. In: S. Sõmiya, N. Yamamoto, H. Hanagida, eds. Science and Technology of Zirconia III. Adv. in Ceramics. Vol. 24. Westerville, OH: Am. Cer. Soc., 1988, pp. 423–429.
33. A. Nisida, S. Fukuda, Y. Kohtoku, K. Terai. Grain size effect on mechanical strength of MgO-ZrO_2 composite ceramics. J. Jap. Cer. Soc. Intl. Ed. 100:203–207, 1992.
34. F.F. Lange, M.M. Hirlinger. Grain growth in two-phase ceramics: Al_2O_3 inclusions in ZrO_2. J. Am. Cer. Soc. 70(11):827–830, 1987.
35. H. Tomaszewski. Effects of Cr_2O_3 additions on the sintering and mechanical properties of Al_2O_3. Cer. Intl. 3(2):60, 1982.
36. J.R. Keski, I.B. Cutler. Effect of manganese oxide on sintering of alumina. J. Am. Cer. Soc. 48(12):653–654, 1965.
37. S.V. Raman. Variations in sintering kinetics and microstructure of alumina with chromium and molybdenum additions. J. Mat. Sci. 22:3161–3166, 1987.
38. T. Nagaoka, M. Yasuoka, K. Hirao, S. Kanzaki, Y. Yamaoka. Effects of CaO addition on sintering and mechanical properties of Al_2O_3. J, Mat. Sci. Let. 15:1815–1817, 1996.
39. J. Fang, A.M. Thompson, M.P. Harmer, H.M. Chan. Effect of yttrium and lanthanum on the final-stage sintering behavior of ultrahigh-purity alumina. J. Am. Cer. Soc. 80(8):2005–2012, 1997.
40. J. Zhao, M.P. Harmer. Sintering of ultra-high-purity alumina doped simultaneously with MgO and FeO. J. Am. Cer. Soc. 70(12):860–866, 1987.
41. E. Sato, C. Carry. Yttria doping and sintering of submicrometer-grained α-Al_2O_3. J. Am. Cer. Soc. 79(8):2156–2160, 1996.
42. G. Wroblewaska. Effect of Titanate Additions on Sintering temperature of Al_2O_3. Cer. Intl. 5(3):120, 1979.

43. G. Wroblewaska. Sintering Kinetics of Al_2O_3 doped with $MgNb_2O_6$. Sintering-Theory and Practice, Mat. Sci. Mon., Elsevier Sci. Pub. Co. Amsterdam 14:165–171, 1982.
44. C. Wanqiu, Z. Yujuan, Z. Yun. Effects of Ta_2O_5 and MgO additives on microstructure and mechanical properties of ultra-pure alumina ceramics. Cer. Intl. 14, 133–140, 1988.
45. L.A. Xue, I.-W. Chen. Low-temperature sintering of alumina with liquid-forming additives. J. Am. Cer. Soc. 74(8):2011–2013, 1991.
46. C.O. McHugh, T.J. Whalen, J. Humenik, Jr. Dispersion-strengthened aluminum oxide. J. Am. Cer. Soc. 49(9):486–491 (1966).
47. D.T. Rankin, J.J. Stiglich, D.R. Petrak, R. Ruh. Hot-pressing and mechanical properties of Al_2O_3 with an Mo-dispersed phase. J. Am. Cer. Soc. 54(6):277–281, 1971.
48. P. Hing. Spatial distribution of tungsten on the physical properties of Al_2O_3-W cermets. Sci. of Cer. 12:87–94, 1984.
49. G. Orange, D. Turpin-Launay, P. Goeuriot, G. Fantozzi, F. Thevenot. Mechanical behavior of a Al_2O_3-AlON composite ceramic material (aluminalon). Sci. of Cer. 12:661–666, 1984.
50. K. Takatori, H. Dori, O. Kamigaito. A preliminary study of alumina sintering by addition of Si_3N_4. J. Mat. Sci. Let. 3:987–999, 1984.
51. R.W. Rice. Fabrication and characterization of hot pressed Al_2O_3. U.S. Naval Research Lab, Report 7111,1970.
52. W.C. Johnson, D.F. Stein, R.W. Rice. Analysis of grain-boundary impurities and fluoride additives in hot-pressed oxides by auger electron spectroscopy. J. Am. Cer. Soc. 57(8):342–344, 1974.
53. S.F. Belov, I.A. Rozdin, V.N. Pilsko, Yu.G. Bushuev, M.S. Igumnov, and N.E. Natanson. Reactions of aluminum oxide with fluorides of certain elements. Izvestiya Akademii Nauk SSSR, Neorganicheskie Materialy 16(6):1026–1028, 1980.
54. A.A. Pirogov, M.M. Mirak'yan, E.N. Leonova, and V.V. Primachenko. Effect of fluoride mineralizers on alumina sintering. Ogncupory (No. 1):37–40, 1970.
55. S.C. Carniglia, J.E. Hove. Fabrication and properties of dense beryllium oxide. J. Nuc. Mats. 4(2):165–176, 1961.
56. R.B. Adams, W.I, Stuart. Effect of adsorbed sulfate and fluoride on the density of hot-pressed beryllium oxide. J. Am. Cer. Soc. 50(12):685, 1967.
57. E.C. Duderstadt, J.F. White. Sontering BeO to variable densities and grain sizes. Am. Cer. Soc. Bul. 44(11):907–911, 1965.
58. N.A. Hill, J.S. O'Neill, D.T. Livey. The properties of BeO containing up to 15 w/o SiC and up to 2 w/o MgO. Proc. Brit. Cer. Soc. (No. 7):221–232, 1967.
59. J. Greenspan. Development of a high-strength beryllia material. Army Materials and Mechanics Research Center Report AMMRC TR 72-16, 1972.
60. R.A. Brown. Sintering of calcium oxide and calcium oxide containing strontium. Am. Cer. Soc. Bul. 44(9):693–695, 1965.
61. R.O. Petersen, I.B. Cutler. Effect of water vapor on the initial sintering of calcia. J. Am. Cer. Soc. 51(1):21–22, 1968.
62. R.W. Rice. CaO: I, fabrication and characterization. J. Am. Cer. Soc. 52(8):420–427, 1969.

63. R.W. Rice.CaO: II, Properties. J. Am. Cer. Soc. 52(8):428–436, 1969.
64. D. Deruto, G. Belleri, L. Barco, V. Longo. Interactions of LiBr with calcite and calcium oxide powders. Cer. Intl. 9(2):53–56, 1983.
65. T.K. Gupta, B.R. Rossing, W.D. Straub. Fabrication of transparent polycrystalline CaO. J. Am. Cer. Soc. 56(6):339, 1973.
66. J.F. Baumard, C. Gault, A. Argoitia. Sintered ceria: a new dense and fine grained ceramic material. J. Less Com. Metals. 127:124–130, 1987.
67. P.-L. Chen, I-W. Chen. Grain growth in ceria: dopant effects, defect mechanism, and solute drag. J. Am. Cer. Soc. 79(7):1793–1800, 1996.
68. H. Yoshida, K. Miura, J.-i. Fujita, T. Inagaki. Effect of gallia addition on the sintering behavior of samaria-doped ceria. J. Am. Cer. Soc. 82(1):219–221, 1999.
69. P.D. Omnby, G.E. Jungquist. Final sintering of Cr_2O_3. J. Am. Cer. Soc. 55(9):433–436, 1972.
70. S.N. Roy, S.R. Saha, S.K. Guha. Sintering kinetics of pure and doped chromium oxide. J. Mat. Sci. 21:3673–3676, 1986.
71. W.D. Callister, M.L. Johnson, I.B. Cutler, R.W. Ure, Jr. Sintering chromium oxide with the aid of TiO_2. J. Am. Cer. Soc. 62(3-4): 208–211, 1979.
72. H. Nagai, K. Ohbatashi. Effect of TiO_2 on the sintering and the electrical conductivity of Cr_2O_3. J. Am. Cer. Soc. 72(3):400–403, 1989.
73. C.J. Malarky, O. Hunter, Jr. Microcracking of Eu_2O_3-Ta_2O_5 bodies. J. Am. Cer. Soc. 61(7-8):315–317, 1978.
74. J. Wright. The effect of calcium oxide on the sintering of ferric oxide compacts. Pwd. Tech. 30:185–194, 1981.
75. P.C. Hayes. The effect of atmosphere on the sintering of Fe_3O_4 and Fe_2O_3 powders at low temperatures. J. Am. Cer. Soc. 63(7-8):387–91, 1980.
76. M.F. Yan. Grain growth in Fe_3O_4. J. Am. Cer. Soc. 63(7-8):443–447, 1980.
77. F.C. Gennari, D.M. Pasquevich. Crystal growth of α-Fe_2O_3 in chlorine atmosphere. J. Mat. Sci. Let. 15,:1847–1850, 1996.
78. H.J.S. Kriek, W.F. Ford, J. White. The effect of additions on the sintering and deadburning of magnesia. Trans. Brit. Cer. Soc. 99:1–34, 1959.
79. R.A. Brown. Sintering in very pure magnesium oxide and magnesium oxide containing vanadium. Am. Cer. Soc. Bul. 44(6):483–487, 1965.
80. P.P. Budnikov, M.A. Matveev, V.K. Yanovskii. Sintering high purity magnesia with additions of hafnium dioxide. Refractories; 190–95, 1965.
81. P.P. Budnikov, M.A. Matveev, V.K. Yanovskii, F.Ya. Kharitonov. Sintering high purity magnesium oxide with additives. Inorg. Mat. 3(5):749–756, 1967.
82. M.N. Chaudhuri, A. Kumar, A.K. Bhadra, G. Banerjee, S.L. Sarkar. Microstructure of sintered natural magnesites with titania addition. Am. Cer. Soc., Bul. 71(3):345–348, 1992.
83. K. Itatani, T. Yamamoto, A. Kishioka, M. Kinoshita. Effect of magnesium bis(dihydrogen orthophosphate) dihydrate on sintering of magnesium oxide. Am. Cer. Soc. Bul. 64(8):1124–1128, 1985.
84. K. Hamano, Y. Akiyama. Effects of boron oxide on sintering of magnesia. Bul. Tokyo Inst. Tech. (No. 126):59–67, 1975.
85. K. Hamano, H. Kamizono, Z. Nakagawa. Effects of minor additives on grain growth of magnesia. Report Res. Lab. Eng. Mats. Tokyo Inst. Tech. (No. 6):47–54, 1981.

86. K. Hamano, M. Miyazaki, Z. Nakagawa. Effects of addition of zirconium oxychloride on sintering of magnesia. Report Res. Lab. Eng. Mats. Tokyo Inst. Tech. (No. 11):73–81, 1986.
87. K. Hamano, T. Yamamoto, Z. Nakagawa. Effects of zirconium sulfate addition on sintering of magnesia. Report Res. Lab. Eng. Mats. Tokyo Inst. Tech. (No. 11):83–91, 1986.
88. H. Watanabe, Z. Nakagawa , M. Hasegawa, K. Hamano. Effects of addition of magnesium basic carbonate, sulfate, fluoride and oxolate on sintering of magnesia. Report Res. Lab. Eng. Mats. Tokyo Inst. Tech. (No. 7):616–621, 1982.
89. R.W. Rice. Production of transparent MgO at moderate temperatures and pressures. (Abstr). Am. Cer. Soc. Bul. 41(4):271, 1962.
90. R.W. Rice. Denser MgO has better properties. Space Aeronautics 109–112, 1963.
91. R.W. Rice. The effect of gaseous impurities on the hot pressing and behavior of MgO, CaO, and Al_2O_3. Fabrication Science 2 Proc. Brit. Cer. Soc. (No. 12):99–123, 1969.
92. R.W. Rice. Fabrication of dense MgO. U.S. Naval Res. Lab. Report 7334, 1971.
93. E. Carnall, Jr. The densification of MgO in the presence of a liquid phase. Mat. Res. Bul. 2:1075–1086, 1967.
94. L.M. Atlas. Effect of some lithium compounds on sintering of MgO. J. Am. Cer. Soc. 40(6):196–99, 1957.
95. G.D. Miles, R.J. Sambell, J. Rutherford, G.W. Stephenson. Fabrication of fully dense transparent polycrystalline magnesia. Trans. Brit. Cer. Soc. 66(7):319–335, 1967.
96. E. Smethurst, D.W. Budworth. The preparation of transparent magnesia bodies. I. By Hot Pressing. Trans. & J. Brit. Cer. Soc. 71(2):45–50, 1972.
97. F.R. Wermuth, W.J. Knapp. Initial sintering of MgO and LiF-doped MgO. J. Am. Cer. Soc. 55(7):401, 1973,
98. M.W. Benecke, N.E. Olson, J.A. Pask. Effect of LiF on hot-pressing of NgO. J. Am. Cer. Soc. 50(7):365–368, 1967.
99. P.E. Hart, R.B. Atkin, J.A. Pask. Densification mechanisms in hot-pressing of magnesia with a fugative liquid. J. Am. Cer. Soc. 53(2):83–86, 1970.
100. F.K. Volynets, L.V. Udalova, L.I. Aranovskil, V.P. Usachev. Compaction kinetics of magnesium oxide with added lithium fluoride at different hot-pressing temperatures. Inorg. Mat. 8:250–253, 1972.
101. F.K. Volynets, L.V. Udalova, L.I. Kleshchinskii, G.N. Vanyukhin. Nature of the “activity” of magnesium oxide doped with lithium fluoride. Inorg. Mat. 10:643–645, 1974.
102. J. Razinger, G.M. Fryer. Effect of the ions of lithium and fluorine on the pressure-sintering of magnesia. Trans. & J. Brit. Cer. Soc. 71:223–230, 1972.
103. M. Shombo, Z.-e Nakagawa, Y. Ohya, K Hamano. Effect of floride addition on crystal growth of MgO. J. Cer. Soc. Jap. Intl. Ed. 97:844–850, 1989.
104. R.W. Rice. Hot-pressing of NgO with NaF. J. Am, Cer. Soc. 54(4):205–207, 1971.
105. M. Banerjee, D. W. Budworth. The preparation of transparent magnesia bodies. II. By sintering. Trans. & J. Brit. Cer. Soc. 71(2):51–53, 1972.
106. T. Ikegami, S.-I. Matsuda, H. Suzuki. Effect of halide dopants on fabrication of transparent MgO. J. Am. Cer. Soc. 57(11):507, 1974.

107. T. Ikegami, M. Kobayashi, Y. Moriyoshi, S.-I. Shirasaki. Characterization of sintered MgO compacts with fluorine. J. Am. Cer. Soc. 63(11-12):640–643, 1980.
108. M.H. Leipold, C.M. Kapadia. Effect of anions on hot-pressing of MgO. J. Am. Cer. Soc. 56(4):200–203, 1973.
109. R. Hanna. Hot Pressing Magnesia. Am Cer. Soc. Bul. 49:548, 1970.
110. S. Bosković, M. Stevanović. Sintering of cobalt-doped nickel. J. Mat. Sci. 10:25–31, 1975.
111. J.A. Varela, O.J. Whittemore, M.J. Ball. Structural evolution during the sintering of SnO_2 and SnO_2 -2m/o CuO. In: G. Kuczynski, Ed. Sintering 85. New York: Plenum Press, 1987, pp. 259–268.
112. V.I. Drozd, E.V. Degtyareva, I.I. Kabakova. Effect of some oxide additives on the sintering of SnO_2. Proizvod. Spets. Ogneuporov (No 6):23–28, 1978.
113. E.V. Degtyareva, B.G. Alapin, V.I. Drozd. Effect of certain oxide additives on the sintering of cassiterite refractories. Proizvod. Spets. Ogneuporov (No 10):40–45, 1978.
114. C.E. Curtis, J.R. Johnson. Properties of thorium oxide ceramics. J. Am. Cer. Soc. 40(2):63–68, 1957.
115. P.J. Jorgensen, W.G. Scmidt. Final stage sintering of ThO_2. J. Am. Cer. Soc. 53(1):24–27, 1970.
116. W.I. Stuart, T.L. Whately, R.B. Adams. Adsorption of fluoride on thoria and its effect on hot-pressed density. J. Aust. Cer. Soc. 8(1):6–9, 1972.
117. G.P. Halbfinger, M. Kolodney. Activated sintering of ThO_2 and ThO_2-Y_2O_3 with NiO. J. Am. Cer. Soc. 55(10):519–524, 1972.
118. C. Greskovich, C.R. O'Clair, M.J. Curran. Preparation of transparent Y_2O_3-doped ThO_2. J. Am. Cer. Soc. 55(6):324–325, 1972.
119. M. Hartmanová, V. Sály, F. Hanic, M. Pisarčik, Ullmann. Microstructure and physical properties of transparent thoria-yttria ceramics. J. Mat. Sci. 26:4313–4317, 1991.
120. C.-J. Chen, J.-M. Wu. Sintering behavior of niobium- and calcium-doped TiO_2 ceramics. Mats. Sci. & Eng. B5:5–15, 1989.
121. K.C. Radford, J.M. Pope. Ammonia as a sintering aid for UO_2. Am Cer. Soc. Bul. 56(2):197–200, 1977.
122. P.J. Jorgensen, R.C. Anderson. Grain-boundary segregation and final-stage sintering of Y_2O_3. J. Am. Cer. Soc. 50(11):553–558, 1967.
123. C. Greskovich, K.N. Woods. Fabrication of transparent ThO_2-Y_2O_3. Am. Cer. Soc. Bul. 52(5):473–478, 1973.
124. W.H. Rhodes. Controlled transient solid second-phase sintering of yttria. J. Am. Cer. Soc. 64(1):13–19, 1981.
125. W.H. Rhodes. Transparent yttria ceramics and method for producing same. U.S. Patent 4,115,134, 9/19/1978.
126. K. Katayama, H. Osawa, T. Akiba, H. Yanagida. Sintering and electrical properties of CaO-doped Y_2O_3. J. Eur. Cer. Soc. 6:39–45, 1990.
127. R.A. Lefever, J. Matsko. Transparent yttrium oxide ceramics. Mat. Res. Bul. 2:865–869, 1967.
128. Y. Moriyoshi, M. Isobe, Y. Hasegawa, W. Komatsu. The fabrication of the translucent ZnO by sintering. J. Mat. Sci, Let. 12:2347–2349.

129a. M. Tronteli, D. Kojar. Sintering and grain growth in doped ZnO. J. Mat. Sci. Let.13:1832–1334, 1978.
129b. J. Luo, H. Wang, Y.-M. Chiang. Origin and silid-state activated sintering in Bi_2O_3-doped ZnO. J. Am. Cer. Soc. 82(4):916–920, 1999.
130a. L.M. Levinson, H.R. Philipp. The physics of metal oxide varistors. J. Appl. Phy. 46(3):1332–1341, 1975.
130b. F.A. Selim, T.K. Gupta, P.L. Hower, W.G. Carlson. Low voltage varistor: device process and defect model. J. Appl. Phy. 51(1):765–768, 1980.
130c. G.S. Snow, S.S. White, R.A. Cooper, J.R. Armijo. Characterization of high field varistors in the system ZnO-CoO-PbO-Bi_2O_3. Am Cer, Soc. Bul. 59(6):617–622, 1980.
130d. D.R. Clarke. Varistor Ceramics. J. Am. Cer. Soc. 82(3):485–502, 1999.
131. J.F. Shackelford, P.S. Nicholson, W.W. Smeltzer. Influence of SiO_2 on sintering of partially stabilized zirconia. Am. Cer. Soc. Bul. 53(12):865–868, 1974.
132. F.-C. Wu, S.-C. Yu. The effect of SiO_2 and Al_2O_3 additives on the sintering of MgO-containing zirconia. Mat. Res. Bul. 23, 1773–1780, 1988.
133. Bi-S. Chiou, W.-Y. Hsu, J.-G. Duh. Fabrication and electrical behavior of liquid phase sintered zirconia. Cer. Intl. 14:7–16, 1988.
134. H.-Y. Lu, J.-S. Bow. Effect of MgO additions on the microstructure development of 3 mol% Y_2O_3-ZrO_2 J. Am. Cer. Soc. Bul. 72(2):228–231, 1989.
135. C.E. Scott, J.S. Reed. Effect of laundering and milling on the sintering behavior of stabilized ZrO_2 powders. Am. Cer. Soc. 58(6):587–590, 1979.
136. W.H. Rhodes, P.L. Berneburg, J.E. Niesse. Development of transparent spinel ($MgAl_2O_4$). AVCO Corp. Final Report AVSD-0509-70-RR for Army Materials and Mechanics Research Center Contract DAAG-46-69—C-0113, 1970.
137. W.J. McDonough, R.W. Rice. Unpublished results at the U. S. Naval Research Lab. Ca., 1975.
138. K.W. White, G.P. Kelkar. Fracture mechanisms of a coarse-grained, transparent $MgAl_2O_4$ at elevated temperatures. J. Am. Cer. Soc. 75(12):3440–3444, 1992.
139. D.W. Roy, J.L. Hastert. Application of transparent polycrystalline body with high ultraviolet transmittance. U.S. Patent 4,983,555, 1/18/1991.
140. F.F. Lange , D.R. Clarke. Morphological changes of an intergranular thin film in a polycrystalline spinel. J. Am. Cer. Soc. 65(10):502–506, 1982.
141. C.A.M. Mulder, G. De With. Translucent $Y_3Al_5O_{12}$ ceramics: electron microscopy characterization. Solid State Ionics. 16:81–86, 1985.
142. G. De With, J.E.D. Parren. Translucent $Y_3Al_5O_{12}$ ceramics: mechanical properties. Solid State Ionics. 16:87–94, 1985.
143. M. Hashiba, Y. Nurishi, T. Hibino. Promotion of $ZnAl_2O_4$ formation by AlF_3. J. Mat. Sci. 23:570–576, 1988.
144. M. Hashiba, Y. Nurishi, T. Hibino. Microstructural observation and reaction mechanism of $ZnAl_2O_4$ formation in the presence of various fluorides. J. Mat. Sci. 23:905–913, 1988.
145. C. Bauden, J.S. Moya. Influence of titanium dioxide on the sintering and microstructural evolution of mullite. F. Am. Cer. Soc. 67(7):C-134–136, 1984.
146. T. Mitamura, H. Kobayashi, N, Ishibashi, T. Akiba. Effects of rare earth oxide addition on the sintering of mullite. J. Cer. Soc. Jpn. Intl. Ed. 99:339, 1991.

147. T. Mori, N. Kosugi, Y. Ishikawa, Y. Kubota. Microstructure and mechanical properties of mullite sintered bodies containing Y_2O_3. J. Cer. Soc. Jpn. Intl. Ed. 98:1320, 1990.
148. C.-S. Hwang, D.-Y. Fang. Effects of Y_2O_3 addition on the sinterability and microstructure of mullite (Part 1)-phase transformation and stability. J. Cer. Soc. Jap. Intl. Ed. 100:1141, 1992.
149. D.-Y. Fang, C.-S. Hwang. Effects of Y_2O_3 addition on the sinterability and microstructure of mullite (Part 2)-phase transformation and stability. J. Cer. Soc. Jap. Intl. Ed. 101:322, 1993.
150. K. Hayashi, K. Tanaka, I. Uei. Effects of titania addition and firing atmosphere on sintering of zircon bodies. Taikabutsu 28(219): 138–143, 1976.
151. J. De Andres, S. De Aza, J.S. Moya. Effect of TiO_2 additions on the sintering behavior of zircon powders. In: W. Bunk, H. Hausner, eds. Ceramic Materials and Components for Engines. Proc. 2nd Intl. Symp., Verlag Deutsche Keramische Gesellschaft, Bad Honnef, Ger., 1986, pp. 299–306.
152. H. Kim. Effect of minor additives on sintering of zircon. J. Korean Cer. Soc. 15(1):3–8, 1978.
153. R.H. Arendt. Liquid-phase sintering of magnetically isotropic and anisotropic compacts of $BaFe_{12}O_{19}$ and $SrFe_{12}O_{19}$. J. Appl. Phy. 44(7):3300–3305, 1973.
154. E. Lucchini, S. Meriani, G. Slokar. Sintering of glass bonded ceramic barium hexaferrite magnetic powders. J. Mat. Sci. 18:1331–1334, 1983.
155. T.-F. Fang, J.B. Hwang, F.S. Shiau. The role of silica in sintering barium ferrite. J. Mat. Sci. Let. 8:1386–1388, 1989.
156. S. Beseničar, T. Kosmač, and M. Drofenik. The influence of ZrO_2 additions on magnetic and mechanical properties of Sr ferrites. Brit Cer. Trans. & J. 86:44–46, 1987.
157. G. Bandyopadhyay, R.M. Fulrath. Effect of NiO and $NiFe_2O_4$ on the processing and properties of lithium ferrite spinel. Am. Cer. Soc. Bul. 54(5):506–509, 1975.
158. P. Peshev, M. Pecheva. A study of sintering and magnetic parameters of spinel lithium ferrite, II. Lithium ferrite containing Bi_2O_3. Mat. Res. Bul. 15:1199–1205, 1980.
159. H.T. Kim, H.B. Im. Sintering behavior of lithium ferrites containing Nb_2O_5 and V_2O_5. J. Mat. Sci. 22:1235–1239, 1987.
160. N. Rezlescu, E. Rezlescu. Effects of addition of Na_2O, Sb_2O_3, CaO, and ZrO_2 on the properties of lithium zinc ferrite. J. Am. Cer. Soc. 79(8):2105–2108, 1996.
161. Y. Bando, Y. Ikeda, T. Akashi, T. Tadak. Role of CaO and SiO_2 in sintering of manganese zinc ferrite. In: H. Housner, ed. Modern Developments in Powder Metallurgy. Vol 4. New York: Plenum Press, 1971, pp. 339–48.
162. F.J.C.M. Toolenaar. Silica-induced exaggerated grain growth in MnZn Ferrite. J. Mat. Sci. 23:3144–3150, 1988.
163. G.C. Jain, B.K. Das, N.C. Goel. Effect of MoO_3 addition on the grain growth kinetics of a manganese zinc ferrite. J. Am. Cer. Soc. 62:79–85, 1979.
164. M. Drofenik, A. Znidaršič, D. Makovec. Influence of the addition of Bi_2O_3 on the grain growth and magnetic permeability of MnZn ferrites. J. Am. Cer. Soc. 81(11):2841–2848, 1998.
165. M. Drofenik, S. Besaniar, D. Kolar. Influence of BaO additions on the microstructure and magnetic properties of Co-Ni-Zn ferrites. Am Cer. Soc. Bul. 63(5):699–701, 1987.

166. F. Mahloojchi, F.R. Sale. Effect of CuO on the sintering of MgO-based soft ferrites. Cer. Intl. 15:51–64, 1989.
167. S. Shimada, K. Kodaira, T. Matsushita, K. Ichijo, R. Maeda. Effect of CdO additive on sintering of $LiNBO_3$. Cer. Intl. 4(4):167, 1978.
168. Z.-G. Ye, R. Von Der Mühll, J. Ravez, and P. Hagenmuller. Dielectric, piezoelectric, and pyroelectric studies of $LiTaO_3$-derived ceramics sintered at 900°C following the addition of (LiF + MgF_2). J. Mat. Res. 3:112, 1988.
169. M.N. Swalam , A.M. Gadalla. Effect of additions on the sinterability of barium titanate. Trans, & J. Brit. Cer. Soc. 74(5):165–169, 1975.
170. W.-Y. Howing, C. McCutcheon. Electrical properties of semiconducting $BaTiO_3$ by liquid-phase sintering. Am. Cer. Soc. Bul. 62(2):231–233, 1983.
171. I. Burn. Flux-sintered $BaTiO_3$ Dielectrics. J. Mat. Sci. 17:1398–1408, 1982.
172. Sarkar, M. L. Sharma. Liquid phase sintering of $BaTiO_3$ by boric oxide (B_2O_3) and lead borate (PbB_2O_4) glasses and its effect on dielectric strength and dielectric constant. Mat. Res. Bul. 24:773–779, 1989.
173. B.E. Walker, Jr., R.W. Rice, R.C. Pohanka, J.R. Spann. Densification and strength of $BaTiO_3$ with LiF and MgO additions. Am. Cer. Soc. Bul. 55(3): 274–278, 1976.
174. B.E. Walker, Jr., J.R. Spann, R.W. Rice. High strength barium titanate ceramic bodies. U.S. Patent 3,753,911, 8/21/1973.
175. B.E. Walker, Jr., J.R. Spann, R.W. Rice. Strengthened and high density $BaTiO_3$. U.S. Patent 3.862,046, 1/21/1975.
176. J.M. Haussonne, G. Desgardin, PH. Bajolet, B. Raveau. Barium titanate perovskite sintered with lithium fluoride. J. Am. Cer. Soc. 66(11):801–807, 1983.
177. G. Desgardin,, I. Mey, B. Raveau, J.M. Haussonne. $BaLiF_3$—A new sintering agent for $BaTiO_3$-based capicitors. Am. Cer. Soc. Bul. 64(4):564–570, 1985.
178. A. Portin, J. Ravez, J.P. Bonnet. Liquid-phase sintering of barium titanate with lithium fluoride. J. Mat. Res. 2(4):485–488, 1987.
179. J.-N. Lin, T.-B. Wu. Wetting reaction between lithium flouride and barium titanate. J. Am. Cer. Soc. 72(9):1709–1712, 1989.
180. J.P. Guha, H.U. Anderson. Reaction during sintering of barium titanate with lithium fluoride. J. Am. Cer. Soc. 69(8):C-193–194, 1986.
181. I. Ueda. Effects of additives on piezoelectric and related properties of $PbTiO_3$. Jap. J. Appl. Phy. 11(4):450–462, 1972.
182. D.E. Wittmer, R.C. Buchannan. Low-temperature densification of lead zirconate-titanate with vanadium pentoxide additive. J. Am. Cer. Soc. 64(8):485–493, 1981.
183. J.-S. Chen, R.-J. Young, T.-B. Wu. Densification and microstructural development of $SrTiO_3$ sintered with V_2O_5. J. Am. Cer. Soc. 63(10):C-260–264, 1987.
184. S.-Y. Cheng, S.-Li Fu, C.-C. Wei. Sintering of $SrTiO_3$ with Li_2CO_3 addition. Cer, Intl. 15:231–236, 1989.
185. M.J. Laurent, G. Desgardin, B. Raveau, J.M. Houssonne, J. Lostec. Sintering of strontium titanate in the presence of lithium salts in a reducing atmosphere. J. Mat. Sci. 23:4481–4486, 1988.
186. J.R. Clarke, D.A. Hirschfeld. The sintering effects of time, temperature, and ZnO additions on CMZP ceramics. Cer. Eng. Sci. Proc. 18 (4):121–128, 1997.

187. V. Ya. Shlyuko, V.V. Morozov, A.V. Besov, L.V. Chernyak, L.V. Guzenko. Effects of yttrium and aluminum oxide additions on the sintering behavior and properties of lanthium hexaboride. Powder. Met. (7):544, 197.
188. Y. Murata, B.R. Miccioli. Inhibition of grain growth in niobium boride. Am. Cer. Soc. Bul. 50(2):182–184, 1971.
189. I.M. Low, R. McPherson. Fabrication of new zirconium boride ceramics. J. Mat. Sci. Let. 8:1281–1283, 1989.
190. E.S. Kang, C.W. Jang, C.H. Lee, C.H. Kim. Effect of iron and boron carbide on the densification and mechanical properties of titanium diboride ceramics. J. Am. Cer. Soc. 72(10):1868–1872, 1989.
191. J. Matsushita, H. Naggashima, H. Saito. Preparation and mechanical properties of TiB_2 composites containing Ni and C. J. Cer. Soc. Jap. Intl. Ed. 99:1047, 1991.
192. B. Champagne, S. Dallaire. Structure and properties of TiB_2-based materials produced by the reaction of ferrotitanium and boron powders. Cer. Eng. & Sci. Proc. 5(7-8):702, 1984.
193. K.B. Shim, M.J. Edirisinghe, B. Ralph. Use of nickel and iron additions as sintering aids for titanium diboride. Trans. Brit. Cer. Soc. 95(1):15–22, 1996.
194. M.K. Ferber, P.F. Becher, C.B. Finch. Effect of microstructure on the properties of TiB_2 ceramics. J. Am. Cer. Soc. 66(1):C-2–4, 1983.
195. M.-A. Einarsrud, E. Hagen, G. Pettersen, T. Grande. Pressureless sintering of titanium diboride with nickel, nickel boride, and iron additives. J. Am. Cer. Soc. 80(12):3013–3020, 1987.
196. P. Angelini, J. Bentley, C.B. Finch, P.S. Sklad. Microstructure of TiB_2 liquid phase sintered with NiAl. Proc. An. Meeting, Electron Microscopy Soc. Am. 42:62–63, 1983.
197. T. Watanabe. Ni•P Binder for TiB_2-Based Cerments. Am. Cer. Soc. Bul. 59(4):485–486, 1980.
198. T. Watanabe, S. Kouno. Mechanical properties of TiB_2-CoB-metal boride alloys. Am. Cer. Soc, Bul. 61(9):970–973, 1982.
199. J. Matsushita, H. Nagashima, H. Saito. Sintering behavior of TiB_2 with Cr_3C_2 additive. J. Cer. Soc. Jap. Intl. Ed. 98:452, 1990.
200. S. Torizuka, K. Sato, H. Nishio, T. Kishi. Effect of SiC on interfacial reaction and sintering mechanism of TiB_2. J. Am. Cer. Soc. 78(6):1606–1610, 1995.
201. F. Thévenot. Boron carbide—a comprehensive review. J. Eur. Cer. Soc. 6:205–225, 1990.
202. C.H. Lee, C.H. Kim. Pressureless sintering and related reaction phenomena of Al_2O_3-doped B_4C. J. Mat. Sci. 27:6335–6340, 1993.
203. Y. Kanno, K. Kawase, K. Nakano. Additive effect on sintering of boron carbide. J. Jap. Cer. Soc. 95(11):1137, 1987.
204. L.S. Sigl. Processing and mechanical properties of boron carbide sintered with TiC. J. Eur. Cer. Soc. 18:1521–1529, 1998.
205. J.S. Nadeau, Very high pressure hot pressing of silicon carbide. Am. Cer. Soc. Bul. 52(2):170–174, 1973.
206. R.A. Alliegro, L.B. Ciffin, J.R. Tinklepaugh. Pressure-sintered silicon carbide. J. Am. Cer. Soc. 39(11):386–389, 1956.

207. A.K. Misra. Thermochemical analysis of the silicon carbide-alumina reaction with reference to liquid-phase sintering of silicon carbide. J. Am. Cer. Soc. 74(2):345–351, 1991.
208. F.F. Lange. Hot-pressing behavior of silicon carbide powders with additions of aluminum axide. J. Mat. Sci.10:314–320, 1975.
209. M.A. Mulla, V.D. Krstic. Pressureless sintering of βSiC with Al_2O_3 additions. J. Mat. Sci. 29:934–938, 1994.
210. M.A. Mulla, V.D. Krstic. Low-temperature pressureless sintering of β-silicon carbide with aluminum oxide and yttrium oxide additions. Am. Cer. Soc. Bul. 70(3):439–442, 1991.
211. E. Lidén, E. Carlström, L. Eklund, B. Nyberg, R. Carlsson. Homogeneous distribution of sintering additives in liquid-phase sintered silicon carbide. J. Am. Cer. Soc. 78(7):1761–1768, 1995.
212. J.-Y. Kim, Y.-W. Kim, M. Mitomo, G.-D. Zhan, J.-G. Lee. J. Am. Cer. Soc. 82(2):441–444, 1999.
213. Z. Chen. Effects of gadolinia and alumina addition on the densification and toughening of silicon carbide. J. Am. Cer. Soc. 79(2):530–532, 1996.
214. S. Prochazaka. The role of boron and carbon in the sintering of silicon carbide. In: P. Popper, ed. Special Ceramics 6, Stoke on Trent: Brit. Cer. Soc.1974, p. 171.
215. S. Prochazaka. Effect of boron and carbon on sintering of SiC. J. Am. Cer. Soc. 58(1-2):72, 1975.
216. S. Prochazaka, W.S. Coblenz. Silicon carbide-boron carbide sintered body. U.S. Patent 4,081,284, 3/28/1978.
217. F.F. Lange, T.K. Gupta. Sintering of SiC with boron compounds. J. Am. Cer. Soc. 59(11-12):537–538, 1976.
218. J. Hojo, K. Miyachi, Y. Okabe, A. Kato. Effect of chemical composition on the sinterability of ultrafine SiC powders. J. Am. Cer. Soc. 66(7):C-114, 1983.
219. S. Shinozaki, R.M. Williams, B.N. Juterbock, W.T. Donlon, J. Hangas, C.R. Peters. Microstructurrial developments in pressureless-sintered β-SiC materials with Al, B, and C additions. Am. Cer. Soc. Bul. 64(10):1389–1393, 1985.
220. Y. Zhou, H. Tanaka, S. Otani, Y. Bando. Low-temperature pressureless sintering of α-SiC with Al_4C_3-B_4C-C Additions. J. Am. Cer. Soc. 82(8):1959–1964, 1999.
221. H. Tanaka, Y. Inomata, K. Hara, H. Hasegawa. Normal sintering of Al-doped β-SiC. J. Mat. Sci. Let. 4:315–317, 1985.
222a. B.-W. Lin, M. Imai, T. Yano, T. Iseki. Hot-pressing of β-SiC Powder with Al-B-C additives. J. Am. Cer. Soc. 69(4):C-67–68, 1986.
222b. J.J. Cao, W.J. Moberly, L.C. De Jonghe, C.J. Gilbert, R.O. Ritchie. In situ toughened silicon carbide with Al-B-C additions. J. Am. Cer. Soc. 79(2):461–469, 1996.
223. G. Passing, R. Riedel, G. Petzow. Influence of pyridine-borane on the sintering behavior and properties of α- silicon carbide. J. Am. Cer. Soc. 74(3):642–645,1991.
224. W. Van Rijwijk , D.J. Shanefield. Effects of carbon as a sintering aid in silicon carbide. J. Am. Cer. Soc. 73(1):148–149, 1990.
225. R. Ruh, A. Zangvil. Composition and properties of hot-pressed SiC-AlN solid solutions., J. Am. Cer. Soc. 65(5):260–265, 1982.
226. V.N. Klimenko, V.A. Maslyuk, I.D. Radomysel'skii. Activated sintering of chromium-carbide-nickel alloys. Inorg. Mat. 17: 768–771, 1978.

227. R.K. Viswanadham, D.J. Rowcliffe, J. Gurland, eds. Science of Hard Materials. New York: Plenum Press, 1983.
228. B.S. Terry, O.S. Chinyamakobvu. Effects of cooling rate and heat treatment on the microstructure of iron-base titanium carbide composites. J. Mat. Sci. 27, 5666–5670, 1992.
229. N.G. Zaripov, R.R. Kabirov, V.N. Bloshenko. Structural pecularities of cermets design based on titanium carbide. Part 1. Influence of chemical composition on the ductile-brittle transition temperature, microstructure and properties of cermets. J. Mat. Sci. 31:5227–5230, 1996.
230. M. Komac, D. Lange. The influence of MoC_x and MbC_x additions on microstructure and mechanical properties of TiC based cemented carbides. Intl. J. Pwd. Met. & Pwd. Tech. 18(4):313, 1982.
231. B.L. Eyre, A.F. Bartlett. The effects of nickel on the structure of sintered uranium carbide. Proc. Brit. Cer. Soc. No. 7:Nuclear and Engineering Ceramics, 1967, p. 127.
232. J.P. Mathers, R.W. Rice. Unpublished work at the U.S. Naval Res. Lab., 1979–1982.
233. P. Ettmayer. Hardmetals and cermets. In: R.A. Huggins, J.A. Giordmaine, and J.B. Wachtman, Jr., eds. Annual Review of Materials Science. 19. Palo Alto: Annual Reviews Inc., 1989, pp. 45–64.
234. L. Froschauer, R.M. Fulrath. Direct observation of liquid-phase sintering in the system tungsten carbide-cobalt. J. Mat. Sci. 11:142–149, 1976.
235. P. Goeuriot, F. Thevenot, N. Bouaoudja, G. Fantozzi. Boron as sintering additive in cemented WC-Co (or Ni) alloys. Cer. Intl. 13:99–103, 1987.
236. R. Cooper, S.A. Manktelow, F. Wong, L.E. Collins. The sintering characteristics and properties of hard metal with Ni-Cr binders. Mats. Sci. & Eng. A 105/106:269–273, 1988.
237. G.S. Upadhyaya, S.K. Bhaumik. Sintering of submicron WC-10 wt. % Co hard metals containing nickel and iron. Mats. Sci. & Eng. A 105/106:249–256, 1988.
238. E.A. Almond, B. Roebuck. Identification of optimum binder phase compositions for improved WC hard metals. Mats. Sci. & Eng. A 105/106:237–248, 1988.
239. H.D. Stromberg, D.R. Stephens. Sintering of diamond at 1800–1900°C and 60–65 kbar. Am. Cer. Soc. Bul. 49(12):1030–1032, 1970.
240. H.T. Hall. Sintered diamonds: a synthetic carbonado. Science 168:168–169, 8/1970.
241. H. Katzman, W.F. Libby. Sintered diamond compacts with a cobalt binder. Science 172:1132–1133, 1971.
242. Y. Notsu, T. Nakajima, N. Kawai. Sintering of diamond with cobalt. Mat. Res. Bul. 12:1079–1085, 1977.
243. M. Akaishi, H. Kanada, Y. Sato, N. Setaka, T. Ohsawa, O. Fukunaga. Sintering behavior of the diamond-cobalt system at high temperature and pressure. J. Mat. Sci. 17:193–198, 1982.
244. M. Akaishi, S. Yamoka, J. Tanaka, T. Ohsawa, O. Fukunaga. Synthesis of sintered diamond with high electrical resistivity and hardness. J. Am. Cer. Soc. 70(10):C-237–239, 1987.
245. M. Akaishi, Y. Sato, N. Setaka, M. Tsutsumi, T. Ohsawa, O. Fukunaga. Effect of additive graphite on sintering of diamond. Am. Cer. Soc. Bul. 62(6):6890–6894, 1983.

246. M. Akaishi, T. Ohsawa, S. Yamaoka. Synthesis of fine-grained polycrystalline diamond compact and its microstructure. J. Am. Cer. Soc. 74(1):5–10, 1991.
247. S. Tajima, H. Itoh, S. Naka. Effects of added diamond powder on the reaction sintering behavior of diamond in the graphite-nickel system. J. Mat. Sci. 24:849–853, 1989.
248. M. Trontelj, D. Kolar. Pressureless sintering of aluminum nitride to digh density. J. Mat. Sci. 8:136–137, 1973.
249. M. Trontelj, D. Kolar. On the mechanism of reactive sintering of AlN. In: P. Popper, ed. Special Ceramics. 6. Stoke on Trent: Brit, Cer. Soc., 1974, pp. 39–50.
250. K. Komeya, H. Inoue. Trans. & J. Brit. Cer. Soc. 70(3):107, 1971.
251. K. Komeya, H. Inoue. Role of Y_2O_3 and SiO_2 additions in sintering of AlN. J. Am. Cer. Soc. 57(9):411–412, 1974.
252. T. Sakai, M. Iwata. Effect of oxygen on sintering of AlN. J. Mat. Sci. 12:1659–1665, 1977.
253. S. Prochazka, C.F. Bobik. Sintering aluminum nitride. In: G. Kuczynski, ed. Mat. Sci. Res., Vol. 13. Sintering Precesses. New York: Plenum Press, 1980, pp. 321–332.
254. N. Kuramoto, H. Taniguchi, I. Aso. Development of translucent aluminum nitride ceramics. Am Cer. Soc. Bul. 68(4):883–887, 1989.
255. T.B. Troczynski, P.S. Nicholson. Effect of additives on the presureless sintering of aluminum nitride between 1500 and 1800°C. J. Am. Cer. Soc. 72(8):1488–1491, 1989.
256. N.S. VanDamme, S.M. Richard, S R. Winzer. Liquid-phase sintering of aluminum nitride by europimum oxide additives. J. Am. Cer. Soc. 72(8):1409–1414, 1989.
257. U. Neumann, T. Reetz, C. Rüssel. The influence of the anions of calcium-containing sintering aids for aluminum nitride. J. Eur. Cer. Soc. 12:117–122, 1993.
258. W. Miao, Y. Wu, H. Zhou. Low-temperature sintering of aluminim nitride substrates with high thermal conductivity. J. Mat. Sci. Let. 16:1245–1246, 1997. (See also: Y. Liu, H. Zhou, L. Qiao, Y. Wu. Low-temperature sintering of aluminum nitride with YF_3-CaF_2 binary additive. ibid. 703–704, 1999.
259. T. Nakamatsu, F. Pomar, K. Ishizaki. The effect of carbon coating of AlN powder on sintering behavior and thermal conductivity. J. Mat. Sci. 34:1553–1556, 1999.
260. K. Watari, H.J. Hwang, M. Toriyama, S. Kanzaki. Low-temperature sintering and high thermal conductivity of $YLiO_2$-doped AlN ceramics. J. Am. Cer. Soc. 79(7):1979–1981, 1996.
261. K.A. Schwetz, H. Knoch, A. Lipp. Sintering of aluminum nitride with low oxide addition. In: F.L. Riley, ed. Progress in Nitrogen Ceramics. Boston: Martinus Nijhoff Pub., 1983, pp. 245–252.
262. S. Nakahata, K. Sogabe, T. Matsuura, and A. Yamakawa. One role of titanium compound particles in aluminum nitride sintered body. J. Mat. Sci. 32:1873–1876, 1997.
263. U. Neumann, T. Reetz, C. Rüssel. The influence of the anions of calcium-containing sintering aids for aluminum nitride. J. Eur. Cer. Soc. 12:117–122, 1993.
264. E.G. Kendall. Intermetallic materials: carbides, borides, beryllides, nitrides, and silicides. In: J.E. Hove, W.C, Riley, eds. Ceramics for Advanced Technologies. New York: John Wiley & Sons, 1965, pp. 143–183.

265. J.H. Steele, R. Engel. Microstructural characterization of commercial hot-pressed boron nitride. Adv. Cer. Mat. 3(5):452–456, 1988.
266. M. Habáček, M. Ueki, T. Sató. Orientation and growth of frains in copper-activated hot-pressing hexagonal boron nitride. J. Am. Cer. Soc. 79(1):283–285, 1996.
267. J.C. Walmsley, A.R. Lang. A transmission electron microscope study of a cubic boron nitride-based compact material with AlN and AlB_2 binder phases. J. Mat. Sci. 22:4093–4102, 1987.
268. K. Shintani, M. Ueki, Y. Fujimura. Microstructure and mechanical properties of sintered cubic boron nitride. J. Mat. Sci. Let. 6:987–989, 1987.
269. B.K. Agarwala, B.P. Singh, S.K. Singhal. Synthesis and characterization of polycrystalline sintered compacts of cubic boron nitride. J. Mat. Sci. 21:1765–1768, 1986.
270. M.M. Bindal, R.K. Nayar, S.K. Singdhal, A. Dhar, R. Chopra. High-pressure sintering of cubic boron nitride. J. Mat. Sci. 21:4347–4351, 1986.
271. B.P. Singh. Characterization of wurtzitic cubic boron nitride compacts. J. Mat. Sci. 22:495–498, 1987.
272. K. Tsukuma, M. Shimada, M. Koizumi. Thermal conductivity and microhardness of Si_3N_4 with and without additives. Am. Cer. Soc. Bul. 60(9):910–912, 1981.
273. I. Tanaka, G. Pezzotti, T. Okamoto, Y. Miyamoto. Dense silicon nitride without additives: sintering and high temperature behaviors. Cer. Eng. & Sci. Proc. 10(7-8):817–822, 198.
274. D.J. Godfrey. The effects of impurities, additions and surface preparation on the strength of silicon nitride. In: R.W. Davidge, ed. Mechanical Properties of Ceramics (2). Proc. Brit. Soc., No 25. 1975, pp. 325–337.
275. G.G. Deeley, J.M. Herbert, N.C. Moore. Dense silicon nitride. Pwd. Met. No.8:145, 1961.
276. G.E. Gazza. Effect of yttria additions on hot-pressed Si_3N_4. Am. Cer. Soc. Bul. 54(9):778–781, 1975.
277. A. Tsuge, H. Kudo, K. Komeya. Reaction of Si_3N_4 and Y_2O_3 in hot-pressing. J. Am. Cer. Soc. 57(6):259–270, 1974.
278. R.W. Rice, W.J. McDonough. Hot-pressing Si_3N_4 with Zr-based additions. J. Am. Cer. Soc. 58(5-6):264, 1975.
279. S. Dutta, B. Buzek. Microstructure, strength, and oxidation of a 10 wt.% zyttrite-Si_3N_4 Ceramic. J. Am. Cer. Soc. 67(2):89–92, 1984.
280. J.T. Smith, C.L. Quackenbush. Phase effects in Si_3N_4 containing Y_2O_3 or CeO_2: I, strength. Am. Cer. Soc. Bul. 59(5):529–537, 1980.
281. F.F. Lange. Ce_2O_3-Y_2O_3-SiO_2 materials: phase relations and strength. Am. Cer. Soc. Bul. 59(2):239–249, 1980
282. A. Tsuge, K. Nishida. High strength hot-pressed Si_3N_4 with concurrent $Y_{2>}O_3$ and Al_2O_3 additions. Am. Cer. Soc. Bul. 57(4):424–427, 1978.
283. A.J. Pyzik, D.F. Carroll. Technology of self-reinforced silicon nitride. In: Annu. Rev. Mat. Sci. 24, eds. Pala Alto: An. Revs. Inc. 1994, pp. 189–214.
284. A. Kuzukévičs, K. Ishizaki. Sintering of silicon nitride with $YAlO_3$ additive. J. Am. Cer. Soc. 76(9):2373–2375, 1993.
285. C.J. Hwang, R.A Newman. Silicon nitride ceramics with celsian as an additive. J. Mat. Sci. 31:150–156, 1996.

286. A.D. Stalios, J. Luyten, C.D. Hemsley, F.L. Riley. The interaction of iron during the hot-pressing of silicon nitride. J. Eur. Cer. Soc. 7:75–81, 1991.
287. P. Popper. Sintering of silicon nitride, a review. In: F.L. Riley, ed. Progress in Nitrogen Ceramics. Boston: Martinus Nijhoff Pub., 1983, pp. 187–210.
288. I.T. Ostapenko, N.F. Kartsev, R.V. Tarasov, V.P. Podtykan. Influence of CaF_2 on compaction kinetics of Si_3N_4 + 10 wt. % MgO during hot pressing. Iorg. Mat. 16:587–591, 1979.
289. J.R. Oswald, F.L. Riley, R.J. Brook. Accelerated densification of silicon nitride using a fluoride flux. Br. Cer. Trans. & J. 86:81–84, 1987.
290. Z.-K. Huang, A. Rosenflanz, I.-W. Chen. Pressureless sintering of Si_3N_4 ceramic using AlN and rare-earth oxides. J. Am. Cer. Soc. 80(5):1256–1262, 1997.
291. J.A. Palm, C.D. Greskovitch. Thermomechanical properties of hot-pressed $Si_{2.9}Be_{0.1}N_{3.8}O_{0.2}$ Ceramic. Am. Cer. Soc. Bull. 59(4):447–452, 1980.
292. N. Ucida, M. Koizumi, M. Shamada. Fabrication of Si_3N_4 ceramics with metal nitride additives by isostatic hot-pressing. J. Am. Cer. Soc. 68(2):C-38–40, 1985.
293. O. Abe. Sintering of silicon nitride with alkaline-earth nitrides. Cer. Intl. 16:53–60, 1990.
294. R.W. Rice. Ceramic fracture mode-intergranular vs. transgranular fracture. In: J.R. Varner, V.D. Frechette, G.D. Quinn, eds. Ceramic Transactions. 64. Fractography of Glasses and Ceramics III. Westerville, OH: Am. Cer. Soc., 1–53, 1996.
295. R.W. Rice. Mechanical properties of ceramics and composites, grain and particle effects. New York: Marcel Dekker, Inc., 1999.
296. R.W. Rice, K.R. McKinney, C.Cm. Wu, S.W. Freiman, W.J. McDonough. Fracture energy of Si_3N_4. J. Mat. Sci. 20: 1392–1406, 1985.
297. C. Greskovitch, G.D. Quinn. Thermomechanical properties of a new composition of sintered Si_3N_4. Am. Cer. Soc. Bull. 63(9):1165–1170, 1984.
298. T. Yamada, M. Shimada, M. Koizumi. Fabrication and characterization of titanium nitride by high pressure hot pressing. Am. Cer. Soc. Bul. 59(6):611–616, 1980.
299. M. Moriyama, K. Kamata, Y. Kobayashi. Mechanical and electrical properties of hot-pressed TiN ceramics without additives. J. Cer. Soc. Jap. Intl. Ed. 99:275, 1991.
300. M. Fukuhara, T. Mitsuda, Y. Katsumaura, A. Fukawa. Sinterability and properties of $Ti(N_{1-x}, O_x)$y-(V, Ta)C-Ni sintered alloys having a golden colour. J. Mat. Sci. 20:710–717, 1985.
301. M. Moriyama, H. Aoki, Y. Kobayashi, K. Kamata. The mechanical properties of hot-pressed TiN ceramics with various additives. J. Cer. Soc. Jap. Intl. Ed. 101:271, 1993.
302. A. Kamiya, K. Nakano. Effect of aluminum addition on TiN hot-pressed sintering. J. Mat. Sci. Let. 14:1789–1791, 1995.
303. O.V. Pshenichnaya, P.S. Kislyi. Influence of ceramic additives on Rrecrystallization of molybdenum disilicide. Inorg. Mat. 15:64–66, 1979.
304. Y. Suzuki, P.E.D. Morgan, K. Niihara. Improvement in mechanical properties of powder-processed $MoSi_2$ by the addition of Sc_2O_3 and Y_2O_3. J. Am. Cer. Soc. 81(12):3141–3149, 1998.
305. K. Uematsu, K. Sawada, Z. Kato, N. Uchida, K. Saito. Effect of additives on the hot pressing of zinc sulphide. J. Mat. Sci. Let. 7:473–474, 1988.

306. W. Stadlbauer, W. Kladnig, G. Gritzner. Al_2O_3-TiB_2 composite ceramics. J. Mat. Sci. Let. 8:1217–1220, 1989.
307. R.A. Cutler, A.C. Hurford, A.V. Virkar. Presureless-sintered Al_2O_3-TiC composites. Mats. Sci. & Eng. A105/106:183–192, 1988.
308. Ki-W. Chae, D.-Y. Kim, B.-C. Kim, K.-B. Kim. Effect of Y_2O_3 additions on the densification of an Al_2O_3-TiC composite. J. Am. Cer. Soc. 76(7):1857–1860, 1993.
309. Ki-W. Chae, D.-Y. Kim, K. Niihara. Sintering of Al_2O_3-TiC composite in the presence of liquid phase. J. Am. Cer. Soc. 78(1):257–259, 1995.
310. D.M. Karpinos, E.F. Mikhashchuck, R.A. Amirov, U.Sh. Shayakhmetov. Physiochemical processes occurring in nitride and oxide-nitride composites with phosphate binders during heating. Pwd Met. 21:388–391, 1982.
311. T. Ekström. Alumina ceramics with particle inclusions. J. Eur. Cer. Soc. 11:487–496, 1993.
312. S. Torizuka, J. Harada, H. Yamamoto, H. Nishio, A. Chino, Y. Ishibashi. Effects of SiC addition on the mechanical properties and sinterability of TiB_2-(2 mol%Y_2O_3-ZrO_2) composite. J. Cer. Soc. Jap. Intl. Ed. 100:685, 1992.
313. G.J. Zhang, Z.Z. Jin, X.E. Yue. Effects of Ni addition on the mechanical properties of TiB_2/SiC composites prepared by reactive hot pressing (RPH). J. Mat. Sci. 32:2093–2097, 1997.
314. S. Hayashi, Y. Kobayashi, H. Saito. TiB_2-B_4C composites presureless-sintered using Ni and C as densification aids. J. Cer. Soc. Jap. Intl. Ed. 101:149, 1993.
315. J.Y. Dai, D.X. Li, H.Q. Ye, G.J. Zhang. Study of $Ni_{31}Si_{12}$ intergranular phase in Ti(CN)-TiB_2-Ni ceramics. J. Mat. Sci. Let. 13:790–792, 1994.
316. K.-S. Cho, Y.-W. Kim, H.-J. Choi, J.-G. Lee. SiC-TiC and SiC-TiB_2 composites densified by liquid-phase sintering. J. Mat. Sci. 31:6223–6228, 1996.
317. K.-S. Cho, H.-J Choi, J.-G. Lee, Y.W. Kim. In situ-toughened SiC-TiB_2 composites. J. Mat. Sci. 1998.
318. B.-W. Lin, T. Iseki. Effect of thermal residual stress on mechanical properties of SiC/TiC composites. Brit, Cer. Trans. J. 91:1–5, 1992.
319. H. Endo, M. Ueki, H. Kubo. Microstructure and mechanical properties of hot pressed SiC-TiC composites. J. Mat. Sci. 26:3769–3774, 1991.
320. R.-R. Le, W.-C. Wei. Fabrication, microstructure, and properties of SiC-AlN ceramic alloys. Cer. Eng. & Sci. Proc. 11(7-8):1094–1121, 1990.
321. Yu-B. Pan, J.-H. Qui, M. Morita, S.-H. Tan, D. Jiang. The mechanical properties and microstructure of SiC-AlN particulate composite. J. Mat. Sci. 33:1233–1237, 1998.
322. K.S. Mazdiyasni, R. Ruh, E.E. Hermes. Phase characterization and properties of AlN-BN composites. Am. Cer. Soc. Bul. 64(8):1149–1154, 1985.
323. D.M. Karpinos, E.P. Mikhashchuk, R.A. Amirov, U.Sh. Shayakhmetov. Physicochemical proceses occurring in nitride and oxide-nitride composites with phosphate binders during heating. Sov. Pwd. Met. & Met. Cers. 21:388–391, 1982.
324. T. Mah, M.G. Mendiratta, H.A. Lipsitt. Fracture toughness and strength of Si_3N_4-TiC composites. Am. Cer. Soc. Bul. 60(11):1229–1240, 1981.
325. K.S. Mazdiyasni, R. Ruh. High/low modulus Si_3N_4-Bn composite for improved electrical and thermal shock behavior. J. Am. Cer. Soc. 64(7):415–419, 1981.
326. H.R. Zhuang, W.L. Li, J.W. Feng, Z.K. Huang, D.S. Yan. Si_3N_4-AlN Polytypoid Composites by GPS. J. Eur. Cer. Soc. 7:329–333, 1991.

327. H. Tanaka, P. Greil, G. Petzow. Sintering and strength of silicon nitride-silicon carbide composite. Intl. J. High Temp. Cer. 1:107–118,1985.
328. J.-Y. Kim, T. Iseki, T. Yano. Pressureless sintering of dense Si_3N_4 and Si_3N_4/SiC composites with nitrate additives. J. Am. Cer. Soc. 79(10):2744–2746, 1996.
329. D.-S. Cheong, K.-T. Hwang, C.-S. Kim. High-temperature strength and microstructureal analysis in Si_3N_4/20-vol% -SiC nanocomposites. J. Am. Cer. Soc. 82(4):981–996, 1999.
330. H. Park, H.-W. Kim, H.-Ee Kim. Oxidation and strength retention of monolithic Si_3N_4 and nanocomposite Si_3N_4-SiC with Yb_2O_3 as a sintering aid. J. Am. Cer. Soc. 81(8):2130–2134, 1998.
331. G. Hillinger, V. Hlavacek. Direct synthesis and sintering of silicon nitride/titanium nitride composite. J. Am. Cer. Soc. 78(2):495–496, 1995.
332. J.J. Petrovic, M.I. Pena, H.H. Kung. Fabrication and microstructures of $MoSi_2$ reinforced-Si_3N_4 matrix composites. J. Am. Cer. Soc. 80(5):1111–1116, 1997.
333. G.-J. Zhang, X.-M. Yue, T. Watanabe. Addition effects of aluminum and in situ formation of alumina in $MoSi_2$. J. Mat. Sci. 34:997–1001, 1999.
334. J.-M. Ting. Sintering of silicon carbide/molybdenum disilicide composites using boron oxide as an additive. J. Am. Cer. Soc. 77(10):2751–2752, 1994.
335. Z. Zakhariev, D. Radev. Properties of polycrystalline boron carbide in the presence of W_2B_5 without pressing. J. Mat. Sci. Let. 7:695–696, 1988.
336. K. Shobu, T. Watanabe, K. Tsuji. Effects of Mo_2B_5 addition to $MoSi_2$ ceramics. J. Cer. Soc. Jap. Intl Ed. 97:1309–1312, 1989.
337. C. Greskovich. The effect of small amounts of B and Sn on the sintering of silicon. J. Mat. Sci. 16, 613–619, 1981.
338. C.-S. Hwang, Z.-e Nakagawa, K. Hamano. Microstructure and mechanical properties of TiO_2-added alumina ceramics. J. Jap Cer. Soc. 94(8):761, 1986.
339. R.W. Rice. Strength and fracture of hot-pressed MgO. Proc. Brit. Cer. Soc. No. 20:329–363, 1972.
340. J.D. Hodge, R.S. Gordon. Grain growth and creep in polycrystalline magnesium oxide fabricated with and without LiF additive. Cer. Intl. 4(1):17, 1978.
341. F.K. Volynets, G.N. Dronova, L.V. Udalova. Influence of lithium fluoride on the electrical conductivity of magnesium oxide doped with lithium fluoride. Inorg. Mat. 8:343–344, 1972.
342. L.V. Udalova, L.A. Kiseleva, I.V. Kurova. General features of compaction of powders of certain lithium fluoride-doped powders. Inorg. Mat. 16:1347–1352, 1980.
343. S. Ordoñez, I. Iturriza, F. Casrto. The influence of amount and type of additive on α-β Si_3N_4 transformation. J. Mat. Sci. 34:147–153, 1999.

6

Other General Densification and Fabrication Methods

6.1 INTRODUCTION

While powder consolidation and pressureless sintering dominate the production of many ceramic products, there are other densification and fabrication methods that are important. These include several fabrication methods that generally have broad applicability, or the potential for it, and those that are more specialized, such as fabrication of fibers or designed porosity. Though the division between these two areas is sometimes uncertain, those processes deemed falling more in the latter category are addressed in the following chapter. Those generally broader applicable processes addressed in this chapter include pressure sintering of powders via hot pressing or hot isostatic pressing, both of which are based on powder processing and use of compaction pressure during sintering, not just before sintering, and have considerable production use—especially hot pressing. There is also press forging of powder compacts, which has had some laboratory demonstration, and press forging of single crystals to polycrystalline bodies, which has had some production use. Another processing method that is also typically based on consolidation of powders, commonly, but not universally used for producing ceramic composites, is reaction processing of powder constituents with themselves or a gaseous media, or an added, or induced, liquid phase.

Other important densification and fabrication methods deviate significantly from traditional powder-based processing. These include polymer pyrolysis, that is the conversion of a preceramic polymer to a ceramic body, which is important for ceramic fiber processing, but also has other possible applications. Chemical vapor deposition (CVD) is another important process used for substantial industrial production of dense monolithic ceramics and also having substantial potential for ceramic composites and some porous bodies. Another major alternate fabrication method for ceramics and to some extent for ceramic composites is melt processing, which has several important technical and industrial manifestations. Melt processing produces the largest ceramic bodies of polycrystalline refractories and glass-based bodies (e.g. for telescope mirrors), as well as the largest volume of ceramic (glass) products.

The basic aspects of these processes are described in this chapter in the general order listed above. Limitations, various manifestations, and opportunities for further use are addressed. Readers should also note that some aspects of these processes are discussed in the following chapter in conjunction with specialized fabrication approaches, for example for ceramic fibers and designed porous bodies.

6.2 HOT PRESSING

6.2.1 Practice and Results

Hot pressing is basically a direct extension of normal sintering where powder is consolidated at or very near room temperature then sintered at higher temperatures without pressure, in that hot pressing uses application of uniaxial compaction at elevated temperatures to enhance sintering (1–6). The primary advantages of hot pressing are easier, faster achievement of at or very near theoretical density at lower temperature than sintering, e.g., by 100–200°C less. This in turn results in better properties: for example, transparency, higher conductivity, improved mechanical properties, and often increased reliability. The disadvantages are mainly higher costs, commonly due primarily to limited shape capabilities requiring more, commonly expensive, machining for many applications, as well as some other cost factors (discussed below). There are also some material limitations due to reaction or reduction. However, despite these issues, hot pressing has progressed over the last 40 years or so from an almost exclusively laboratory method to one with substantial industrial production and provides opportunities for further development and use, as discussed below.

A hot press consists of a press frame and a furnace around the die that provides sufficient insulation to the press system, with management of heat losses up the top (and typically the only moving) ram managed by selection of materials for sections of the ram. Hot pressing has typically been done by filling the die cavity with loose powder without any binder (but possibly with densification or other additives, Chap. 5), often followed by cold pressing in the die before hot pressing. In limited cases, a green body from other die pressing or alternate consolidation has been inserted in the hot pressing die, which is one area for further development of hot pressing as discussed further below—but for now, hot pressing of a loose powder fill or cold-pressed powder with no organic binders is addressed. Such powder is hot pressed by heating the die and its powder content to the hot pressing temperature with the uniaxial hot pressing pressure being applied, often in a graduated fashion at the hot pressing temperature or somewhat below it. Details of the pressure application depend on the material and powder, and possibly the amount of powder, driven mainly by adequate outgassing of the powder, as discussed further below. The system is held at the hot pressing temperature and pressure for periods of a few minutes to a few hours, commonly 0.5 to 2 hr depending on material and powder characteristics. The uniaxial hot pressing pressure is released at the hot pressing temperature when desired, typically full, densification has been achieved (indicated by ram travel) or on initial cooling since maintaining pressure during cooling can cause cracking. Hot pressing temperatures are typically inversely proportional to the hot pressing pressures and to some extent to the time at hot pressing temperature. Hot pressing pressures are typically a function of pressing temperatures, allowed die materials and sizes, but can become press limited in hot pressing large parts as discussed further below.

Die material selection and preparation is a critical factor in hot pressing, since the material selected is a major factor in the temperature, pressure, material, and atmosphere capabilities of the hot pressing. The basic capabilities of the die material are important, as are its compatibility with both the material to be hot pressed and the atmosphere in which this can be suitably done, but can be extended some via effects of interface or coating materials on the die body or die components, that is, rams and spacers (Fig. 6.1). The predominant choice for die materials are various graphites due to basic temperature capabilities (to ~ 3000°C), reasonable compatibilities with many ceramics (and extension of this via use of die liners and coatings), and acceptable to good properties for much hot pressing. The large die sizes available, and their reasonable costs (due in part to use of conventional machining) for many graphites are also factors in their use. Various grades of commercial graphites provide selection from a range of performance and cost factors, ranging from coarser microstructures (often anisotropic) to finer microstructures, isotropic (e.g., POCO) graphites. These graphites differ in thermal expansion (which must be adequately accounted for in die design) and in performance and cost, with the former being more moderate and latter, higher performance and cost. The bulk of hot pressing, especially industrially, is done with lower cost graphites, which allow use at pressures of the order of 35 MPa in laboratory pressing, but is commonly used at pressures of the order of 15 MPa industrially. Higher performance, e.g., isotropic, graphites can be used to pressures of 70 or more MPa. However, op-

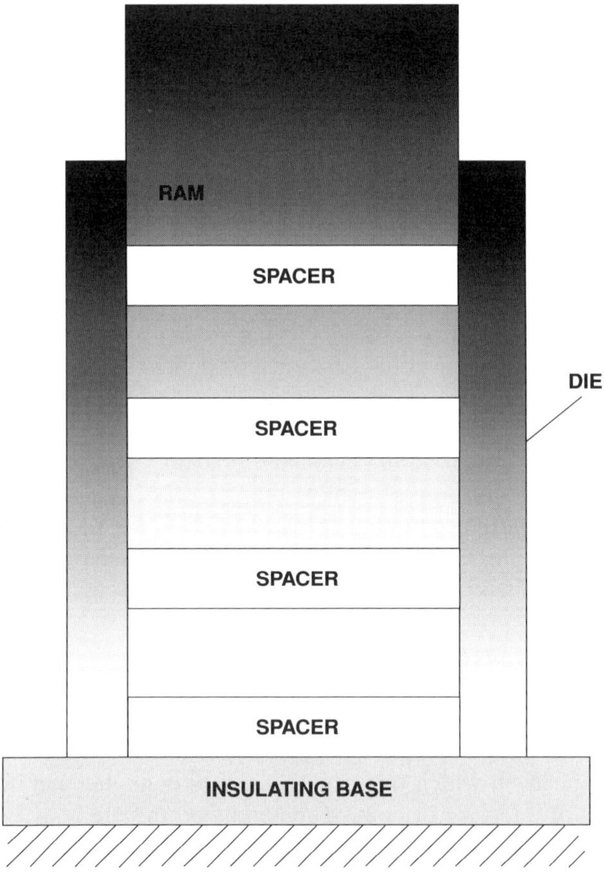

FIGURE 6.1 Schematic of a hot pressing die with three parts and associated spacers. Note that an insulating spacer (not shown) may also be used on the top ram. Press frame and heating system not shown.

eration of graphites near their higher pressure/stress capabilities is more demanding in pressure train alignment to avoid die, and especially ram failure and attendant losses. (Other factors such as permeability for binder burnout for some graphite hot pressing dies are discussed below.) More recently, some, mainly smaller, dies have been made using carbon-carbon composite sleeves with a suitable graphite liner. These can be operated at higher pressures, give longer life at normal pressures, or some combination of these. Though dependent on several factors as noted above, representative hot pressing temperatures for some common oxides are shown in Table 6.1. Note that corresponding tem-

TABLE 6.1 Representative Hot Pressing Temperatures for Common Oxide Ceramics[a]

Single oxide	Pressing temperature (°C)	Mixed oxide	Pressing temperature (°C)
Al_2O_3	1300–1500	$BaTiO_3$	1100–1200
BeO	1600–1800	PZT	1000
CaO	1200	$PbTiO_3$	1200
CaO (+ LiF)	1000	Mn-Zn Ferrite	1350
MgO	1200–1400	$MgAl_2O_3$	1400–1600
MgO (+LiF)	1100	$MgAl_2O_3$ (+LiF)	1200–1300
TiO_2	1200–1300		

[a]Specific temperatures will vary with the specific powder, actual pressure, and time at temperature but representative temperatures are shown as an approximate guide, e.g., for pressures of the order of 35 MPa and times of 0.5 hr for common powders.

peratures for common refractory borides, carbides, and nitrides, with typical additives are commonly in the range to 1700–1900°C.

Other die materials used have been refractory metals such as W or Mo (or alloys such as TZM) and some ceramics, such as Al_2O_3 or SiC, mainly made via hot pressing [2,3, 6,7]. These are restricted by both fabrication limitations and costs to smaller dies, for example, to cavities of 5 -cm diameter. In using ceramic and, especially, metal dies, cautions such as looser ram-die clearances or use of coatings to inhibit or prevent ram-die sintering or welding are often needed. Another limitation is creep of the die; even small amounts of additives or impurities that enhance creep, such as SiO_2-based impurities, e.g., even at the 0.1% level can seriously limit temperature/pressure capabilities—that is, 99% alumina may not be adequate. SiC dies have been commercially produced (by hot pressing) for commercial production of ferrite components in air or other atmospheres (since hot pressing in graphite dies is not acceptable due to reduction of the ferrites). Various high-pressure systems, including those used for making synthetic diamonds (which use special systems consisting of massive metal dies with small volumes for product, internal heater, and electrical insulation), have also been applied to hot pressing dense ceramics. While bodies of NiO, Cr_2O_3, and Al_2O_3 at or near theoretical density with grain sizes < 1μm have been obtained in massive metal dies at temperatures of ~ 800–1200°C with pressures of ≤700 MPa [8], they often have strengths much below those expected for their fine grain size. This is attributed to retention of anion species at grain boundaries [3,9], as discussed in Section 8.2.1.

Before proceeding with more description of conventional hot pressing, two other related processes for densifying metals, that have had some application to ceramics, should be noted. The first is high-rate mechanical compaction by a metal ram accelerated to high speeds by expanding gas, such as a Dynapak system[1,5]. This was applied to some ceramics, particularly UO_2-PuO_2 nuclear fuel materials, which were sealed in a metal can, evacuated while heating to 1200°C in a separate furnace, sealed off, and rapidly loaded in the massive metal, high-pressure die, then impacted, giving a transient high consolidation pressure, 1750 MPa (250,000 psi). In this case the fuel went from 55% to 99% of theoretical density. Cracking from rapid cooling was not a particular problem since the densified body was crushed to produce dense grain for swaging in metal tubes for fuel rods. Tests with other ceramics were generally less successful due to incomplete densification, cracking, or both: for example, trials with alumina, thoria-alumina composite, and thoria gave ~85, 90, and 95% of theoretical density, respectively, at ~ 1200°C with ~ $^1/_2$ million psi impact pressure. The other related process is explosive compaction [5,6], which is normally done at room temperature, and can be done at elevated temperatures, but still generally results in severe cracking in most ceramics; though cracking is reduced some at higher temperatures, it can still be a problem. The limitations of both processes have constrained their use and development.

Graphite and metal dies are generally used in neutral or nonoxidizing conditions, but are sometimes used in an atmosphere open to the air, especially graphite dies, with some surface insulation to limit oxidation rates of the outer surfaces of the die and ram, and hence retain reasonable die lives. Graphite dies used open to the air maintain a reducing atmosphere determined by the CO-CO_2 balance in and around the die as determined by operating conditions such as temperature. Such use of metal dies exposed to oxidative atmospheres is often restricted by possible detrimental internal oxidation and bonding of dies and rams. Again, coatings or interface layers may maintain reasonable die lives and performance (i.e., avoiding oxidative bonding of die, ram, or components with one another).

Having selected the material to be pressed and the dies to be used, and hence the allowable temperature and pressure ranges, specific pressing temperature and related parameters of pressure and time are selected based on factors such as prior experience, trial pressings, or both. Such information can be systemized to give a probable pressing temperature for a specific material-powder system by one of two methods [1,4,10]. The first, referred to as the *isothermal technique* requires several hot pressing runs in which the time to reach the target density, e.g., 99% or more of theoretical density, at the selected pressure is plotted versus the pressing temperature of each hot pressing run. This generally results in the time to reach the target density with the selected pressure decreasing substantially and linearly with increasing pressing temperature followed by much less decrease in time as temperature is further increased. The target pressing temperature is that where the two linear curves intersect (Fig. 6.2A). The second technique, referred to as the *climbing temperature program*, requires fewer trial pressings and is done by heating the body, under the selected pressure, slowly through the regions of densification until densification essentially ceases

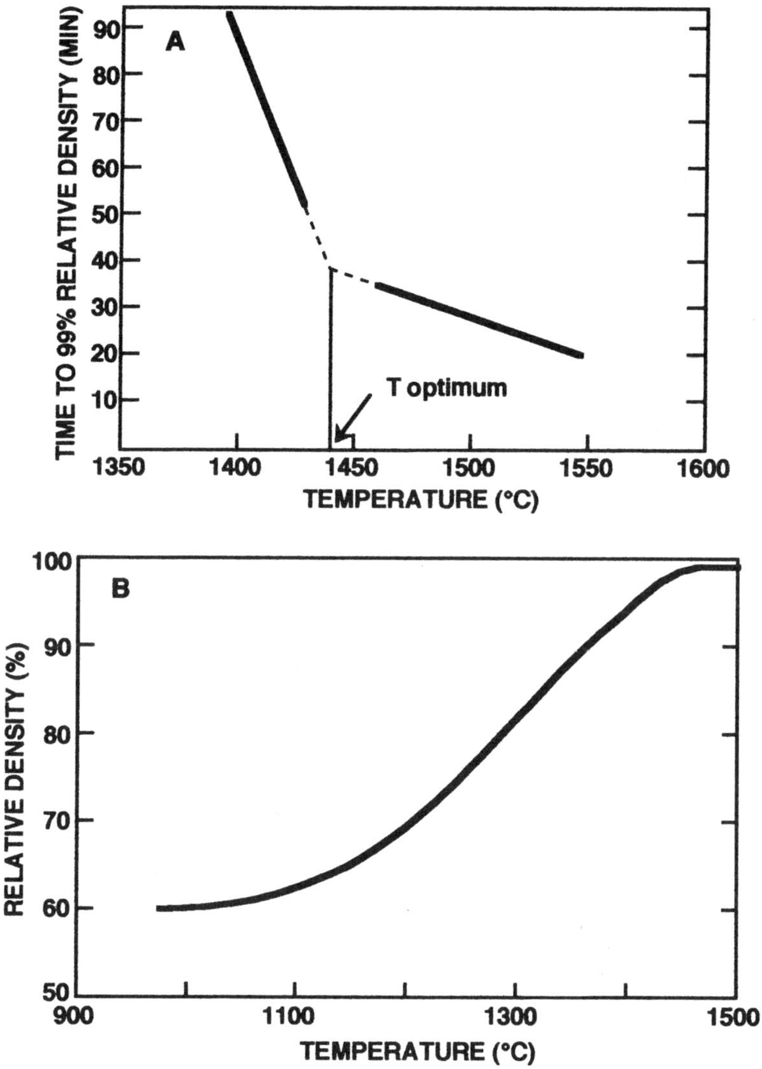

FIGURE 6.2 Schematic of data analysis to determine target hot pressing temperatures: (A) isothermal and (B) rising temperature.

(Fig. 6.2B). A few trial runs should map out the temperature and pressure parameters for pressing the selected material and powder. A newer approach becoming available is based on universal hot pressing curves analogous to universal sintering curves developed by Johnson [11].

An issue in much hot pressing is whether to do so under vacuum or to evacuate an enclosed hot pressing system, then backfill with a selected, usually, nonoxidizing atmosphere. This is most extensively done in laboratory hot pressing, and while used in some industrial hot pressing, is much less common since near-zero porosity bodies generally can be achieved without vacuum hot pressing. This results from gas in the powder compacts being expelled, at least in small-to medium-sized bodies, by thermal expansion as well as by mechanical pressure, which also reduces sizes of residual gas-filled pores, whose pressure is greatly reduced upon cooling from hot pressing. The extent to which residual gas and related pore gradients start to occur in the interior of large hot-pressed bodies appears unknown. Vacuum hot pressing may thus be more important for larger bodies, bodies with larger initial pores, and for other reasons—for example, to remove other sources of outgassing such as binders (discussed below).

The issue of heating for hot pressing should also be noted. Most laboratory hot presses use furnaces with resistive heating elements, which are typically carbon/graphite or refractory metal elements, thus also requiring nonoxidizing atmospheres. On the other hand, most industrial hot presses use inductive heating with coupling to a graphite suceptor or more commonly directly to the graphite dies. Such inductive heating also favors use of nonoxidizing atmospheres, but as noted above, can often be done with some system exposure to the air atmosphere. The differences in rates and uniformity of heating differ significantly between the two heating systems, with induction heating allowing faster heating rates, hence shorter hot pressing cycles. Though not directly studied, it is likely that inductive heating, especially via direct coupling to the die, is more uniform. However, it should be noted that alternative heating systems have shown promise for at least some hot pressing (and some sintering) that deserve attention as discussed in the next section.

Determining actual component temperatures in hot pressing is a challenging problem since thermocouples often become inoperative, and where operative, typically must be some distance from the actual part(s). Optical pyrometer temperature measurements, while improved in some ways, have similar or greater uncertainties in their specific proximity to a component in the die. Thus, hot pressing "temperatures" for one size and shape component versus another of the same material may differ substantially due to differences in thermal environments and temperature sensing in the dies, and especially from one hot press to another. The latter differences can be 100–200°C, so operating parameters for a specific component, and especially a specific hot press, are typically refined by operational trials.

Hot-pressed ceramics are briefly characterized in terms of three aspects: (1) the materials, (2) their microstructures and properties, and (3) their numbers, shapes, and sizes. Hot-pressed ceramics consist of an extensive and diverse array of materials including a variety of oxide and nonoxide ceramics, many of these, especially nonoxides, with densification additives (Chap. 5). Oxides and other compounds that are subject to reduction or other stoichiometric changes in common hot pressing environments may require post pressing annealing or other treatment to return them to suitable stoichiometry, but opportunities for this are limited, especially for larger parts. While pressureless sintering of several important refractory nonoxides has progressed substantially, there is still important impetus for hot pressing these and other nonoxides. The increased interest in ceramic composites with dispersed particulate, platelet, whisker, or fiber phases has further added to the use of hot pressing since these are more challenging to densify by pressureless sintering, especially the latter three (Sec. 8.2.3). Thus, all of the commercial production of composites of SiC whiskers in alumina matrices is by hot pressing, and most fiber composites have been made by hot pressing.

Turning to microstructure, hot pressing commonly gives bodies at or near zero porosity,:for example, frequently hot pressing gives translucent to transparent bodies for suitable dielectric materials, with finer, often more equiaxed, grain structures than typically obtained from pressureless sintering the same material to within a few percent of the same density. Thc typically resultant higher transparency, and mechanical, electrical, thermal, and related performance is, in fact,a primary driving force for hot pressing. However, though dependent on material, powder, and pressing conditions, it should also be noted that hot pressing often results in some measurable degree of anisotropy in properties due to some crystalline preferred orientation, anisotropy of residual porosity, or both. Varius studies of anisotropy of hot-pressed silicon nitride bodies shows there is commonly some anisotropy in commercial hot-pressed bodies and that this arises from effects of both grain orientation and anisotropy of residual pores [9]. An earlier reference for some of the more comprehensive data on the crystallographic orientation aspect is the report of Iwasaki and coworkers [13], and Rice[9] has summarized some more recent results.

The almost universal output of hot pressing is one or a few pieces of very basic shapes such as cylindrical or prismatic rods or plates. Where more than one of these is produced in a given hot pressing run, it is primarily by use of spacers to allow two to four other identical parts to be produced above or below one part (Fig. 6.1). This requires that the die be large enough and have sufficiently uniform heating along its length and diameter, and that the powder used has reasonable pour/tap densities. On the other hand, hot pressing can produce sizable parts—$75 \times 40 \times \geq 0.1$ cm and $45 \times 45 \times 20$ cm—and parts to nearly a meter in diameter are seen as feasible. (R. Palicka, Cercom, Inc., personal communications, 2000). However, such large parts require very slow cooling to avoid cracking, and may impose maximum dimensions—for example 2-–3 m to avoid cracking. They also present challenges in filling such large size die cavities uniformly and reproducibly. Hot pressing of rods has been shown to be feasible in a semicontinuous fashion by periodic addition of powder and pressing it onto a rod in steps[15,16], but this is probably of limited practicality for most purposes. However, further development to extend the technical capabilities and lower costs of hot pressing are feasible as outlined below, after first considering approximate limits on the size of hot pressed parts.

Another basic size limit of hot-pressed parts arises from impacts of lateral part dimensions being ultimately limited first by capital costs (press size) and second by operating costs. The force needed to supply a fixed pressing pressure increases as the square of the part area being pressed; for example it reaches 1600 tons at a meter square part size for 14 MPa pressing pressure (~2000 psi), that is a sizeable press and hydraulic system. Further, the sizes of the cross-members (beams) of the press sustaining such forces significantly increase with part lateral dimensions. Though I or truss beams are likely candidates for cross members, consideration of a solid rectangular beam is illustrative of increased cross member bulk with increasing cross member length for pressing larger parts. Thus, in order to keep the same center beam deflection and pressing pressure as the scale of a press increases, the width of a solid beam or its thickness (or both) must increase as the beam length is increased to press larger parts. For example, if the beam length is doubled, increasing its thickness to keep the same deflection and pressing pressure results in a minimum of about a six fold increase in beam volume, hence mass. Part heights are limited first by the compactibility of the powder and heights of dies and of uniform heating, as well as by die-wall friction effects and resultant densification gradients (see Sec. 4.2.1), which are functions of material and part aspect ratio. Part dimensions are also ultimately limited by heat-transfer limitations within the part, possibly in some cases by heating and inadequate temperature uniformity for suitably uniform densification, but more commonly by cooling stresses, especially when they lead to cracking, which is also material and part shape dependent. Note also that large hot-pressed parts require powder loading systems that can either load the powder uniformly in the die in a practical fashion or suitably load one or more powder preforms in the die, as well as unload the parts. The mass of large parts becomes a factor in the engineering for such part handling; for example typical ceramic parts of 1 m^2 x 1 cm dimensions will weigh 200–400 pounds.

Consider now part shape: Some deviations from simple rods or plates is already feasible, including some pressing of parts with some simple holes or cavities. Also, hot pressing of silicon nitride turbine vanes with a two-rather than a three-dimensional variation of shape have been shown to be feasible for net shape pressing and at costs less than for injection molding and sintering of limited numbers—for example, a few thousand per year where there is still a substantial penalty for each vane for the injection molding die costs. More complex shapes, such as hemispherical domes [17] or ogive radomes, have been hot pressed with specialized tooling, and some potential has been shown for hot pressing more complex shapes by using a refractory powder to apply a quasi-isostatic pressure from the die rams and constraint of the die wall [18]. However, such advances (discussed below) are mainly applicable to specialized, niche, production, rather than lowcost, largescale production.

6.2.2 Extending Practical Capabilities of Hot Pressing

Consider now increasing the number of parts hot pressed per unit time, which can be done by reducing the pressing cycle time, by increasing the number of parts in a given pressing, or both. Simple engineering, which of course depends on both the part size and the die size feasible, can do quite a bit of this given the sizes of dies feasible for pressing parts approaching a meter in diameter and at least half a meter in height. A horizontal hot press was built to accommodate insertion and removal of dies into and from a horizontal chain of dies in the pressing train[19]. Typically there would be three loaded dies in the train at any one time, one at the input side being heated for pressing, the middle die being in the hot zone for actual hot pressing, and the die at the output end cooling to be removed when the middle die is done pressing and a new die is inserted in the input side of the train, moving the two remaining dies to their next station in the press. While used some and showing potential for increased output, this semicontinuous hot press presented limitations due to its horizontal nature, requiring the same or very similar size dies and powder loads, and presenting issues of control of pressure on the die being heated up and its heating, as well as possible problems of the part(s) in the die being cooled.

More promise in reducing the times for a pressing cycle for a given hot press, which consist of a heating, a pressing, and a cooling stage, is seen by moving heating and die assemblies through a vertically oriented press frame such that the primary or only time each heating and die assembly spends in the press frame is for actual heating and pressing portions of the hot pressing cycle. Thus, one heating and die assembly can be being heated while one is cooling, another is being loaded and another unloaded, with such assemblies moved in and out of the press frame. Such movement of heating and die assemblies has been accommodated by having two of them on metal wheels being moved back and forth in and out of the press frame, or having two or more assemblies in a lazy Susan arrangement. This approach avoids the limitations of the horizontal semicontinuous hot press noted above.

While the above use of two or more heating and die assemblies for a given press frame can reduce costs of hot-pressed parts, the most significant cost reductions should result from using large dies and heating-pressing systems such that substantial numbers of parts are pressed simultaneously in one die in one pressing cycle. This requires that parts not only be commonly stacked vertically (using spacers, Fig. 6.1) but that multiple stacks arranged in horizontal arrays be used, such as using multicavity dies. Some use of such dies has been made, but much more extensive use requires a fundamental shift in die loading procedures, which also has other benefits. As noted above, the standard procedure for loading powder into dies has been to simply pour powder into the die cavity. However, this is labor intensive, and has poor reproducibility of die fill needed to obtain uniform powder filling and consolidation in parallel stacks of powder allotments for components. Such powder loading of multicavity dies may also introduce defects, as shown by efforts of an industrial developer of ceramic bearing components (who demonstrated success in such multicavity die pressing, as have others). However, greater rates of failure of silicon nitride bearing balls from multicavity hot pressing of individual ball blanks was found versus from balls made from blanks machined from large silicon nitride billets. This higher ball failure rate was traced to greater inclusion of graphite particles introduced in the powder filling stage for multicavity pressing due to the large number of graphite edges from which graphite particles could readily be abraded and entrained by powder loading. The potential solution to this problem, namely of using ball-blank preforms to be loaded into the die was not considered, and the hot pressing of individual ball blanks was abandoned. However, subsequent developments in other aspects of hot pressing indicate substantial potential for hot pressing of green-formed parts as outlined as follows.

The first of two examples of significant use and potential of green forming of parts for hot pressing, not only demonstrated green forming of sophisticated electronic ceramic parts, but also showed a significant new opportunity for hot pressing. This was the conception and demonstration by Rice and coworkers [20--22] that first electronic substrates and subsequently multilayer electronic packages made by green forming, with cofireable metalization, could be successfully and advantageously densified by hot pressing (Fig. 6.3). Thus, they showed that conventional as well as large multichip module electronic packages could be hot pressed to high quality—better than by sintering—with significant advantages. These included higher densities, such as transparent AlN, but more importantly with much greater flatness and lateral dimensional control than in pressureless sintering. Such dimensional advantages arises since there is no lateral shrinkage in hot pressing (it is all in the axial pressing direction) versus the nearly isotropic (20% linear) shrinkage in sintering. This lack of lateral shrinkage is a major advantage in this packaging (and some other) applications both for customers that require a high degree of dimensional control, as well as having the ability to surface metallize with thin film metallization. The high density of

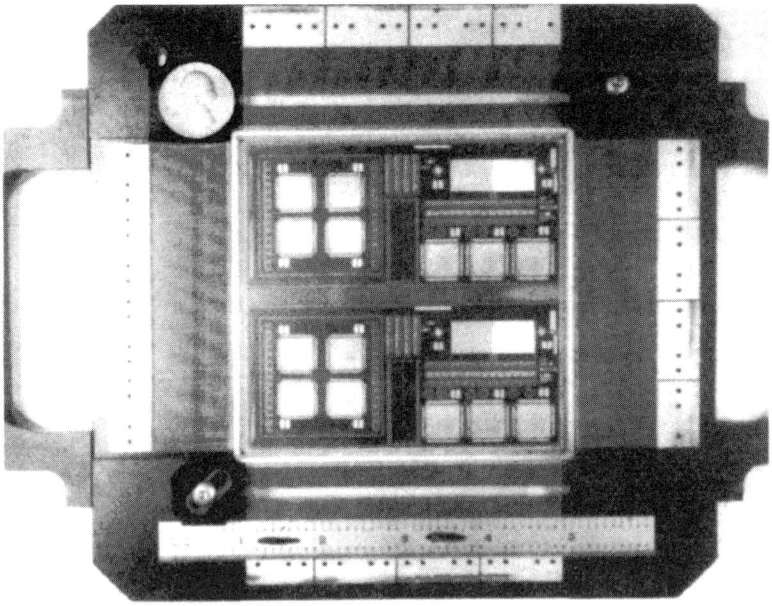

FIGURE 6.3 Multilayer AlN electronic package for multichip modules (probably the largest ever) made by hot pressing. Package is shown with metallization in a frame for populating the package with components. (From Ref. 20, published with permission of the American Ceramic Soc.)

the hot-pressed parts and their flatness also lend itself well to thin film processing. The lack of lateral shrinkage in hot pressing also allowed formation of hybrid electronic packages, for example, of low K dielectric with Cu metalization to be hot pressed onto previously hot pressed AlN or other bases.

The above technical advances were achieved by using conventional, binder-based, green forming (tape forming), screen printing, and lamination techniques, then burning out the binder in the hot pressing die in the heating stage for hot pressing. Initially a binder system based on polyethylene and mineral oil that yielded good tape by hot extrusion was used, since this is probably one of the cleanest burning binder systems (especially after solvent removal of the mineral oil, which phase separates from the polyethylene on cooling from extrusion). Subsequently, conventional binder and tape casting for other, (alumina) packages were successfully adapted for hot pressing. Conventional binders for screen-printed metallizations were used in either case. Graphites with greater permeability for enhanced binder burnout were found to be advantageous for die components.

The above development of hot pressing ceramic electronic substrates and multilayer packages was judged to be economically viable based in part on a substantial number of parts potentially being hot pressed simultaneously in a given pressing (with successful demonstrations of hot pressing 3–6 sizable packages at once). However, among other important factors was that substrates and most multilayer packages are, or can be made to appear as, flat plates for hot pressing, eliminating machining costs that are a common major cost disadvantage of hot-pressed versus pressureless sintered parts. Further, the general cost disadvantage of hot pressing relative to sintering can be limited by lower temperatures/shorter densification cycles, especially for high temperature materials such as AlN (where avoiding the need of overpressure processing atmospheres or powders for burying components needed in most of its sintering is also an advantage). Also, specific to the electronics application, the higher costs often allowable are an advantage for this specific application. The economic viability of such hot pressing is supported by the technology having been licensed to CEPCO (Coors Tek, Inc., Golden, CO.), who has used it in some commercial production.

Green forming of ceramic bodies for hot pressing has also been successfully used for making large bodies, e.g. 40 cm square by several cm thickness (and some modest shapes beyond a simple plate). While use of binders with some bodies needs further development, they have been successfully used in a number of cases. Though some binder burnout of some medium size parts has been done in the hot-press dies, it may often be best to have a separate binder burn out step, though this can pose handling problems.

The above two uses of binder-based green forming of components have potentially significant ramifications for the economics of hot pressing, besides adding new product opportunities such as packages. Thus, die loading via preforming can be made more efficient using existing techniques widely available for sintering. It can also increase the number of parts for a given pressing where higher green densities can be obtained, for example, with finer powders (whose frequent voluminous character is often a limitation in hot pressing (Fig. 4.1). However, even larger potential may exist for using green body formation for making attached arrays of green preforms for multiple cavity hot pressing; thus, for example, molding arrays of parts (such as bearing ball preforms) attached by thin rods of green material such that the array can be dropped into a multipart die would eliminate the added graphite contamination noted earlier, and possibly allow automation of the hot pressing process.

Besides issues of inductive heating of a die directly and potentially more uniformly, or by heating a die via a suceptor (giving more flexibility in die size with a given inductive coil) or direct coupling to the die versus heating the die via a conventional surrounding resistively heated furnace, there are other variations in heating that may have potential for special or more general application. These commonly entail either or both of two options of resistive heating of the graphite dies, powder compacts of conductive particles, or both. One of the earlier reports of such heating is that of Glaser and Ivanick[23], who investigated direct current resistive heating of the graphite dies and TiC powder compacts in making small (2.5 × 1 × 0.5 cm) TiC samples with or without metal binder. They reported reaching very near theoretical density without metal binders by hot pressing TiC powder (2 μm particle size) at pressures of only 1–12 MPa (applied well before the maximum temperature of 3000°C) with a maximum hold time at temperature of 30 sec. Flexure strengths of resultant specimens of ~ 870 MPa indicate high-quality, fine-grain bodies achieved due to the short time at high temperature. While their paper is not clear on what portion of the resistance heating may have come from the powder compact versus the die, it clearly indicated promising hot pressing cycles with such resistive heating. Others have demonstrated successful hot pressing using resistive die heating; for example,. Jackson and Palmer [24] report such successful hot pressing of somewhat smaller samples of eight common carbides, four common borides, and four common oxides at modest pressure (21 MPa), typical temperatures, and short times. Thus, pressing WC at 2000°C and ZrB_2 at 1900°C for 5 and 3 min, respectively, gave dense bodies with grain sizes near the starting particle sizes. The Army Research Laboratory in Watertown, MA, (circa the 70s) had a sizable hot press (apparently mainly for metals) using resistive heating of the die (and presumably the powder compact). In view of the limited resistive heating likely in oxides (e.g., alumina), their heating probably came from the die. This author and colleagues attempted some hot pressing trials where the powders to be hot pressed were conductive, but were insulated from the graphite dies (via a BN sleeve), so only the powder was resistively heated. Such trials were unsuccessful since densification of the conductive powder greatly decreased as the powder partially densified and increased in electrical conductivity. Use of a higher current welding power supply instead of a conventional fixed voltage system did not solve the problem, supporting a focus on die heating. However, efforts to develop hot pressing with resistive heating mainly or exclusively of the die, despite some very promising short cycle times and properties, was generally neglected for a long period of time. Whether this was due to engineering issues of power leads to the dies or problems of practical handling of different die sizes and configurations is not clear.

Recently, there has been further investigation and development of at least laboratory systems using more sophisticated power supplies, apparently providing combinations of variable direct or alternating current and possibly pulses of these, e.g., of higher frequency. While the details of such heating system design and use as well as the mechanism(s) for its success are uncertain, there have been claims of very promising results, again associated with very short times at maximum temperature (2–5 min), and thus promising overall hot pressing cycle times [25–27]. Some units with at least modest (for example, mechanical pump) vacuum capability have been also offered for sale. In some cases, retention of fine grain sizes has been observed, while in others, substantial grain growth occurred, posing some issues. However, there are other methods of grain growth control, especially use of second phases as in composites. Possibilities of plasma generation in the pores between particles, and resultant cleaning/activation of particle surfaces have been suggested as a mechanism for the pressing benefits, as also noted for some novel heating methods for pressureless sintering (Sec. 8.2.2). Thus, alternate heating systems may have important benefits, but face issues such as scaling to larger sizes and flexibility— for different die sizes, shapes, and fills, different materials and powders, and use with binders.

Finally, note that an important extension of hot pressing is as an important tool in reactive processing of ceramics, as discussed in Section 6.5. Also note that press forging and some other deformation forming processes discussed in the next section are extensions of hot pressing.

6.3 PRESS FORGING AND OTHER DEFORMATION FORMING PROCESSES

A natural extension of hot pressing is press forging, which basically entails pressing a body at elevated temperatures such that it has little or no lateral constraint, at least initially, other than normal viscous and pressing surface resistance to vertical and lateral flow. This is commonly done in an oversized die to provide ram alignment as well as possible shape and size definition in the later state of deformation. There are two basic manifestations of this, the first being forging of powder-based bodies to both densify and shape them, which was discovered by Spriggs and coworkers [28,29] as a result of a die failure during a normal hot pressing run. The die failure was not observed until after the pressing run, but had been accompanied by considerable densification and plastic flow of the ceramic powder compact. Subsequent investigation showed that not only reasonable shaping, but also possible enhanced densification was possible, during press forging. However, as discussed in Chapter 1, forging rates achieved so far indicate that the process is likely to at best be restricted to special niche applications. (Recently Kim, an coworkers [29] have reported that strains of 1000% have been achieved with a nanograin composite of ZrO_2 with 30% each of spinel and alumina in tensile elongations at higher strain rates, e.g., 0.4 sec^{-1} at 1650°C, which could aid practicality.) Considerable research subsequently followed initial observations based on demonstrated possibilities of significant preferred grain orientation and resultant, often beneficial, effects on properties, such as electrical and magnetic ones [30–33]. However, green body formation and seeding techniques often do as good or better job with these and are generally more versatile and practical techniques.

The other manifestation of press forging, and one that has had some industrial use, is of converting single crystals into polycrystalline bodies. This has been commercially applied to making IR windows of halide materials, e.g., of KCl and CaF_2, since it offers two advantages over the as grown single-crystal form. Press forging of a single crystal and converting it into a polycrystalline body provides greater strength and more isotropic properties. It also has a practical shape advantage in that most optical windows are thin discs with diameters several times their thickness, while single-crystal growth typically produces boules of limited diameter and greater length. There are cost and other limitations on growing crystals of large diameter, so press forging which typically compresses rods into plates while also providing some property advantages is attractive. Such processing thus offers advantages that offset the moderate cost of forging compared with those of growth of larger diameter crystals.

The above press forging of single crystals of ceramics and related materials has various roots including those in metals forming. More specific roots are in deformation of NaCl to recrystallize it, e.g. as in the review of Schmidt and Boas [34], and more specifically in Stokes and Li's studies of NaCl deformation using extrusion of single crystals to produce high-quality polycrystalline specimens for test [35]. Day and Stokes [36] also recrystallized MgO crystal specimens by tensile straining them 60% at 1800°C then annealing at 2100°C; and Rice and coworkers [37] had demonstrated hot extrusion of MgO crystals and their recrystallization to polycrystalline bodies. Rice [3,37,38] demonstrated press forging and recrystallization of not only MgO crystals but also other cubic crystals of CaO, spinel, and TiC (but that noncubic sapphire and ruby deformed very anisotropically and did not recrystallize). Height reductions of 50–60% were readily achieved in the pressure range of 20–35 MPa in times of the order of 0.5 hr at 1850–2000°C (Fig. 6.4), with height reductions of 70–80% seen as feasible. Subsequently Becher and Rice [39] demonstrated press forging of KCl and its benefits for application as high-power laser windows, which has been used in commercial production of halide windows.

Whilc press forging has its advantages, other hot working was also of interest, particularly extrusion, which, as noted above, was demonstrated for both NaCl [34,35] and MgO crystals [3,30,37] before press forging was demonstrated. Hot extrusion of polycrystalline MgO, as rods with area reduction ratios of 10, ram speeds of 5 cm/sec at temperatures of 2000°C, was also demonstrated [37,38]. This was done in thick wall, coextrudable cans of Mo-based alloys or of W. Extrusion of some other cubic ceramics was also demonstrated, along with methods of starting with compacted (rather than dense) powder billets that were simultaneously densified by pressures generated in the initial stage of extrusion. Both some shaping, (extrusion of round pieces versus slabs) as well as possibilities of lowering extrusion temperatures and broadening the scope of extrudable materials via high-temperature hydrostatic extrusion and "fluid to fluid extrusion" were demonstrated. However, unless some unusual applications were to

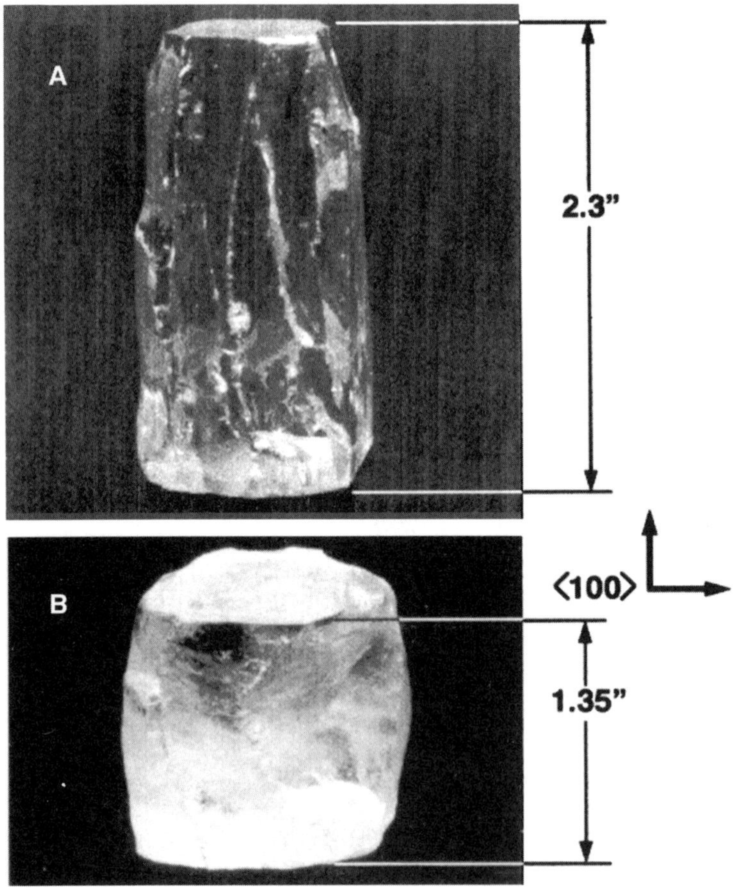

FIGURE 6.4 Press-forged MgO crystal: (A) before press forging and (B) after partial forging. (From Refs. 30 and 38. Published with permission of J. American Ceramic Society.)

arise, refractory ceramic extrusion technology remains only partially developed and unused because of high costs and limited yields and benefits relative to other fabrication. Coextruding glasses and metals was demonstrated by Hunt [38] as a possible method of making glass-to-metal seals, but was not successful in finding practical application.

Turning to forming and hot working methods that may also entail densification, hot rolling has been considered primarily as a ceramic densification and forming method, e.g. as a method of continuous hot pressing, rather than as a means of hot working. Earlier experiments focused on placing ceramic sleeves on metal rolls to allow higher temperatures (via a furnace attached to the rolling mill) and avoiding metal contamination of the ceramic [38]. Trials were unsuccessful since the unheated rolls limited temperatures to 1000–1100°C and gave unfavorable temperature gradients. Thus, substantial glass phase had to be added to obtain sufficient flow of ceramic-glass mixtures for some limited densification.

Greater, but still limited, success was achieved in hot rolling MgO with LiF additions since this yields a more "fluid" system and at lower temperatures, for example, 1000–1100°C [3,38,40]. This development started with a suggestion of Hunt to densify powders in metal tubes by swaging the powder-filled steel tubes first at room, then at elevated temperatures. However, it was quickly recognized that filling powders in steel tubes, then densifying the powder first by cold (room) temperature rolling (which readily yields very high green densities, Sec. 4.5), then rolling the tube and partially densified powder at elevated temperatures, was more promising. Some success was achieved in limited rolling trials in a system with alumina-sleeve-insulated metal rolls in a resistive furnace to heat the rolls and the part to be rolled. Somewhat more success was obtained with a prototype custom rolling mill consisting of larger, but less massive, lower pressure rolls and two metal strip heaters that were fed between the tube containing the powder and the bottom and top rolls (which were partially thermally insulated from the heaters by sheets of asbestos). MgO slabs ≥99% of theoretical density a few millimeters in thickness and 2–3 cm in width and several centimeters long were obtained in hot rolling at 1000°C at pressures of 20 MPa at speeds of 2.5 cm/sec., but there were major problems of cracking and outgassing. Thus, during hot rolling of previously cold-rolled powder, the filled tubes generally bloated back up to about their original diameters prior to cold rolling, despite being evacuated at temperatures of 500°C, then sealed prior to cold rolling. Drilling holes in the sealed, cold-rolled tubes just prior to hot rolling allowed outgassing and prevented tube bloating within a few centimeters of the drilled holes. Some progress on these problems was made; for example giving some translucent-transparent, crack-free pieces a few centimeters in lateral dimensions. However, a basic problem was seen as the limited temperatures achievable in the powder to be rolled, due to having to heat the rolls substantially. This along with costs and problems of the cans and their removal were reasons for ceasing development.

Later interest in self-propagating high-temperature synthesis not entailing gaseous reactants or reaction products, where highly exothermic reactions can generate a reaction front that, upon ignition of the reaction, moves through a compact of solid reactants, along a bar of the reactants, was seen as a possible basic solution to the basic heating problem for hot rolling noted above [41]. Thus, filling a metal tube with reactant powders and cold rolling the tube and powder, then igniting the reaction at one end of the tube so the reaction and associated high-temperature reaction front propagates along the length of the reactant powder compact in the tube was seen as a possible way of providing desired heating for hot rolling to simultaneously densify the reactant products. Such heating would have the potential of heating not only primarily the material to be consolidated, but to do so primarily in the moving reaction zone, where consolidation would be focused, and potentially aided, by transient liquid phases that are often generated. Further, rolling speeds might be controlled by a feedback system to match the rolling speed to the reaction speed since this should minimize rolling pressures, which can be sensed and used as the feedback control. Trials showed some promise, a number of challenges, and some possible requirements, such as that the roll diameter probably needs to be on the scale of the size of the hot reaction zone, that is, substantially smaller than typical rolls in laboratory metal rolling mills. Such smaller rolls could also be backed up with more massive rolls. Thus, while there are possibilities to try, substantial challenges and uncertainty remains for hot rolling as a continuous hot pressing method.

Various approaches have been considered for placement of green or partially sintered bodies in a powder bed, or their encapsulation in a glass, to act as pressure transmitting media such that placement of the part to be densified with the pressure transmitting powder or the glass encapsulation in a hot pressing die to turn the hot press into a quasi-hot isopress [6,42–44]. While general hot isopressing is discussed in the next section, some of these approaches are discussed here since they use hot presses, and one [27] uses resistive heating of the graphite die and the powder transmitting media. Development and possible commercialization of these concepts for quasi-isostatic densification of ceramic (or metal) powder preforms surrounded by a pressure transmitting carbon or other ceramic powder in a die cavity, where the part is pressed by applying axial pressure to the pressure transmitting powder have been pursued. In order to cut cycle time (and limit heating costs) some part preforms and pressure transmitting powder were heated outside of the hot press die into which they were rapidly loaded and pressed, then unloaded from the die with only partial cooling. Actual times at temperature for densification were of the order of seconds due to the high pressures used (0.8–1 GPa), indicating use of massive metal dies that limit exposure to higher temperatures and times at temperature. This approach may be more successful with metals because of their greater densification by plastic flow and consolidation at lower temperatures, as well as to some lower temperature ceramics, such as high temperature-superconductors, to which the process was applied. However, clear commercial success has apparently not been achieved, due to temperature and pressure limitations and pressure variations in the powder bed.

As noted above, glasses have been considered for pressure transmitting media for high-temperature pressing. Thus, some have reported using glasses for this at temperatures in the range of > 1000°C to about 1600°C [44]. A more recent use of glass as a pressure transmitting media is in the rapid omnidirectional compaction (ROC) process [45]. This entails encapsulating a ceramic, metal, or composite powder preform in a glass such that the preform and encapsulating glass form a disc or cylinder "billet" that fits into the cavity in a massive metal die. Heating the "billet" to the temperature desired and rapidly loading the hot "billet" into the die cavity and pressing at 800 MPa for a few seconds before hot ejection allows substantial densification of a number of materials. However, use of metal dies, while allowing high pressures, restrict temperatures significantly, to 600–700°C, limiting applicability; it is seen as particularly useful for densifying ceramic-metal composites.

Use of glass as an encapsulant in some of the above methods raises two issues concerning processing with glass. First is the extent to which glass is extruded into the powder compact, for example, into surface pores. While this is typically limited if pores are small and limited in number, it can be a desired end in some cases, mainly with more porosity. Thus, pressing parts encapsulated in glass to extrude glass into the body to form a glass-crystalline composite may be useful in some cases. The other issue is processing of composites of a crystalline and a glass phase, which can have some practical ramifications. Thus, for example, Clark and Reed [46] investigated low-pressure (5 MPa) forging of glass-bonded abrasive wheels, showing some potential for this. An important example of glass-crystalline ceramic composites that are hot pressed are glass-bonded mica bodies used for various electrical applications [47]. A variety of shapes and sizes of parts are pressed, apparently in the range of 430–800°C, including irregular tubes weighing 4.5 kg and sheets to 70 x 50 × 2.5 cm. Parts can be molded to tolerances as close as 10-15 μm. These capabilities of size, shape, and tolerance reflect advantages that may be achievable in a number of cases of applying glass forming techniques to glass-crystalline ceramic composites. In both of the above cited cases of hot pressing or forging of glass-ceramic composites, annealing after pressing was needed to relieve forming-densification stresses.

6.4 HOT ISOSTATIC PRESSING (HIPING)

Extending hot consolidation from the typical uniaxial densification of hot pressing to triaxial hot pressing—hot isostatic pressing—is a logical step, as is the extension of uniaxial die cold pressing to cold isopressing (CIPing), but is more costly. Some earlier HIPing was conducted in a CIP by selecting suitable metal sheets of it to be formed and welded to form a larger "metal bag" (with power feed-throughs) that contained a pressure transmitting and electrically and thermally insulating ceramic powder [48]. A powder compact inside another sealed metal can was placed in the center of this larger metal bag with a surrounding, but noncontacting, resistive heater. Thus, the heater provided the heating for HIPing, while the ceramic powder in the metal bag transmitted the hydrostatic pressure from the CIP environment outside the outer metal bag and insulated the CIP media and vessel from the internal heating of the part in the inner can. There were also trials using glass as a pressure "fluid" for HIPing, as discussed in the previous section [48]. While such approaches had some success and offered some potential safety advantages over HIP units using gases as the pressuration media, the latter were developed commercially and are the basis of the HIPing industry today.

Some benefits can arise from HIPing a porous preform in a HIP with no canning or encapsulation, since closed pores are reduced in size, number, or both occurs [49], but no significant effect occurs on open pores since there is no net pressure across them to cause them to shrink. However, HIPing is predominately used to achieve high property levels, which typically requires achieving at, or near zero porosity, which cannot be done without an impervious envelope around the specimen so the HIPing pressure is applied equally to open and closed pores in the body to be pressed. Earlier HIPing was commonly conducted by sealing a powder compact in a refractory metal can that would apply the HIP pressure to the encapsulated body. Such cans typically had an umbilical cord that allowed high-temperature evacuation in a separate furnace before being sealed off and moved, after cooling, to the HIP. However, such metal canning was very cumbersome and expensive, which lead to development of two alternative approaches. The first was glass encapsulation of porous compacts via use of a coating of glass powder, such that the selected glass powder ideally sinters to seal off the surface of the compact after the compact is suitably outgassed (either of adsorbed gas species or of gases from binder burnout) [50]. The glass, which commonly allows good preservation of powder preform shape, is removed after HIPing, usually by chemical dissolution in strong acids. The second alternative is to separately sinter the preform to closed porosity, in which case it can generally be HIPed to, or near, theoretical density without any can or encapsulent. There are, however, various pros and cons to these two process, as discussed below, after briefly reviewing the capability of HIP units today.

As HIPing has become more widely used for research and development, as well as some production, especially of metal parts, costs of HIP units have decreased and capabilities have increased [51,52]. Temperature capabilities to ≥2000°C with pressures in the range of 100 to ≥400 MPa are available with hot zones of 5 to ≥100 cm diameter and lengths of 10 to ≥250 cm commonly available with more extreme parameters feasible. Costs increase as temperature, pressure, or size (mainly diameter) increase. Most units operate with nitrogen or argon as the pressurizing gas, but specialty units can be obtained that can operate with oxygen gas (with added cost and some operational restrictions), and use of some other gases may be feasible. Because of the interests in sinter-HIPing and especially outgassing prior to sealing of glass encapsulation via sintering, HIP units that can be operated with a vacuum at lower temperatures then switched over to HIPing are available so the glass sealing operation does not need to be made in a separate facility. While some HIP units can also sinter prior to HIPing, it is generally more efficient to sinter in a separate facility.

As with hot pressing, higher pressure and longer time at temperature and pressure lower HIPing temperatures. Thus, typical HIPing temperatures will be below those for hot pressing (Table 6.1), by 50–100°C. HIP production of metal parts is much more advanced because of the much larger and more diverse metal components markets as well as the lower temperatures required, allow larger more economical HIP units. However, some ceramic HIP production exists and more is likely to occur. Thus, ceramic bearing components are HIPed, as are some speciality wear parts, and engine components are clearly candidates. Reaction processing (see next section) and some other speciality processing, for example, of graded bodies [52], may be practical. Component size is a serious limitation for some possible applications such as IRdomes, radomes, and other windows. However, construction of special HIP units is feasible for such specific components that make one or a few parts at a time, but with little excess of gas volume, temperature, or pressure capabilities, as indicated by Huffadine and coworkers [53], so facility and operation costs may be lower than with a general purpose HIP. While such possible custom HIP units may aid production of larger parts, HIPing is likely to limit sizes substantially more that pressureless sintering and hot pressing.

Besides these trade-off between specialized and general purpose HIP units, there are other important trade-offs, especially between sinter-HIP and glass encapsulated HIPing. Thus, in some cases HIPing may drive excess glass encapsulent into a porous preform, which may be negligible in many cases, but may be a problem, or an advantage in other different cases. Also, the typical glass encapsulent removal by dissolution in strong acids may be deleterious to ceramic parts of some compositions. This may simply require machining off some surface material in some cases, but may present basic limitations in other cases. On the other hand, there are also issues for sinter-HIPing, a basic one being that open pores near the surface will generally remain open, thus, possibly requiring some surface machining. Another issue is incompatibility between the composition being HIPed and the HIP environment, especially the gas used; for example, HIPing compositions sensitive to stoichiometry, such as some oxides, may be adversely affected. Effects may vary from minor to major—actions varying from none to some surface machining or annealing may be required, to more serious limitations. The latter may be solved in some cases by special HIP units designed for use with oxygen, but these are more expensive and limited in capabilities. Other problems may arise in certain materials, such as some desintering of some Si_3N_4 bodies due to dissolution of nitrogen from the grain boundary phase and associated with use of BN crucibles [54]. These types of problems may be sporadic in location on a part and in time of occurrence, but can be serious. Also, outgassing effects can occur [55], and large processing pores may not be removed, leaving still serious defects [56]. More recently, Andrews and coworkers [57] reported that commercial Si_3N_4 with Al_2O_3, Y_2O_3, and Nd_2O_3 gas pressure sintered at > 1600°C had a clearly visible surface reaction layer 1 mm deep in rods 11 mm diameter, though possibly more extreme in extent was similar to other observations. This darker area had randomly distributed small black spots and larger white snowflakelike features that were often the strength limiting flaws, giving different strength behavior for the surface versus interior material for both test specimens and prototype valves machined from sintered blanks. More generally, nonuniformities may occur in HIPing densification due to combinations of gradients of densification associated with part geometry and thermal variations of the HIP heating system, as well as perturbations of this due to HIP loading and support structure for the components being HIPed [58]. This is directly analogous to issues of the loading ceramic components in furnaces for firing.

6.5 REACTION PROCESSING

Chemical reaction is an integral part of many ceramic processes. It is usually involved in preparation of ceramic powders, especially those tailored to high purity, fine and controlled particle size, or both. The reactions of concern here are those involving one or more particle species reacting with other such species or with other body constituents in a fluid state, with the reaction occurring in conjunction with densification during fabrication. Other reaction processing such as chemical vapor deposition (CVD) where all reactants are in the vapor state are discussed in the next section. Preparation of ceramics via polymer pyrolysis [59–61] partially falls in this area, but is addressed in Section 6.6. Reactions considered in this section are generally considered in the order of increasing reactivity, as indicated by their increasing exothermic character.

An earlier reaction process long used for making many SiC bodies is the infiltration molten Si into a compact of fine carbon or carbon and is the infiltration of SiC grain [1,63]. This process, developed by K. Taylor of Carborundum Co., Niagara Falls, NY (Carborundum named the product KT SiC in honor of him) used relatively coarse raw materials, giving modest properties, room temperature strengths of 100–200 MPa but versatile fabrication, yielding a variety of sizes, shapes and character of resultant bodies. Subsequently Popper and others refined this (RBSC) process [64], especially via use of finer powders to give much better properties, e.g., room temperature strengths of > 500 MPa. The resultant bodies, composites of SiC and residual Si, are of moderate cost, but their use has been limited by the excess Si, and by the later development of dense SiC via sintering or hot pressing with much smaller amounts of different more compatible additives (Sec. 5.5).

Another older reaction processes in use is that to make reaction sintered or bonded silicon nitride (RSSN or RBSN) by the in situ nitriding of silicon metal powder compacts [65,66]. This has been used for some time to make Si_3N_4 bodies of reasonable strength at moderate cost, and was of particular interest before ways of densifying Si_3N_4 powder were discovered. The reaction process yields not only bodies of good integrity, but also of good dimensional control despite the large intrinsic volume increase of Si on nitriding to form Si_3N_4 since much of the Si to Si_3N_4 conversion occurs in the vapor state within the pores of the Si powder compact (Sec. 3.2). Thus, Si vaporizes from sharp or other fine protrusions of the Si particles into the pores where reaction occurs with the N_2 (often with some H_2) or NH_3 and deposits out Si_3N_4 onto the pore walls. This vapor-phase reaction in the pores of the Si-powder compact avoids the incompatible product-precursor stresses that would occur if the reaction occurred on the surface and via diffusion into the bulk of the Si. This also allows the Si_3N_4 product dimensions to be within 0.5% those of the green Si compact, making it a desirable net-shape process, reproducing a variety of shapes obtainable by various powder consolidation/forming processes. The nitriding process is aided by additives, such as Cr and, especially, Fe (Sect. 3.2); but temperature control is needed to keep the reaction from becoming exothermic and melting the Si, since surface tension of the melt forms large Si agglomerates that cannot be fully nitrided and severely limit strengths. Despite the above advantages of properly nitriding Si, use of the process has been limited since additive sintering of Si_3N_4 yields two times the strength of RSSN due to the near-zero porosity of the latter versus the 20% porosity of good RSSN. However, it was later discovered that sintering additives for Si_3N_4 could be incorporated in the Si compact, then used to densify the resultant nitrided body by conventional additive-based sintering [66]. While this nitriding/sintering process involves two stages of "sintering", (reaction sintering and actual sintering), it has some cost advantages and has been used in production of some components, such as some cutting tools.

Another earlier reaction process is that of hot pressing some oxides directly from uncalcined precursors, such as carbonates and hydroxides, and effectively combining the decomposition of the precursor with the densification of the oxide. Thus, $Mg(OH)_2$ yielded dense transparent MgO when hot pressed at 35 MPa and 1000°C, while hydroxides of thorium and aluminum gave theoretically dense oxides at 1300 and 1500°C, respectively, and trials indicated similar effects with $Ca(OH)_2$ and $La(OH)_3$ [67]. Subsequent study of making MgO by this technique showed that full densification could be achieved at lower temperatures with some increases in pressure, but that hydroxide precursors always left considerable contained water and often considerable preferred grain orientation, while similar densification of $MgCO_3$ precursors did not show preferred orientation and had much less carbonate retention [67,68]. However, later work showed that much MgO hot pressed from calcined hydroxide, carbonate, or bicarbonate precursors also retained some hydroxide, bicarbonate, or both which often caused clouding, blistering, or gross bloating of specimens on exposure to elevated temperatures [69]. These problems—which may also vary some with storage times and conditions, especially for calcined fine MgO powders, and should significantly increase with hot-pressed body size—while more extreme for some powders, are a broad production issue (Sec. 8.2.1).

Turn now to a broad class of reaction processing that has been used for some time and applied to a variety of compound ceramics—those containing two or more metals or metalloids—widely used for mixed-oxide compounds, such as $MgAl_2O_4$ and mullite. Whenever both compounds to make the desired compound are sufficiently stable under handling (storage) and processing conditions, there are two options: (1) calcine the mixed powders to form the desired compound in powder form and consolidate and sinter the compound powder or (2) react the mixed and consolidated powders during the densification process. A variety of factors impact this choice, but frequently there are cases where the second choice is made, to avoid a separate powder-calcining step, as well as achieve as much densification before extensive reaction has occurred. The latter is often desired for mixed-oxide compounds, such as mullite, that are more difficult to densify than their constituents. Such reaction processing is closely related to that often used for making some composites as discussed below.

Consider now some related reaction processes for producing ceramic-metal composites, especially those involving reaction of molten Al with oxygen or nitrogen [70–74]. The original discovery was that bulk alumina bodies with several percent residual Al could be produced by oxidation of molten Al, especially with some additions of Mg, Si, or both (Sec. 3.5), giving substantial alumina growth rates at temperatures of 1100–1300°C. The process has considerable potential for producing larger and more complex shaped parts via use of molds and growth inhibitors (to shape part boundaries). Like any process, this Lanxide process produces a range of compositions and microstructures that determine its properties, which along with cost aspects determine its utility. The alumina produced had reasonable properties consistent with its moderate to larger grain size, some toughening at low to moderate temperatures due to the residual Al, and somewhat better relative strength at elevated temperatures, apparently due to Si-free grain boundaries.

The above alumina-Al composite processing was further developed in several respects. One modification was to replace the residual Al by an aluminide that was more corrosion-and wear-resistant (but probably reducing the modest toughening from the Al). A more significant advance was to use the process to grow a matrix through a preform of different composition, (of ceramic particulates, platelets, or fibers) to produce a range of ceramic composites [75]. Both the basic composition and the range of composites were expanded by the demonstration that the process could also produce AlN-based bodies [74], as well as some other composites, some being more metal matrix rather than ceramic matrix composites [76,77]. Examples of the latter are composites made from molten Zr infiltration of B_4C preforms to produce composites of Zr with ZrC grains and ZrB_2 platelets having high strengths (450-900 MPa), toughness (11–23 $MPa{\bullet}m^{1/2}$) and Weibull moduli [21–68]. There has been some commercial production of products via the above processes, but much or all of this has ceased, reflecting challenges to new materials technologies. Recently, Sandhage and coworkers [78] have shown that such exchange reactions to produce metal-ceramic composites can often have processing temperatures greatly reduced by limited additions of another metal. Thus, they showed that limited additions of Cu to the Zr reactant allowed formation of a reactive liquid to infiltrate at much lower temperatures (≤1300°C) to react with a WC preform to yield in situ a dense W-ZrC-based composite of interest.

There are a variety of other composite fabrication methods, some of which use molten metal infiltration of dense ceramic preforms, for example, of Al into Al_2O_3 [79]. Others introduce the metal as solid particulates mixed with the ceramic particulates in various powder forming and densifying processes, often with the goal of achieving low or zero shrinkage on densification due to metal particle oxidation eliminating much of the compact void space. Of particularly popular in this regard has been processing of Al-Al_2O_3 compacts, where there have been both past industrial as well as more recent academic investigations. Industrial tests showed that aluminum-alumina powder mixtures could be formulated and formed by conventional alumina fabrication methods that gave essentially "zero shrinkage" from green body to fired body. While generally lower and more variable strengths obtained via this zero shrinkage route versus normal commercial alumina production could probably be improved with further development, part of the problems appeared to arise from more basic dimensional effects. While the green versus fired part dimensions were nearly or actually the same, that is, being zero shrinkage, the intermediate stages of firing gave very different results, as shown by dilatometer tests over the firing cycle. Thus, initial body thermal expansion rapidly increased over that of alumina due first to substantially higher expansions of the binder and then that of the Al particles, with the latter extending this high expansion to higher temperatures, then further increasing it at and beyond the melting of the Al particles, until they were consumed and the body expansion subsided to that of alumina. These resultant increases then decreases in thermal expansion of the body, and gradients of them as a function of body size, shape, and stage of the process were seen as potential basic problems with such processing on an industrial scale. This, along with some possible storage-handling-safety issues for Al powder and some possible cost increases, lead to ceasing further development.

Substantial recent investigation of processing of alumina-based bodies using Al-alumina powder mixtures (named reaction bonded Al_2O_3 designated as RBAO), have shown refinement and extension of the above process, particularly by Claussen and coworkers and more recently, Messing and coworkers [80–82] which deserve attention. Thus, use of finer Al powders and additives to enhance oxidation (similar to the Lanxide process above) and processing such as intensive attritor milling to achieve substantial oxidation before firing (by forming hydroxides) and during earlier firing stages, have given strengths more competitive with those of commercial aluminas: e.g. 350 MPa. Another important extension is to use the RBAO process for fabrication of ceramic composites [83,84], as well as combining it with molten Al infiltration to increase the Al content while the body is still porous [85].

Other metals have been investigated for oxidation to oxides; for example this author showed a number of years ago that Ti and Zr sponge particles could be oxidized to coherent pieces of the respective oxides with the Ti giving small centimeter scale TiO_2 test bars with strengths ≥70 MPa. However, other more serious and extensive investigations have been made, of which those of Sandhage and colleagues are particularly noteworthy [86-90]. They recognized that some metals, especially heavier alkaline earth metals such as Ba, have oxides that have somewhat less volume than the metal so there is some shrinkage of the metal on oxidation, which occurs at low temperatures, e.g., ~ 300°C. The smaller oxide than metal volume is attributed to the size of the Ba++ ion in the oxide being substantially smaller than the Ba atom in the metal. Thus, mixing with other metals that expand on oxidation and possibly some oxide product can yield a net-zero shrinkage (though again dimensional changes with intermediate reactions my be important). Applications to producing various Ba containing materials such as aluminates, ferrites, and titanates, as well as 1-2-3 superconductors have given encouraging results. Further, Ba is quite ductile, such that laminates with alternate layers of Ba with Ti particles and of Ag or Pd have been formed by metal rolling operations, then the $BaTiO_3$ formed by oxidation at 300°C and annealing at 900°C (i.e., below the melting point of Ag), indicating feasibility of making electronic ceramic devices such as multilayer capacitors.

The above bodies made by various oxidations of metals, which have been recently reviewed by Sandhage and Claussen [91], are a further addition to the diverse methods of fabricating ceramics and composites. However, much more evaluation, especially of scale-up tests, is needed to better access their potential, as is the case with any process, and often more so with composite fabrication, especially via reaction processes as discussed further below. Particularly pertinent to the above processes are interactions of body size and shape and reaction exotherms on controlling internal temperatures (e.g., as shown by melting problems in making RSSN) and of outgassing from reactive metals and handling costs for them.

Consider now a large number of reactions that inherently produce ceramic composites, primarily particulate composites that are of interest. A good example is the reaction of zircon, alumina, and silica powders to yield mullite with dispersed zirconia:

$$ZrSiO_4 + Al_2O_3 + SiO_2 \Rightarrow 3Al_2O_3{\bullet}2SiO_2 + ZrO_2 \qquad (6.1)$$

The potential advantages of this reaction is that it uses potentially lower cost raw materials and offers the possibility of substantially densifying the compact of the reactants before much reaction occurs since the formation of mullite usually retards densification, especially by pressureless sintering. This reaction also has the more general potential advantage of most reaction processing to produce ceramic composites, that, since all phases are nucleated during the processing, there is the potential of controlling grain growth and hence grain sizes of the resultant constituent phases. Such reaction processing of ceramic composites has been reviewed by Rice [92], who shows that densification of the reactants by pressureless sintering prior to much if any reaction of the constituents is advantageous, since the reaction and resultant added porosity can complicate the normal sintering process. Because of this and a general densification advantage, hot pressing or HIPing are often preferred for much ceramic composite processing. Other reactions that fall in this category of reaction processing are shown in Table 6.2. Another example is making composites of an alumina or mullite matrix with significant dispersion of BN flake particles. This was originally done by mixing BN powders with alumina or alumina + silica powder and hot pressing, giving reasonable properties [93a,93b]. However, Coblenz and Lewis [94] conceived of instead using reactions between Si_3N_4 and $B_2O_3 + Al_2O_3$ or AlN and $B_2O_3 + SiO_2$ to yield mullite with BN that was not only more uniform (Fig. 6.5) and superior in performance, but was also somewhat lower cost due to both Si_3N_4 and AlN being lower cost than BN.

Another subgroup of reactions for producing either single-or mixed-compounds, or composite ceramic products are those that can be sufficiently exothermic such that a green compact (e.g., a bar) of the reactants, once ignited locally,

TABLE 6.2 Reaction Hot Pressed Ceramic Composite Data[a]

Reaction	Vol. % Nonoxide	Density (gm/cc)[b]	HV (1 kg) (GPa)	Costs ($/lb)[c]
$4Al + 3SiO_2 + 3C \rightarrow 2Al_2O_3 + 3SiC$	43	3.67	26	1.11/4.83
$4Al + 3TiO_2 + 3C \rightarrow 2Al_2O_3 + 3TiC$	42	4.29	22	1.58/6.87
$10Al + 3TiO_2 + 3B_2O_3 \rightarrow 5Al_2O_3 + 3TiB_2$	27	4.14	22	1.69/7.97
$8Al + 3SiO_2 + 2B_2O_3 + 4C \rightarrow 4Al_2O_3 + 3SiC + B_4C$	37	3.62	19.9	1.37/6.69
$6Mg_3Si_4O_{10}(OH)_2 + 36Al + 25C + 2B_2O_3 \rightarrow 18MgAl_2O_4 + 24SiC + B_4C + 6H_2O$	31	3.45	15	0.91/4.55
$(La_2O_3 \bullet 6B_2O_3) + 14Al \rightarrow 7Al_2O_3 + 2LaB_6$	35	4.09	21.5	3.55/9.46
$Si_{3>}N_4 + 4Al + 3C \rightarrow (4AlN \bullet 3SiC)$	100	3.24	25.4	3.21/9.41

[a]Compiled from data of Rice and coworkers [98,100].
[b]Theoretical density of solid product.
[c]Raw materials costs. Top figure is for the raw materials for reaction hot pressing. Bottom figure is for powders to produce the same product by directly hot pressing of powder mixtures of the same ceramic composite compositions.

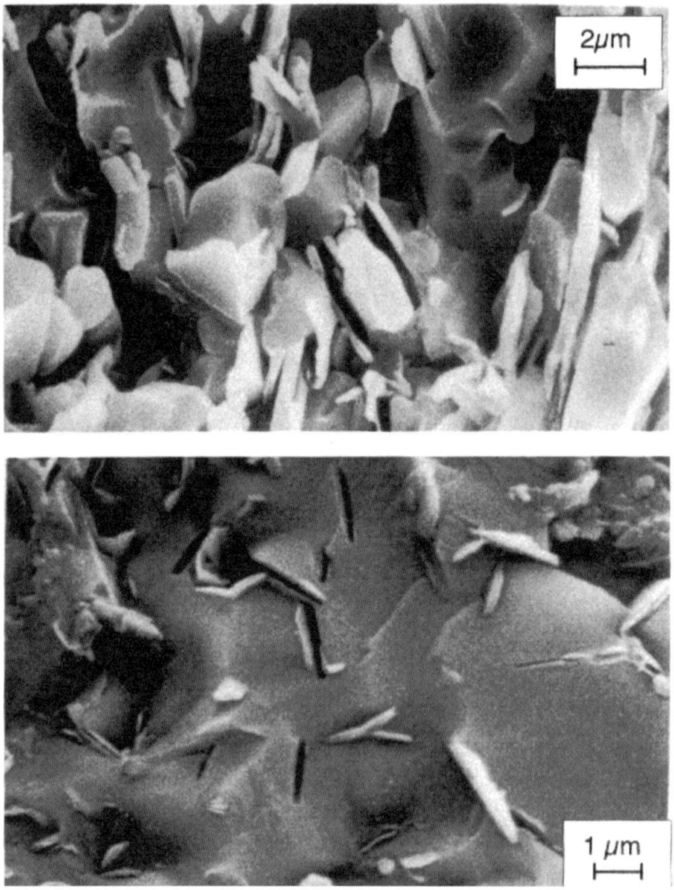

FIGURE 6.5 Al_2O_3-30% BN composites: (A) hot pressed from a mixture of Al_2O_3 and BN; (B) reaction hot pressed. Note the more uniform microstructure in B.

(on one end), the reaction propagates along the bar to complete the reaction of the body without any other thermal energy other than the modest amount used for reaction ignition [95]. Such reactions are referred to as self-propagating high-temperature synthesis (SHS), high temperature referring to the fact that the adiabatic temperatures from the reaction can be quite high (Table 6.3) [95,96]. Such reactions have attracted much attention, due to the possibility of achieving densification with such little input of thermal energy for very refractory compounds. Many of these reactions, especially some of the most vigorous ones, are between elemental reactants, but there are also a substantial number that are reactions be-

TABLE 6.3 Intrinsic Volume (ΔV) and Density Changes in Forming of Ceramic and Intermetallic Products

Products[a]	ΔV (%)	ρ_P/ρ_R[b]	Q_R + (J/kg × 10^6)[c]	T_{ad}(K)[d]
$MoSi_2$	-40.6	1.39	0.28	1900
SiC	-28.4	1.39	1.73	1800
$TiSi_2$	-27.5		0.40	1800
TiC	-24.4	1.32	3.08	3210
WC	-23.8	1.31	0.18	3060
TiB_2	-23.3	1.29	4.03	3190
TiSi	-22.9		2.01	2000
VC	-21.0	1.27	1.93	1620
$ZrSi_2$	-20.9	1.26	1.03	2100
ZrB_2	-20.4		3.12	3310
VB_2	-19.2	1.24	2.81	2670
NbB_2	-19.1	1.24	1.53	3270
NbC	-17.4		1.54	1840
ZrC	-17.0		1.95	3690
Cr_3C_2	-16.9		0.70	
TaC	-15.1	1.18	0.50	4270
CaB_2	-14.8	1.17		
W_2C	-9.7	1.11	0.12	3070
B_4C	-8.3	1.09	0.05	1200
AlB_{12}	-6.3	1.07		
BaB_{12}	-4.2			
Al_4C_3	-1.7	1.02	1.04	600

[a]Single ceramic products from elemental reactions, P_p and P_R theoretical densities respectively of the product and the reactants, Q_R heat of reaction, and T_{ad} adiabatic temperature. After Rice and McDonough (96). Published with permission of the American Ceramic Society.

tween compounds, (Table 6.3), and some are reactions between combinations of compounds and elements.

An important factor in using these reactions, especially very vigorous ones, is controlling them to keep them from propagating as discussed below. This can be done not only by selection of the reaction, but also by control of the microstructure of the reactant compact since its microstructural factors play an important role in its reaction, with higher reaction propagation velocities corresponding to more vigorous, less controllable reactions [97]. Thus, compact porosity impacts reaction propagation, with reaction velocities for a given reaction often being a maximum at intermediate levels of compact porosity (e.g., 40%), but may depend on the character of the porosity. Increasing the particle sizes of the reactant particles also decreases reaction velocities, as does increasing content of particles of reaction products or other dispersed particles that are inert to, hence not involved in the reaction, thus diluting the reactants. Finally, ignition of reactions can be inhibited by contact of reactant compact surfaces with higher thermally conducting environments, for example by contact with graphite tooling versus contact with a gaseous atmosphere.

Much of the earlier attention to processing using vigorous reactions was motivated by the possibility of obtaining dense compacts of very refractory ceramic bodies by simply compacting the reactants then igniting the reaction, for example, with a small coil of high temperature resistively heated wire or a torch flame or laser beam. Further, it was proposed that this could be done with major cost reductions since it required so much less energy than normal ceramic processing, and that it could often result in unique compositions, microstructures, or both due to the very transient nature of such reactions. However, these expectations were generally poorly founded due to neglecting some basic and practical factors[95,98]. Thus, extraction of many elements, especially some of those for the most vigorous reactions to produce very refractory products such as TiB_2, ZrB_2, TiC, and ZrC, is generally expensive, making many such SHS processing routes fairly to highly expensive. Further, eliminating the energy costs for densification (the gas or electric bills for sintering or hot pressing) generally eliminates $< 5\%$ of typical ceramic production costs, which, while not negligible, is not a major savings. Elimination, or substantial reduction, of furnaces for densification would be a more significant savings, but is generally not feasible, as discussed below.

Limitations in two other basic factors of obtaining low porosity and sound, that is uncracked, ceramic bodies by SHS were also not fully recognized [98,99]. Porosity issues are discussed here, and avoiding cracking is discussed below. Besides the porosity of the compact of powder reactants that must be removed for most applications, there are two other important sources of porosity that require additional densification. The first is extrinsic generation of porosity due to outgassing of adsorbed species on compact powder surfaces, which can be very rapid due to rapid heating to high temperatures from the reaction of compact constituents. This generally scales with the temperatures reached in the compact, is a serious problem for many reactions, and can be exacerbated by the propagation of the reaction (discussed below) as well as increasing sizes of bodies being fabricated. Such extrinsic porosity generation can in some cases result in at least some minor explosions. The other basic source of porosity is that intrinsically generated by the reaction itself—the exothermic nature of the reactions basically arises from the reaction products having stronger atomic bonds, hence higher densities, than the reactants. This increase in solid density of the reaction products in the reaction compact is accommodated by intrinsic generation of porosity, that is typically substantial, and generally increases with the energy of the reaction (Fig. 6.6) [96]. The combination of the three porosities that must be eliminated to produce a dense body—the initial porosity of the compact of the reactants and intrinsically and extrinsically generated porosity—pose challenges

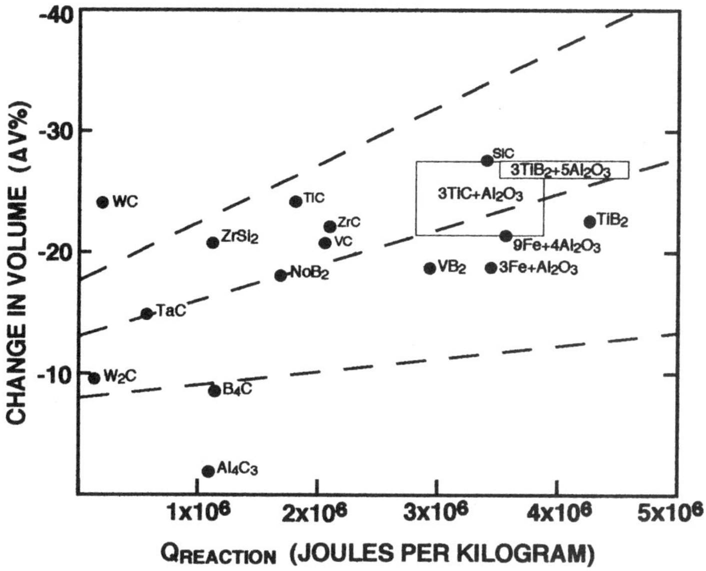

FIGURE 6.6 Plot of the change in solid volume as a function of the enthalpy of the reaction for more vigorous reactions. (From Ref. 96, Published with permission of the American Ceramic Society.)

to producing quality bodies (but may in some cases be useful for making porous bodies, e.g., foam, Sec. 7.3.2). Thus, while some less vigorous reactions that produce some intermetallic products have yielded dense billets, mainly via sintering (in some cases aided by plastic flow feasible in some of the reactants and some products), the most common approach to solving the problem eliminating most or all porosities in ceramic products, especially from more vigorous reactions, has been the application of pressure during reaction. While pressure may be applied by hot rolling (Sec. 6.3) or HIPing, it has been most commonly been applied by hot pressing—making the process reaction hot pressing.

Reactive hot pressing can yield dense bodies from reactant compacts of some of the most vigorous reactants, but also shows that elimination of all heating is generally not feasible, and that propagating reactions are generally undesirable for hot pressing, and probably for HIPing. Thus, it has been shown that nearly dense TiC can be produced by reactive hot pressing of compacts of Ti and C, at least when the reaction is ignited so it propagates essentially axially in the graphite die [99]. However, use of an unheated die resulted in thermal stress cracking of even modest size discs, e.g., 3–4 cm diameter. Cracking in such small parts was eliminated by heating the die to 1000°C, but more heating would be needed for larger parts, thus seriously restricting reductions in heating with such SHS processing. Such restrictions are even more constrained by indicated needs to avoid propagating reactions, at least for larger parts.

As is so often the case, scaling up the size of bodies often reveals or exacerbates important processing issues, as was the case in scaling up reactive hot pressing using reactions in Table 6.2 [100,101]. While hot pressing discs 2–3 cm diameter and < 1 cm thick gave uniform dense bodies, scaling to 5–7 cm diameter by 2 cm thick gave discs with serious gross problems—having seriously rumpled and sometimes cracked top surfaces and macropores or porous areas with diameters of the order of half the disc diameter and thickness. Tests and evaluation revealed that these areas were the result of the reactions initiating primarily from the disc periphery and secondarily from the top and bottom of the disc with reaction propagation primarily on radial and axial directions, respectively [101]. The secondary ignition from the top and bottom surfaces densified those surfaces, sealing off much escape of gases released by the reaction exotherm, while the radial ignition sealed off the cylindrical periphery. However, radial propagation of the reaction allowed for little densification since in the early stages, the bulk of the unreacted interior resisted consolidation of the reacting material, while in the late stages of radial reaction propagation the densified outer peripheral area resisted consolidation. Changing the temperature gradients (by reducing heat losses through the pressing rams) so most ignition of the reactions occurred in the axial versus the radial direction, substantially reduced, but didn't eliminate, the above gross problems. This and subsequent tests showed the solution to achieving large, quality hot-pressed bodies was to make the reactions nonpropagating, via use of coarser reactant particle sizes and modest dilution of the reactants with product particles. Thus, eliminating reaction propagation—having a normal diffusion reaction—allowed successful scaling of the reaction hot pressing to produce billets 15 cm square and 3–5 cm thick for ballistic testing. While it might be argued that achieving sufficient axial reaction propagation might have been sufficient for successful scaling, this is considered doubtful, and would probably have required hot pressing one body at a time, rather than a few to several at a time (which would be more economical). Thus, use of propagating reactions for producing bulk bodies by reactive hot pressing or closely related fabrication is seen as something to be avoided. The advantage of using the reaction processing route can often be that of lower raw materials costs as illustrated in Table 6.2.

Another potential of reaction processing, especially with pressure consolidation such as hot pressing, is obtaining a finer, possibly more homogeneous microstructure, since all grains are nucleated by the reactions, rather than simply growing from the starting powder particles. Thus, if there is sufficiently limited temperature and time during densification, mutual inhibition of growth of particles of one phase by those of other phase(s), or both, finer microstructures may be attained. Comparison of reaction hot pressed versus conventional hot pressing of mixed oxide-nonoxide composites has shown promising results for the former versus the latter bodies despite there being more experience to guide the conventional processing [100]. However, again requirements for improvement were noted, since reaction hot-pressed composites often failed from isolated larger grains or clusters of them. It was suggested that these resulted from heterogeneities in the reactant compact as well as local accumulation of transient liquid phases during the reaction, for example of Al and B_2O_3, with the two being interactive. It was also noted that local excesses of transient liquid phases can often be inhibited by coating all particle or those that will melt with a reactant that does not melt. Thus, promising results have been obtained by using polyfurfural alcohol, a polymerizable liquid that can be used as an infiltrant or part of the binder, as a moderate cost source of carbon.

There are several processes that depend on reaction in a liquid or gaseous state that can have limited to extensive applicability to fabrication of ceramics and ceramic composites. Starting with the former: Electrolytic deposition of metal oxides or their hydroxide precursors from water solution of metal ions [102] has some applicability, especially for electronic applications, but is limited to thin depositions, for example, 1–10 μm, such as coatings. Similarly, electrodeposition of other materials, especially nonoxide materials, from various molten salts (Chap. 3) has some applicability since substantially greater deposition should be feasible, but the use of molten salt baths is a serious limitation.

Much, if not all, of the use of preceramic polymers (Sec. 6.5) also falls in this category via their preparation, polymerization, or both [59–61]. This is by far most developed for carbon bodies, as illustrated in Fig. 6.7. There is again applicability for thin layers, (coatings), and bodies with small cross-sectional dimensions, especially fibers (Sec. 7.2.1), but also some potential for bulk monolithic or composite ceramics. Costs, large shrinkages and related issues, for example, of stress, limit the use of polymeric ceramic precursors for producing bulk bodies. While there are some possible means of extending fabrication of some ceramics by polymer pyrolysis via some CVD processing as discussed below, much application of such processing for both monolithic and composite ceramics appears to be use of preceramic polymers as part or all of the binder for green body fabrications (Sec. 4.3). The frequent requirement of spray-drying systems with organic solvents is a definite limitation for such fabrication, but such spray drying is done where its costs can be justified. In the case of composites, especially fiber composites, use of preceramic polymers as the matrix source is promising (Sec. 8.2.3).

Turning to vapor-phase reactions, there are important uses of physically generated vapors—by evaporation or sputtering of—metals,and reacting the metal vapor with a gas—methane to form carbides or nitrogen to form nitrides. Such processing is again limited to thin layers or coatings, commonly for wear applications. A good example is arc vaporization of Ti and its reaction in the vapor state to form TiN, TiC, or combination coatings on consumer and industrial drill bits for wear resistance and bathroom fixtures.

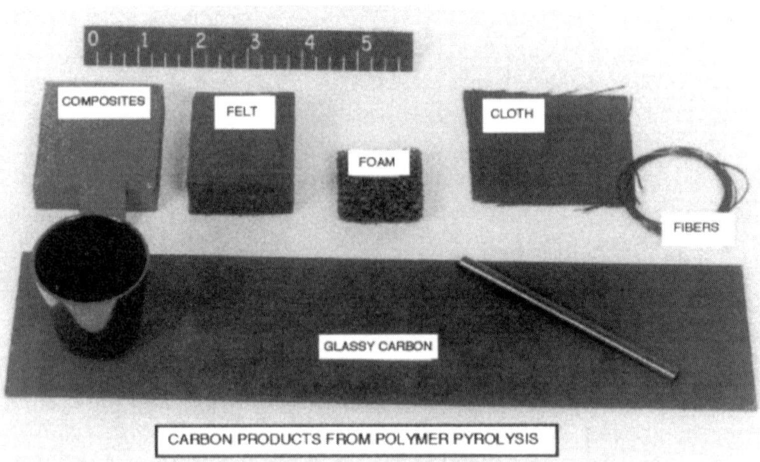

FIGURE 6.7 Examples of carbon bodies and items made via polymer pyrolysis ranging from graphite fibers and cloth (top right), carbon-carbon composite (top left) felt, and foam (center), along with bulk glassy carbon bodies (plate, rod, and crucible, bottom). Scale in inches. (From Ref. 61. Published with permission of Plenum Publishing Corp.)

The most important vapor-phase reaction process, and the focus of this section, is that of chemical vapor deposition (CVD) [103–108] and an important and growing subprocess of chemical vapor infiltration (CVI) of a porous preform, e.g., of fibers. These processes entail either of two closely related reactions of suitably selected and heated gaseous compounds. The simplest and a widely practiced manifestation is decomposition of a single compound to a metal or occasionally a ceramic (e.g., SiC or BN, from methyltrichlorosilane and borazine, respectively) with the deposition of these products as a dense, coherent solid on a heated, "mold," surface. The other and also widely used manifestation is the decomposition of more than one, commonly two, gaseous compounds in the same reactor such that a solid product of the reaction of atomic species resulting from the decomposition and reaction deposits out as a dense, coherent solid on a heated, "mold," surface. This latter manifestation can produce a wide variety of ceramic and other (e.g., semiconductor) compounds, as well as a variety of "alloys" or composites, and thus complements the use of a single decomposing gaseous species that produces a few compounds and many metals and related elements such as B and Si. In either case, the heated "mold" surface can be of simple or fairly complex shape to produce correspondingly shaped parts (Fig. 6.8).

The common source of most metal and related cation species are halide compounds, commonly chlorides (many of which are solids, not gases, at room and

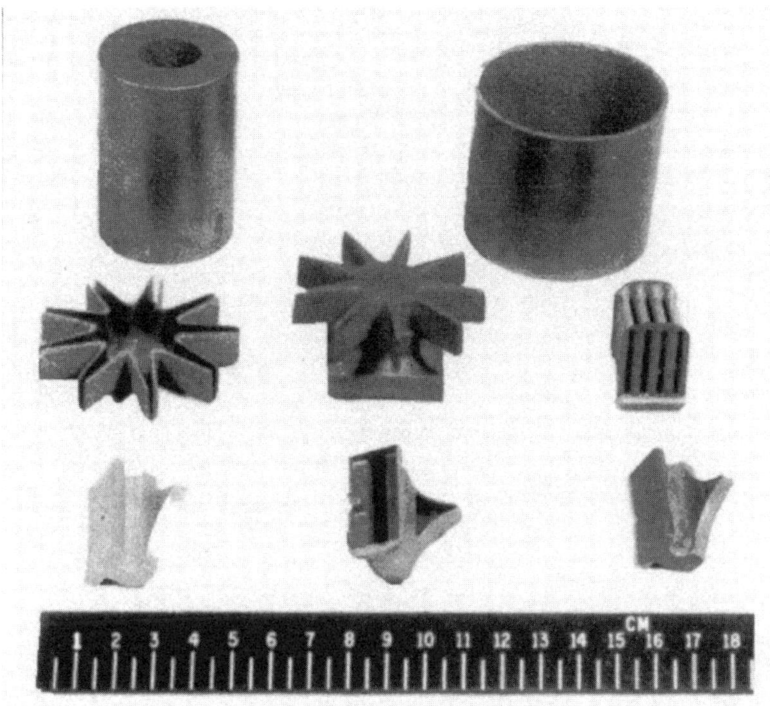

FIGURE 6.8 Examples of CVD SiC bodies. Top row: thick and thin wall cylinders, middle row; carbon mold for a simple prototype turbine rotor, the SiC rotor, and a small, thin wall heat exchanger; and bottom row: metal prototype, carbon mold for CVD, and SiC CVD part. (Samples courtesy of R. Engdahl of Synterials. From Ref. 61. Published with permission of Plenum Publishing Corp.)

modest temperatures, but are readily vaporized well below CVD temperatures, by passing a halogen gas over heated particles of the metal). Such halides are typically the low-cost sources for many metals and are widely used with suitable dilution and carrier gases. While, O_2 and N_2 can be used to form oxides and nitrides, more often other gases are used, e.g., H_2O or CO_2 for oxides and ammonia for nitrides, along with methane for carbides, and boron halides for borides (and B and BN). Ceramics are typically produced with such reactions at temperatures of 900–1500°C, though these may be reduced by plasma assistance of the reactions. Other gases are also used, especially metalorganic ones for metals and semiconductors which are extensively used in the electronics industry despite their frequent toxicity and high cost since they are used in small amounts and allow processing at much lower temperatures [107,108]. Such gases can generally be reacted at temperatures of a few to several hundred degrees Centigrade. Reactions with either type of precursor are typically done at pressures substantially below one atmosphere. However, the above noted reactants, other than most metalorganic ones, are quite suitable for most metal and ceramic processing, and offer low to modest cost sources for CVD. Such lower costs, and potential diversity and quality of CVD materials that may be produced, and the potential for near net shape fabrication give CVD substantial potential, only a portion of which has been realized. Possibilities for improved products and use to go beyond some of the current limitations are discussed after summarizing key aspects of the status of CVD.

There are substantial applications of CVD. Its largest use is for films in the electronics industry, where substantial process development of control and reproducibility has occurred, including plasma assisted CVD. There is also successful commercial development of first, B, and later, SiC, filaments made by CVD, which have also provided some insight on reproducibility. Some of these process advances can be transferred to CVD production of bulk bodies of particular interest here. However, there are also important successes for CVD of bulk bodies [104]. Earlier ones were for graphite bodies, pyrolitic graphite (PG), hence CVD graphite. These included re-entry nose tips for intercontinental ballistic missiles (ICBMs) (before being replaced by carbon-carbon composites for all-weather capability), as well as nozzles for various rockets, both demonstrating large and complex shapes, and measurable thicknesses (e.g., a centimeter or more). Besides a number of commercial technical applications for PG, it became widely used as the liner in the bowls of tobacco pipes [104], which required substantial volume production at modest cost (for example reduction in costs from a few tens of dollars for similar size custom PG crucibles to of the order of a dollar or less per pipe bowl). Another commercial extension of the PG market was the chemical exfolliation of bulk PG and then the rolling of this material to make Grafolil®, a thick graphite paper. Substantial CVI is used to produce some of the carbon matrix in carbon-carbon composites, for example for high-performance aircraft brakes.

Technological and commercial successes are not restricted to PG. They include the commercial development of ZnSe and ZnS for IR windows, for example, as large plates of 90 × 120 × 2.5 cm [104]. Various size IRdomes of other ceramics such as ZnS and MgO have been demonstrated, from 8 to 20-cm-base diameters. Another important commercial and technological development was that of preforms (i.e., "billets") from which glass optical fibers are drawn for optical communications [109–111]. While there are also fusion or sintering methods, there are at least three variations of CVD processing as a key step in processing such preforms, for example of 2.2-cm dia. by 0.5 to 1-m long (giving a few kilometers of 125-μm diameter fiber, commonly drawn at 1 m/s).

Frequently other uses of CVD for fabricating bulk bodies as opposed to films, coatings, and filaments, though considerable, face limitations due substantially to four partially interrelated factors of moderate deposition rates, microstructures obtained, residual stresses, and limited toughening/strengthening. Thus, deposition rates can be quite high, e.g., millimeters per minute, at higher temperatures and reactant partial pressures, to produce thick parts as noted above and shown in Figs. 6.8 and 6.9. However, most deposition is at much lower, by 1–2 orders of magnitude, since higher deposition rates often result in less desirable microstructures. This includes not only larger, often columnar, grains, but also the occurrence, or larger scale, of colony structure, which is the nucleation and growth of clusters of oriented (often, but not necessarily, elongated) grains that are common for most deposition processes. Such structures, which commonly resemble shiefs of cut grain plants (Fig. 6.10) are often sources of weakness since they act as large grains because of the crystalline missorientation between the body matrix and a colony or two abutting colonies. This behavior of colonies as large grains is enhanced by possible collection of impurities, the occurrence of some limited porosity, or both at colony boundaries.

CVD, like most deposition processes can result in substantial residual stresses that can also be a problem (Fig. 6.8). Such stresses can arise from different sources, with a common one being differences in thermal expansion between the deposit and the substrate ("mold") material. This source of stresses can commonly be minimized by selection of a substrate material with similar expansion as the deposit, which is generally feasible. Other sources of residual stresses appear to include gradients of the degree of preferred grain orientation and of atomic composition across the deposit, but a clear understanding of these is not in hand. Thus, the minimization of such stresses is still primarily empirical on a material/process parameter basis. Deposition of bodies of mixed composition—of solid solutions or composites—should give insight to sources and possible solutions to residual stresses.

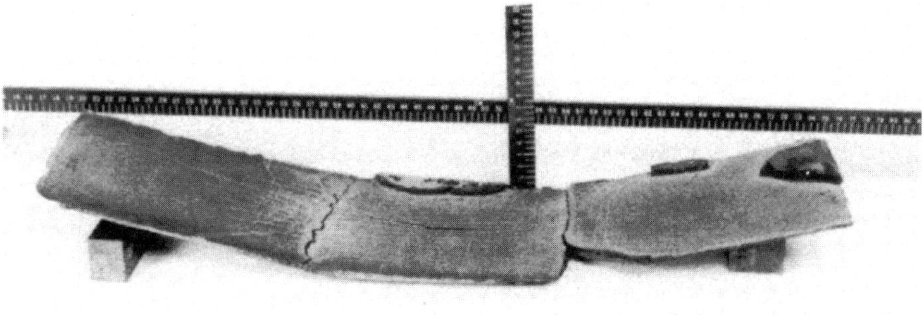

FIGURE 6.9 Photo of a large, thick SiC plate made by CVD, having significant bowing and cracking due to residual stresses. Scale in centimeters (From Ref. 61. Published with permission of Plenum Publishing Corp.)

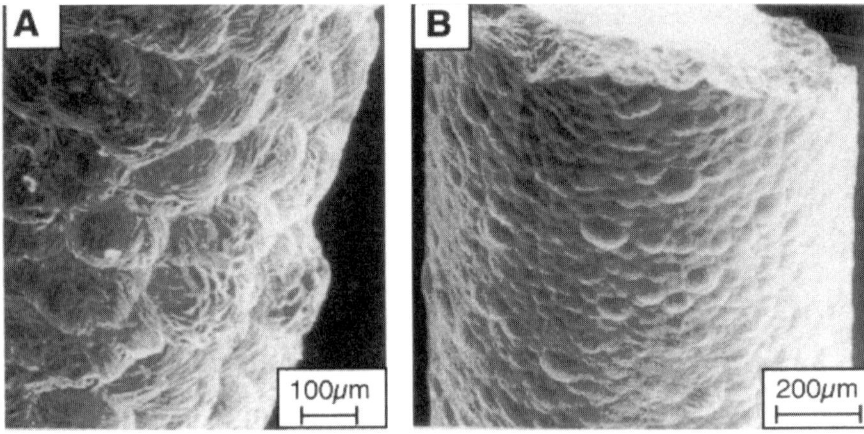

FIGURE 6.10 Examples of CVD colony structure. (A) "Bumpy," botryoidal as-deposited CVD surface on Si_3N_4. (B.) Fracture cross section of a colony showing common, but not universal, elongated grain structure in SiC.

A limitation of CVD, however, has been a focus on producing only pure, single phase materials. Purity of the deposited material is often a strength of CVD; for example, its ability to give dense bodies of materials that otherwise can only be densified with additives, which may give lower temperature toughening, but seriously limit high-temperature performance. However, only limited attempts have been made to purposely make CVD bodies with two or more compounds that can have one or more of several functions, besides possible effects on residual stresses noted above. Where one or more additional phase is formed, the functions may be (1) limiting grain sizes, (2) providing some composite character, particularly toughening, and (3) compositional gradients for both design of materials, as well as developing surface compressive stresses. Where a single phase such as a ternary compound is formed, it expands the material possibilities of CVD, and where solid solutions are formed, compositional gradients may be purposely introduced to provide surface compressive stresses. Thus, instead of thinking only about pure compounds and CVD as the only step in the fabrication process, consider a broader range of compositions and processing. Processing of ternary compounds and especially composites and possible additional steps in the processing are examples, e.g., sintering of porous preforms from CVD, similar to forming optical fiber preforms, and heat treatment to change compositions and microstructures, which could include transient phases, including liquid ones. Thus, to expand CVD and related processing, it should be viewed more from a materials perspective and not restricted to the chemical perspective that was essential to its development, but appears to be restricting its further expansion and diversification.

There are results that indicate the promise of addressing the substantial task of investigating the above possibilities. First, some CVD of ternary ceramic compounds has been made, such as Ti_3SiC_2 [112–116]. Second, some deposition of two different cations can significantly reduce grain sizes and improve mechanical properties [117,118], and similar results should occur for two different anions. Production of ceramic particulate composites, e.g., of Si_3N_4 with dispersed (often nanometer) particles of BN(118) or TiN [118–120], some with particle-matrix epitaxial relations, an amorphous matrix, and transparency have been reported.

The first of four other important factors for more study is the preferred orientation that often occurs (often associated with colony structure, Fig. 6.8). Control of this could give not only more reproducible bodies, but also may allow additional uses. Control of orientation could allow more tailoring of properties, providing oriented grains and properties to better meet some uses. An extreme of this is deposition of coating materials with very anisotropic thermal expansion such that the high crystal expansion directions are oriented parallel with the metal substrate surface on which the deposit is made. Thus, greater compatibility between coating and substrate can be obtained, while giving low coating expansion normal to the substrate (a characteristic often of interest in coatings). Thermal expansion anisotropies of some ceramics that may be used for obtaining more thermal expansion compatibility with metal substrates when the ceramic is deposited with high expansion directions parallel with the metal surface to maintain more constant clearances of coating from other components as temperature varies (Table 6.4). Second, gradation of composition in solid solution or composite bodies can allow grading of thermal expansion and hence generation and tailoring of surface compressive stresses to give greater mechanical reliability (until use temperatures reach fabrication temperatures) should be feasible. Third, CVI has demonstrated important capabilities in production of carbon-carbon composites and shows promise for a number of experimental ceramic fiber composites for specialized, high value-added demanding applications (Sec. 7.5). The demonstration that the residual porosity typically left by CVI is amongst the most benign in limiting mechanical properties [121,122] further aids future possible uses of CVI for ceramic composites. There are also important opportunities for CVI processing of other specialized ceramic bodies, for instance for designed porous structures (Sec. 7.3.2). Fourth, there have been demonstrations of modifications of CVD that results in the reacting gases forming an intermediate liquid phase on the surface of the deposit, that decomposes to the product, for example, W-C or SiC. This liquid intermediate, which is apparently related to a polymer precursor, results in extremely fine, nanoscale, grains and very high as-deposited strengths at room temperature [123–127]. While these results are for small samples with limited scaling, they suggest possibilities, of combining CVD and polymer pyrolysis, that deserve further investigation.

TABLE 6.4 Examples of Thermal Expansion Anistropy of Ceramic Oxides

Material	Melting temp. (°C)	Thermal expanion coeffiencent (10^{-6} °C^{-1}) α a	α b	α c
Al_2O_3	2050	8.6	8.6	9.5
TiO_2	1825	8.3	8.3	10.2
Cr_2O_3	2260	8.1	8.1	6.1
SiO_2 (α quartz)	1720	22	22	12
$LiA1siO_4$ (ß eucryptite)	1400	8.2	8.2	-17.6
$MgTiO_3$	1630	9.4	9.4	12.4
$LiNbO_3$	1410	21.2	21.2	0.1
Al_2TiO_5	1860	-3.0	11.8	21.8
$MgSnB_2O_6$		4.6	4.6	19.0
$MgTi_2O_5$	1650	2.3	10.8	15.9
$MgFeTaO_5$		1.9	9.1	13.7
$CaWO_4$	1580	13.7	13.7	21.5

After Rice [61]. Published with permission of Plenum Publishing Corp.

6.6 MELT PROCESSING

6.6.1 Glasses and Polycrystalline Bodies

Melt processing, though widely neglected by many in the field of ceramics, especially those in high technology ceramics, is both diverse and very important. While many incorrectly feel that melt processing is expensive because of energy costs, such costs are generally low as attested by the modest costs of many melt processed ceramics (and metals). The largest volume of ceramics produced, namely glasses, is via melt processing, which clearly shows its low cost capabilities. Melt processing entails not only the melting technology, but also diverse forming technologies, which results in a diverse array of sizes and shapes [128]. These include large architectural glass pieces and telescope mirrors, e.g., the 200-in. (~ 5 m) diameter Palomar mirror cast as one piece (including the large honeycomb backing, though more recent large mirrors have been made in sections to be joined, Sec. 8.3.3). Some of this technology is used in forming some glass-crystalline ceramic, e.g., mica, composites (Sec. 6.3) [47], indicating other possible extensions. Another important extension is via crystallization of glasses, commonly aided by additives (Sec. 3.5) to improve properties [9].

Much of the crystalline-based ceramic refractories produced, most of which are composites, are made by arc skull melting, that is, melting almost all of the material in a suitably designed water-cooled container, leaving a thin layer of unmelted, and hence noncontaminating, material (i.e., the "skull") against the container (which can be a few meters in diameter). A substantial volume of these materials is solidified in the skull form, then crushed into grain for various refractory and other uses, for abrasives and for thermally conducting, electrical insulation for encapsulated electrical heating elements (Sec. 2.4). (Note that the comminution costs are generally more than the melting costs). The other common use of melt forming for refractories is by tapping the melt in an arc skull melter to cast the melt in molds (generally of graphite) to make refractory bricks and furnace parts, many of much larger size (some weighing well over a ton, being some of the largest ceramic bodies made) and more complex shapes. An important factor in such fusion casting is controlling the microstructure, which is often done by controlling nucleation and growth of grains, and often the crystalline phase in which they occur, both of which are often impacted by the use of additives (Sec. 3.5). Another critical factor is controlling sources of both extrinsic and intrinsic porosity. A clearly extrinsic one is outgassing of raw materials, with entrapment of much of the resultant gas in the melt, which is an important problem not well recognized outside the field of fusion derived ceramics. It is discussed some below and more extensively in Section 8.2.1. Exsolution of gasses dissolved in the melt, released on cooling to and through solidification, can be an important problem, but is generally very material, and, to some extent, process dependent, and hence not addressed further. A clearly intrinsic, and very important, source of porosity in fusion processing is the intrinsic changes in volumes of materials between the melt and solid state.

An important factor in casting any crystalline body from the melt is the change in volume on solidification. While a few materials expand on solidification, e.g., H_2O and Si (which can result in stresses and cracking), most materials shrink on going from the liquid to the solid state, which, if not controlled in the nature of its occurrence, is an important source of porosity. Most cast metals have volume shrinkages of 5–10 v/o on solidification, but the (somewhat limited, mainly oxide) data for ceramics shows that, while some have solidification volume shrinkages similar to metals, they frequently have shrinkages of 10–20 v/o or more—20% for Al_2O_3 [129] and 10–17% for rare earth oxides [130]. (Limited data also suggest that some halides may have even higher solidification shrinkages, e.g., 40 v/o.).

Solidification shrinkage leads to porosity in the solidified body whenever melt can no longer accommodate the solidification shrinkage. This occurs whenever melt to be solidified becomes sufficiently surrounded by solidified material, to inhibit or prevent continued supply of melt to the solidification front (Fig. 6.11). The volume of porosity left on final solidification in such cases is determined by

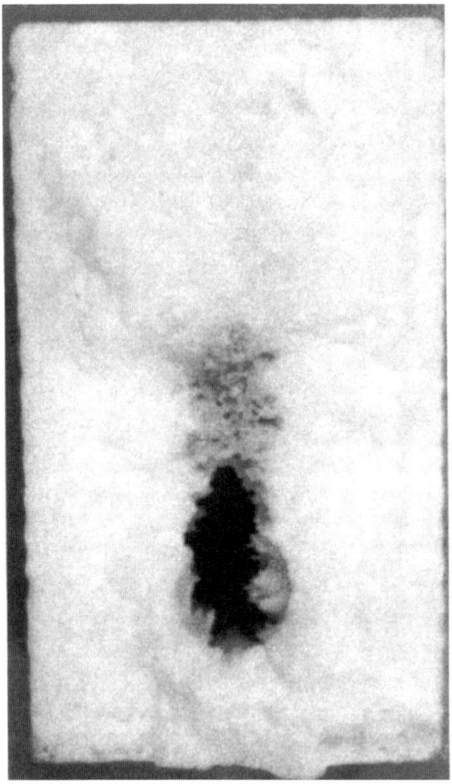

FIGURE 6.11 Cross section of a fused cast refractory brick (~ 5 cm thick) showing the substantial porous area near the back-center of the brick, completely encased in dense solidified material.

the intrinsic solidification shrinkage and the volume of melt involved (as well as any exsolved gas). Whether the entrapped porosity is a single large pore or a porous region depends on local solidification conditions (e.g., a solidification front involving two or more phases of differing melting points is likely to have distributed porosity. Single-phase solidifying is more likely to involve a single larger pore, but this can be impacted by grain morphology.

There are two basic ways in which the effects of porosity generation from solidification can be reduced or eliminated: (1) reduce the effects of the porosity and (2) reduce its amount, preferably eliminate it. Making the porosity more benign can be accomplished to varying extents by making the pores finer in size, more dispersed, and with geometries and locations that are more benign [121], via compositional changes to have phases of different solidification point and different solidification morphologies. However, controlling the location of the porosity within parts to a more benign locations is a particularly important factor, for instance for refractories. Thus, for many refractory applications casting molds are designed so the solidification generated porosity occurs in a more benign area—in the center or toward the back side (opposite from the surface exposed to the use environment) (Fig. 6.11).

The primary approach to limiting or eliminating porosity from solidification shrinkage is directional solidification; i.e., where solidification always occurs at a solid surface that remains in contact with the bulk of the melt so the shrinkage on solidification always occurs at the melt-solid interface and is thus totally taken up by the melt in contact with the solidifying surface. Such directional solidification has been successful in producing fully dense transparent plates for IR optical windows of the cubic materials of CaF_2 or SrF_2 25-cm dia. and 1-cm thick [131] and of $MgAl_2O_4$ 12-cm dia. and a centimeter or more thick [132]. Because of the relative slow cooling to maintain the desired directional solidification, grains were large (e.g., 1–2 cm), but nominally equiaxed. In both cases post solidification annealing was necessary to reduce stresses (revealed by birefringence). However, strengths of test bars were promising since they were similar to those for single crystals of the same material of comparable surface finish. It is expected that such casting can be extended to other cubic materials, and possibly some of limited anisotropy, as well as to larger sizes (of both lateral and thickness dimensions) quite possibly substantially larger ones.

There are other ways to approach or achieve fully directional solidification and obtain finer grain structures; that is, rather than fill a mold with melt, add limited amounts of melt, initially to a mold surface and then to the exposed parts of the part where growth is desired. This has, for example, been done by temporarily dipping simple metal molds of female or male shapes into a pool of skull melted Al_2O_3-ZrO_2 eutectic or Y-PSZ such that a thin layer (~ μm thick, Fig. 6.12) solidifies before and as the mold and part are removed from the melt pool, before being inserted repeatedly until the size "casting" desired is achieved. Though some limited, finer porosity was entrained due to local deviations from directional solidification, strengths of test bars approaching 600 MPa were promising.

The other and more versatile way of directionally and more rapidly solidifying small amounts of melt at a time on a surface is to accelerate molten droplets to splat on the surface where growth of the body is desired. A very practical manifestation of this is the use of melt spraying; i.e., via use of arc (for electrically conductive materials that can be made in wire form), flame (e.g., oxyacetylene), or plasma spraying that are widely used for melt spraying coatings. This listing is in the order of increasing temperature capability and cost, though plasma spraying is still generally of moderate cost and offers significant increases in process control, not only for the spraying parameters including spray environment, but also for substrate preparation, such as cleaning. While melt spraying coatings results in very rapid quench-

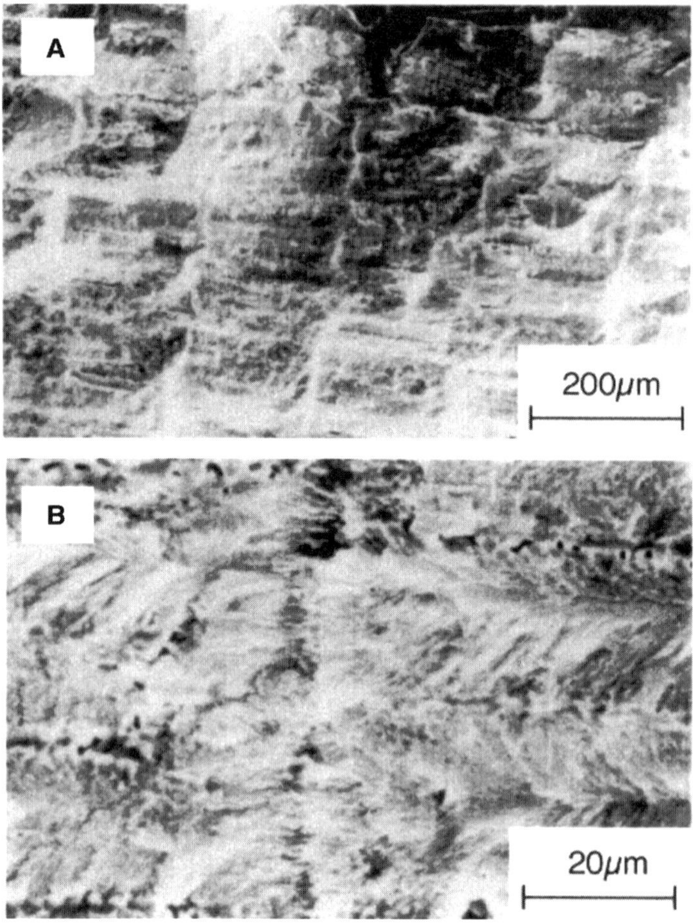

FIGURE 6.12 Dip-cast Al_2O_3-ZrO_2 ceramic and its microstructure. (A) Lower magnification of fracture cross section. Horizontal striations are demarcation of individual dips and resultant directional solidification. (B) Higher magnification showing the eutectic colony structure and boundary (with some porosity) between dip layers.

ing giving both fine microstructures and microcracks that may be beneficial to coating performance, more control of thermal stresses of parts during deposition is necessary to preclude larger scale part cracking. Earlier work showed this to generally be accomplished by heating molds and parts to 1000°C (i.e., as for welding of ceramics, Sec. 8.3.3) [133]. Though resulting in some increase in grain sizes, room temperature flexure strengths of the order of 15 MPa were somewhat promising [134]. However, use of grain growth inhibitors or toughening additives would be expected to give considerable strength improvement. Bodies of substantial sizes and various shapes and materials have been demonstrated by such melt spraying. It has also shown promise for macrocomposites (e.g., of a body with a ferrite core surrounded by a ceramic dielectric [135]. Feasibility of fabricating another macrocomposite consisting of a ZrO_2 cylindrical shell encircled by a metal shell, both substantially thicker than normal coatings, was also shown. The motivation was to form a lower cost ceramic cylinder liner by greatly reducing the major costs of machining such liners, which by the normal fabrication of a free standing ZrO_2 sleeve, machining both its surfaces and those of a metal sleeve into which the ZrO_2 sleeve would be shrunk fit and the metal-ZrO_2 composite sleeve then presses into the cylinder liner. Melt spraying of the metal-ZrO_2 composite sleeve would eliminate most of the most expensive machining, that of the ZrO_2 since its ID surface, being formed on a mandrel should be nearer final dimensions and finish than an as-fired free standing ZrO_2 cylindrical sleeve, and only the outer metal sleeve needs machining rather than both surfaces. While only a brief exploratory program was carried out, potential was demonstrated by fabricating metal-ZrO_2 composite sleeves (~ mm thick, ~ 6 cm in dia and ~ 12 cm in height) that showed > 700 MPa hoop tensile strengths (R.W. Rice, unpublished work circa 1986). Some large freestanding ceramic parts are apparently manufactured by melt spraying; for example, fused silica cylindrical shells ~ 40–50-cm diameters and heights with thicknesses of 1–3 cm for insulators for induction heated furnaces for hot pressing and other applications.

Melt spraying to form bulk bodies, not just coatings, also has other potentials for expansion. One is in the fabrication of bodies with controlled porosity, as demonstrated by cospraying oxide particles with carbon spheres or with selected resin intermediates, followed by burnout to produce porous manganite coatings for fuel cells [136]. Another approach has been to obtain denser coatings (hence also denser bodies) by HIPing following spraying [137], or to use CVD for some infiltration of pores, or surface sealing, or both [138]. Finally, a potentially large area of expansion is that of using melt spraying as a means of forming ceramic matrices in ceramic fiber or particulate composites. There are limited reports of melt spraying to produce ceramic composites; LaPiere and coworkers [139] sprayed alumina-SiC particle composites with resultant strengths of ~ 100 MPa after postdeposit annealing. This can probably be improved significantly by spraying the body at elevated temperatures, as for bulk alumina bodies as noted above. Again, considering other postspraying fabrication steps may be practical and significant.

6.6.2 Single Crystals

Consider next single-crystal growth, which is mainly via directional solidification from the melt, and can be divided into techniques for growing one single crystal at a time versus those that grow many crystals at a time, the latter typically via growing a large ingot of large grains. Consider the latter first. This is typically done via skull melting—the melting of almost all of the material in a suitably designed water-cooled crucible, leaving a thin layer of unmelted, and hence noncontaminating, material (the "skull") against the crucible (which can be up to a few meters in diameter). While in principle applicable to many materials, this process is typically applied to insulating materials, mainly oxides, for which melting is typically achieved using either (typically graphite electrode) arc or high-frequency inductive heating, with the heating choice depending in part on the material to be melted. Thus, oxides that are not susceptible to serious oxygen loss in the proximity of very hot graphite electrodes (consumed in the process) are commonly arc skull melted; e.g., to produce abrasive or refractory grain of Al_2O_3 or MgO refractory grain (Sec. 2.4). Materials that are amenable to inductive heating at high temperatures (which includes a large number of oxides), especially those subject to serious oxygen loss under reducing conditions (e.g., with TiO_2 and ZrO_2) and hence less amenable to arc melting, are inductively skull melted [141]. This method is used on a large industrial scale to produce the large volumes (probably hundreds of tons) of the cubic zirconia crystals for the jewelry trade. Regardless of the heating method, nucleation of grains to be formed in the solidification of the melt occurs from the unmelted material, but subsequent growth often occurs with some preferred orientation in the growth direction, especially with highly directional solidification that is usually imposed. The latter typically results in a substantial columnar character of the grains, e.g., aspect ratios of 2–5, at least in smaller crystals, such as those 1–3-cm diameter grown in skulls < 20-cm diameter (Fig. 6.4A). The other primary control over the grain size is via the melt dimensions, that is, larger diameter melts give larger diameter grains, reaching of the order of 10 cm in large ZrO_2 skull melts for the jewelry trade. A fortuitous aspect of much skull melting is that larger grains of at least some materials such as MgO, CaO, and ZrO_2 tend to separate along grain boundaries (attributed to possible effects of elastic anisotropy [9]). This greatly limits thermal stresses and cracking on cooling ingots after solidification.

While the market for skull-melted ZrO_2 crystals is dominated by the jewelry uses of cubic zirconia, there are technical uses of it, for example, as substrates for superconductor electronics. Further, the same facilities can be used to grow partially stabilized crystals, which have very promising properties at both room and elevated temperatures (e.g., respective strengths of 1.4 GPa and 0.7 GPa [142]. Trial engine-wear components machined from such crystals have proved superior to polycrystalline components, and could potentially be cost competitive with the latter given the economies of scale in both growing and machining of such crystals for the jewelry trade. Further, concepts for significantly strengthening components made from stabilized ZrO_2 crystals used for the jewelry trade have been proposed.

Consider now growth of individual single crystals one at a time, which is a major method of growing single crystals [143]. While there are a number of methods of growth and variations of these, most proceed from a seed crystal of the desired material and selected orientation. There are various ways of providing melt for controlled solidification from the seed. The oldest is Verneuil growth which entails feeding suitable powder particles through a torch—an oxy-acetylene, flame to deposit molten particles on the boule growing from a seed. Such crystals, which generally have more imperfections than crystals from most other growth methods, are still in production, especially for the jewelry trade, often with dopants for various coloration. Boules about 2 cm in diameter and 7 cm in length are common products.

There are various methods of growing larger, more perfect crystals from a molten bath. A major, long established method in laboratory and industrial use is Czochralski growth, which has melt in a crucible drawn up to a seed just contacting the meniscus with the melt surface, with the resultant growing seed slowly retracted above the melt. Such crystal growth by "pulling" from the melt can provide some rough control of the size of the crystals grown via the volume and diameter of melt bath available and speed of growth. Most crystals are nominally cylindrical boules and usually limited in crystal diameters and lengths to respectively less than the diameter and the volume of the crucible, but also depend on the material. Thus, sapphire is grown to diameters of about 15 cm and apparently in some cases to 20 cm with lengths a few times this, with lengths and diameters having inverse trends with each other as limited by crucible volume. Si crystals can be even larger, to 30 cm dia. and 100–300 kg. Crystals grow in only a few preferred growth directions dependent on crystal structure and composition, giving very limited control of the boule shape other than selecting a seed crystal with a crystallographic orientation that yields suitable growth near or along crystal directions of interest, though in some cases crystal orientations can be grown as larger slab shapes, e.g., of sapphire. However, while there are growth constraints as outlined above, some materials and growth conditions yield growth of hollow and other novel crystal shapes and morphologies [144]. While such crystal pulling is applicable to many crystals, sapphire is a major industrial product, as for other growth methods in use. Products such as those of Fig. 6.13 are often machined from boules.

An important development in "pulling" single crystals from the melt to give versatile shapes was the discovery and development of the edge-defined, film-fed growth (EFG) technique [145–149]. This simple, clever method basically consists of having a crucible of melt from which to grow desired crystals and a die, e.g., of the same material as the crucible so it is compatible with the melt under the growing conditions. The die is shaped with channels in it so that when it is held in contact with the melt will be raised by capillary action to the top of the die, where crystal growth proceeds at or slightly above the top of the die where thermal gradients are maintained for crystal growth as in any crystal "pulling" process. As in the latter processes, growth is initiated using a seed crystal, whose orientation, along with the growth conditions, determines the crystal growth axis, again like other crystal "pulling" processes. Thus, the basic, and critical invention of the EFG® process was the use of a die to locally shape the liquid in the immediate vicinity of the crystal growth. This process was invented by Harry LaBelle, a technician on a government contract project to grow sapphire filaments from the melt (he became the president of the company, Saphikon (Milford, WA), set up in the mid- to late sixties to commercially produce crystal products grown by this process).

The EFG® process has been demonstrated with some metallic, intermetallic, and different ceramics (the latter apparently including some orientations of BeO that apparently allow pushing the phase transformation back toward the meniscus and thus suppressing it). Commercial production of sapphire and Si is dominant, where a variety of two-dimensional shapes can be made by growth (mainly with a constant horizontal cross section), and machining (Fig. 6.13). Shapes include filaments, tubes of different cross sections, and tapes (some used for wear resistant windows for bar code readers at supermarket and other checkout stations), some of these shapes with differing crystal orientations. Some variations are feasible—by twisting the axis of growth with sapphire tubes of rectangular cross section, Bourdon gauge tubes can be made. A limitation of the process for some applications has been the occurrence of limited porosity somewhat under the surface, especially at faster product growth rates. However, some reductions of this porosity problem have been made, and advances in machining efficiency make it more practical to machine off the porous layer where it occurrs. Some attempts have been made to make larger, more complex bodies such as IRdomes, but with incomplete success. However, more recently rapid prototyping/solid freeform processing concepts have been applied to the EFG process using a die that can be articulated to build up a dome shape of sapphire in small layers [150]. Such concepts may have considerable potential for EFG and other melt processing (Sec. 6.7.3).

The EFG process makes a diversity of sapphire products, with significant ones being a major source of thin sapphire plates to be laminated to glass sheets to give wear-resistant windows for supermarket checkout bar code scanners and a variety of windows and other components for semiconductor and other processing furnaces. Though some subsurface microporosity still occurs, it can be

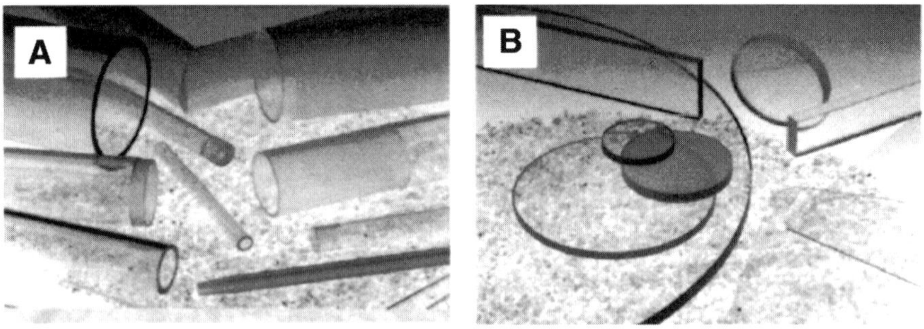

FIGURE 6.13 Examples of sapphire parts made by the EFG® method. (Photos courtesy of J. Locher of Saphikon).

significantly reduced by slower growth rates and use of W instead of Mo dies, but costs of this are generally higher than to simply machine off the porous layer, which is commonly done for flat parts such as for scanner window application. Pieces 30-cm square can be grown and sizes of 60-cm square are seen as feasible [151]. Added possibilities are indicated by the EFG method having been used to grow porous sapphire crystal by wicking melt into a porous W body and directionally solidifying the melt then etching out the W (Sec. 7.3)

The other important method of growing individual single crystals is the heat exchanger method (HEM), which instead solidifies melt in a crucible to a single crystal; that is, in contrast to normal crystal pulling the crucible dimensions and shape are typically those of the resultant crystal. This is done by melting all of the charge in a large crucible, except for a small seed situated at the bottom center (cooled) area of the crucible so that on cooling the melt in the crucible, all of the melt will solidify to become part of the crystal the seed starts. This method was developed by Fred Schmid and Dennis Viechnicki, then of the Army Materials and Mechanics Research Center in Watertown, MA, and has been used for commercial production of crystals, especially sapphire, since the formation of Crystal Systems in 1969, of which Schmid is president.

Because the crystal size is directly determined by the crucible size, not how much liquid can be moved vertically across a meniscus, and large crucibles are feasible via welding of refractory metal sheets, the HEM allows by far the largest crystals to be grown (Fig. 6.14). Thus, sapphire boules up to 34 cm dia. and a substantial fraction of this thick, e.g., weighing 65 kg have been grown [152], and crystal sizes to of the order of 50 cm dia. are seen feasible. Again, sapphire is the main product, grown in four grades ranging from the highest quality sapphire free of light scattering and lattice distortion to that with measurable light scattering and lattice distortion suitable for mechanical, structural, and electronic applications. Because the crucible dimensions perpendicular to the growth direction determine the crystal size, they can in principle be used to some extent to shape the crystal, but this must be balanced against growing more material in a given run and sawing a larger body into bodies of the shape desired. Further, some possibilities of inserting refractory metal sheets into the crucible has been shown to provide some resultant shaping, e.g., of sapphire Irdomes [153,154]. Note that this method is also applicable to directional solidification of polycrystalline ingots, e.g., of 66-cm square, 240-kg Si ingots [155].

Two factors should be noted about the above production of single-crystal components. First, while some important shaping during growth is feasible, machining still plays a large role. Improvements in machining efficiency, hence lowering of costs, e.g., via gang slicing and boring, continues to expand production opportunities. Second, while a primary raw material purity requirement for single crystal growth is for low levels of cation species, especially those of refractory impurities, sources of volatile species are also an important problem.

FIGURE 6.14 Large sapphire boules (~34 cm dia.) grown by the Heat Exchanger Method (HEM). (Photo courtesy of F. Schmid, President, Crystal Systems, Salem, MA.)

Such species, e.g., from adsorbed species and anions left from calcining or powder storage (Sec. 8.2.1), are also a serious problem since they often result in trapped bubbles. In the past, much crystal growth was from previously melted material, from single-crystal scrap from machining, Verneuil boules, as well as purposely melted material for the sole purpose of reducing species causing bubbles. However, suitably dead burned powders have apparently been identified for much of the melt feed, with some remelt feed.

Two other factors should be noted about crystal growth processes in general. First, the above melt growth processes are practical mainly for some materials, such as those that melt congruently and lack destructive phase transformations on cooling. Thus, materials such as a number of nonoxides such as SiC and Si_3N_4 are not amenable to such growth since they do not melt at atmospheric pressures and BeO presents a high-temperature destructive phase transformation (which, as noted earlier can apparently be suppressed in some growth conditions). Further, the above crystal growth processes depending on crucibles, are limited to those materials for which crucible materials can be found that are afforded and compatible with the material to be grown as crystals in the environments in which the crucibles can be used. These factors present limitations for some oxide materials and a variety of nonoxides. However, many of these materials, especially many nonoxides are amenable to some of these techniques such as skull melting, and some to other crystal growth methods, e.g., moving molten zone processes. Further, there are other crystal growth processes. Thus, in particular vapor phase processes that are used to grow some macro crystals, are also widely used to grow miniature single crystals such as platelets and especially whiskers, which are typically grown with the aid of additives (Sec. 3.5). Further, there are some molten flux baths in which some crystals can be grown, again often with additives, well below their melting points, e.g., BeO to avoid its high temperature phase transformation, and there are some possibilities of growing some nonoxide crystals via electrochemical methods, again often with use of additives (Sec. 3.5). Finally, recent developments of solid-state methods of crystal growth, again with an important role of additives, shows promise (Sec. 3.5).

6.6.3 Eutectic Ceramics and Directional Crystallization of Glasses

Consider now primarily melt processing of eutectic materials that entails processing of bodies of both single and polycrystalline character. Many combinations of ceramic compounds and some combinations of a ceramic compound with a refractory metal can be grown by directional solidification to produce an array of eutectic composites. Such composites often consist of single-crystal lamelli of rods or strips of one phase in a given crystallographic orientation in a single-crystal matrix of the major phase with its own single-crystal orientation. Well-grown bodies of such composites often have favorable properties such as hardness, wear resistance, and higher resistance to failure by brittle fracture and especially creep and related high temperature deformation [9]. Typically such properties increase as the size and spacing of the lamelli decreases, i.e, at higher growth rates, which have serious limitations for bulk bodies. Further, directional solidification of such eutectic materials is limited in sizes and to very simple shapes, mainly rod-shaped bodies, often with substantial costs. Another important limitation is frequent breakdown of the desired lamellar crystal structure, by not achieving it uniformly through out the body, instead a global lamellar structure is replaced by a mix of local, possibly distorted lamellar structures of some varying orientation. The latter structure, which is termed a cell structure, i.e. consisting of cells (or grains) of lamellar structure, has lower properties that the homogeneous and global lamellar structure of similar lamellar spacing.

There have been some indications that making bodies by consolidating eutectic particles, e.g., by hot pressing can provide reasonable compromises in properties, while giving much more versatile size and shape processing, as well as lower costs. Thus, Claussen [156] reported that hot pressing bodies from particles derived from comminution of some metal-ceramic eutectics (with colony structures and hence of compromised properties) gave bodies with the eutectic structure preserved and having promising fracture energies, e.g., about an order of magnitude larger than the ceramic matrix alone (strengths were not reported). Krohn and coworkers [157] obtained high toughness, e.g., 15 MPa•m$^{1/2}$, by hot pressing eutectic particles. Rice [92] used particles comminuted from commercially produced, approximately eutectic, Al_2O_3-ZrO_2 abrasive particles, and, though some porosity remained, the retained eutectic structure gave promising toughnesses and strengths, e.g., > 500 MPa. Further comminution to reduce particle size and hence resultant porosity was limited in effectiveness by this beginning to lose the eutectic structure. Also, properties decreased on exposure to elevated temperature oxidizing environments, due to resultant oxidization of the partially reduced ZrO_2, thus allowing it to form monoclinic ZrO_2, since partial stabilization of the abrasive depends on reduction of the ZrO_2 and not on the use of oxide stabilizers. Recently Mah and coworkers [158] showed similar hot pressing of Al_2O_3-YAG eutectic particles gave reasonable strengths of 270 MPa. However, particularly significant are higher strengths of 700 MPa with good toughnesses of 7–8 MPa•m$^{1/2}$ that Homeny and Nick [159] obtained by hot pressing Al_2O_3-ZrO_2 eutectic particles produced via melt quenching droplets using a plasma torch.

The above results for bodies made by consolidating eutectic particles suggest promise for further development, as does the following possible mechanism. As noted above, eutectic mechanical properties tend to scale as the inverse square root of the interlaminar eutectic spacing provided there is no cell structure. When such structure is present it determines mechanical properties generally in inverse relation to the square root of the colony cross-sectional dimensions, which are much larger than the lamellar spacing, thus giving much lower mechanical properties. The approximately single-phase colony, i.e., "grain," boundaries apparently provide a highly preferred path for fracture initiation. Thus, making eutectic particles significantly smaller than the typical colony sizes should give considerably improved mechanical properties as observed. Further, densified bodies of eutectic particles appear to lack nominally single-phase grain boundaries since there is often some joining of lamelli in abutting "grains" (Fig. 6.15). Thus, additional strengthening may be obtained by limiting single-phase fracture paths around eutectic grains, and thus may allow strengths to be greater than those of single-phase polycrystalline bodies with the same grain size and similar elastic properties. Note that quenching, e.g., by splatting of eutectic melt droplets may allow finer structure with higher properties and reasonable costs.

Directional solidification of both single crystals and eutectic structures has a much less used analog in directional crystallization of crystalizable glasses as opposed to normal surface or bulk random crystallization. A particularly illustrative example of the potential of directional crystallization is work of Abe and coworkers [160] in uniaxially crystalizing CaO-P_2O_5 glasses to develop fiberous crystallites parallel with the long axis of bars parallel with the imposed thermal gradient for crystallization. They showed an increase in room temperature flexural strengths of small bars for tension parallel with the bar axes and the fiberous

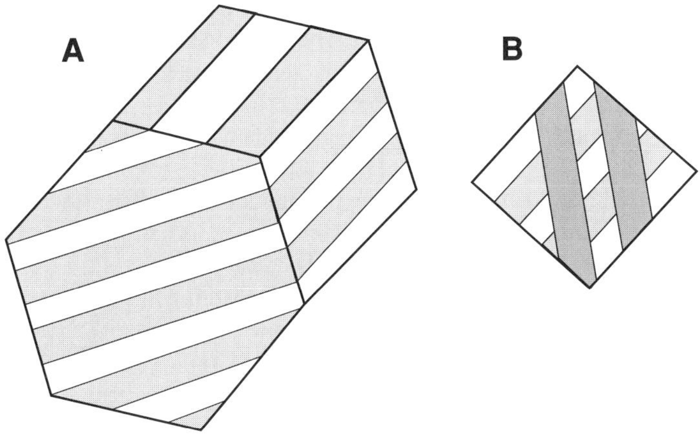

FIGURE 6.15 Schematics to illustrate the complexity of grain boundary structure between consolidated particles of a eutectic structure versus normal grain boundaries. (A) An idealized hexagonal particle with a lamellar eutectic structure. (B) A view of the intersection of another eutectic particle boundary on a section of the boundary shown for the particle in (A). The orientation of this idealized twist boundary is one with limited intersections of the lamellar eutectic structures. Note that such idealized boundaries will consist of various combinations of separate boundary sections between abutting (1) matrix; (2) second (lamellar or rod); and (3) matrix second phases versus boundaries in monolithic ceramics consisting of only the matrix phase. These eutectic boundary sections of such idealized boundaries will generally not be coplanar due to differing crystal structures, misorientations, and resultant surface energies and dihedral angles. Real boundaries between eutectic particles should further accentuate this noncoplanar character of the differing boundary sections due to the irregularities in the topography of actual eutectic particles and the angular differences from differing degrees of diffusion to form boundaries between real particles. Thus, eutectic boundaries have a structure similar in a number of aspects to that of the structure within the eutectic particles, which will not differ greatly in fundamental character for lamellar versus rod eutectics.

crystallites from 300 MPa at a mol ratio of CaO/P_2O_5 of 0.925 to a maximum strength of 640 MPa at a ratio of 0.94, then decreasing to 400 MPa at a molar ratio of 1. Fractures were generally fiberous and noncatastrophic, implying high toughness. These high strengths and indicated toughness are impressive, especially for a material of modest Young's modulus (85 GPa).

6.7 SUMMARY

This chapter has addressed major alternatives to pressureless sintering addressed in Chapters 4 and 5, i.e., other fabrication methods that have fairly wide potential and sometimes extensive, usage as well as substantial potential for further development.

These techniques include pressure sintering methods of hot pressing, hot isostatic pressing (HIPing), and press forging, then various reaction processes ranging from those involving substantial solid, liquid, or mixed-state reactions to those dominated by gas-phase reactions, i.e., CVD and CVI. These were followed by a variety of melt-based processes ranging from polycrystalline and composite bodies to glasses and single crystals.

Hot pressing is well established and has good potential for further growth in general as well as in newer areas, with composites being an important factor. HIPing is also established, but of narrower usage scope because of size, volume, and cost limitations, but has reasonable growth potential. Press forging and related extensions of hot pressing have achieved only very limited application, and are likely to continue to be restricted to specialized applications unless some special application or breakthrough in engineering to make them more practical occurs. Reaction processing, based mainly on liquid and solid-state reactions, has had some past and continuing successes, and much potential for future applications, especially for composites, but realizing these probably depends substantially on technical requirements or opportunities being meshed with further practical development. Gas-phase reaction-based processing, i.e., CVD, which is well established, and developing CVI, have excellent potential for further growth both for monolithic as well as for composite ceramics. Much of this growth probably requires using the underlying chemically oriented technology with a much broader materials perspective, e.g., deposition of two or multiphase bodies to control microstructure, making preforms for final densification by sintering, and using post deposition heat treatment to modify microstructures. Finally, melt processing of glasses and single crystals as well as both polycrystalline and composite bodies is well established, in fact dominates some major areas of application such as glasses and crystals. However, there are also reasonable opportunities for further successes, including more use for higher technology applications.

The first of four other important assessments is that overall a broader perspective on fabrication is needed, both within and between the above fabrication methods as well as traditional sintering-based fabrication. Thus, much reaction processing is better done by hot pressing, and CVI has broad possibilities, not only in composites, but also in various specialty materials (Sec. 7.3.2) and possible some other, e.g., melt sprayed, materials. Also, while some processes such as CVD have excellent capabilities for producing dense materials, they may also have other uses, e.g., as in producing porous preforms for optical fibers, and again in making some designed pore structures.

The second point is to note the general trends in component characteristics of size, shape versatility, properties (i.e., reflecting microstructures achieved) and general cost trends, as summarized in Table 6.5. Third, note that scale-up is again an important issue for all processes. However, for reaction processing scale up can be especially important since effects of exotherms from reactions

TABLE 6.5 Summary of General Component Capabilities of Other Fabrication Trends[a]

Process	Size	Shape	Properties	Costs
Hot pressing	Large	Simple	High	Moderate to high
Press Forging	Moderate	Somewhat versatile	Often high	Often high
HIP	Small to moderate	Reasonably versatility	High	Generally high
Reaction processing	Moderate-large	Modest to versatile	Moderate-high	Often moderate
CVD/CVI	Large	Fairly versatile	Often high	Moderate
Melt	Very large	Simple to moderate	Low-moderate	Low-moderate
Sintering	Small-moderate	Versatile	Moderate-high	Low-moderate

[a]Relative to pressureless sintering.

can change with component sizes and shapes, furnace loading, and heating, and if melting, even local and transient, occurs, it can result in coarser microstructures and phase distribution and serious component property limitations. Fourth, while a single fabrication process is often used in making ceramics, there are increasing opportunities to use two, or possibly more, as indicated by CVD of billet preforms for optical fibers, followed by sintering prior to fiber pulling.

REFERENCES

1. R.M. Fulrath. A critical compilation of ceramic forming methods III. Hot forming processes. Am. Cer. Soc. Bul. 43(12):880–885, 1964.
2. T. Vasilos, R.M. Spriggs. Pressure sintering of ceramics. Progress in Ceramic Science. Vol. 4. New York: Pergamon Press, 1966, pp. 95–132.
3. R.W. Rice. Hot forming of ceramics. In: J.J. Burke, N.L. Reed, V. Weiss, eds. Ultrafine-Grain Ceramics. New York: Syracuse Univ. Press, 1970, pp. 203–250.
4. A. Ezis, J.A. Rubin. Hot pressing. In: S.J. Schnieder, ed. Engineering Materials Handbook. Vol. 4. Ceramics and Glass. Metals Park, OH: ASM, 1991, pp. 186–193.
5. D.J. Bowers. A critical compilation of ceramic forming methods v. miscellaneous forming methods. Am. Cer. Soc. Bul. 44(2):145–150, 1965.
6. R.F. Davis, H. Palmour III, R.L. Porter, eds. Emergent Process Methods for High-Technology Ceramics. Materials Science Research, Vol. 17, Plenum Press, 1984.
7. R.M. Spriggs, L.A. Brissette, M. Rosetti, T. Vasilos. Hot-pressing ceramics in alumina dies. Am. Cer. Soc. Bul. 42(9):477–479, 1963.
8. R.M. Spriggs, T. Vasilos, L.A. Brissette. Grain size effects in polycrystalline ceramics. In: W.W. Kriegel, H. Palmour III, eds. Materials Science Research. The Role of Grain Boundaries and Surfaces in Ceramics. 3. New York: Plenum Press, 1966, pp. 313-53.

9. R.W. Rice. Mechanical Properties of Ceramics and Composites, Grain and Particle Effects. New York: Marcel Dekker, Inc., 2000.
10. P.H. Crayton, J.J. Price. Prediction of effective isothermal hot-pressing temperature. Am. Cer., Soc. Bul. 63(5):715–717, 1984.
11. K. An, D.L. Johnson. The pressure-assisted master sintering surface", to be published.
12a. H. Su, D.L. Johnson. Master sintering surface. J. Am. Cer. Soc. 79(12):3211–3217, 1996.
12b. D.L. Johnson, H. Su. A practical approach to sintering. Am. Cer. Soc. Bul. 76(2):72–76, 1997.
13. H. Iwasaki, S Shikanai, T. Takahashi, A. Tsuge, H. Ohkuma, T. Ishii. Studies of Elastically Structural Anisotropies in Hot Pressed Ceramic Materials. Toshiba Research and Development Center, Report, March 1975.
14. Reference deleted.
15. G.J. Oudemans. Continuous hot-pressing technique. Proc. Brit. Cer. Soc. No. 12, Fabrication Science 2, 3:83–98, 1969.
16. W. Rhodes, J. Thomson, Jr. Continuous hot pressing of $Ba_2Ti_9O_{20}$ and its evaluation. Am. Cer. Soc. Bul. 55(3):308–310, 1976.
17. H. Sheinberg. Hot pressing spinel hemishells with retractable tooling. Am. Cer. Soc. Bul. 58(7):719–721, 1979.
18. F.F. Lange, G.R. Terwilliger. The powder vehicle hot-pressing technique. Am. Cer. Soc. Bul. 52(7):563–565, 1973.
19. E.H. Zehms, J.D. McClelland. Semi-Continuous Hot Pressing. Am. Cer. Soc. Bul. 41(1):10–11, 1963.
20. R.W. Rice, J.H. Enloe, J.W. Lau, E.Y. Luh, L.E. Dolhert. Hot-pressing–a new route to high-performance ceramic multilayer electronic packages. Am. Cer. Soc. Bul. 71(5):751–755, 1992.
21. J.H. Enloe, J.W. Lau, C.B. Lundsager, R.W. Rice. Hot Pressing Dense Ceramic Sheets for Electronic Substrates and for Multilayer Electronic Substrates. U.S. Patent 4,920,640, 5/1/1990.
22. J.H. Enloe, J.W. Lau, R.W. Rice. Electronic Package Comprising Aluminum Nitride and Aluminum Nitride-Borosilicate Galass Composite. U.S. Patent 5,017,434, 5/21/1991.
23. F.W. Glaser, W. Ivanick. Sintered titanium carbide. J. Metals 4:387–390, 1952.
24. J.S. Jackson, P.F. Palmer. Hot pressing refractory hard materials. In: P. Popper, ed. Special Ceramics. New York: Academic Press, 1960, pp. 305–328.
25. Y. Zhou, K. Hirao, M. Toriyama, H. Tanaka. Very rapid densification of nanometer silicon carbide powder by pulse electric current sintering. J. Am. Cer. Soc. 83(3):654–656, 2000.
26a. Materials Modification, Inc., Fairfax, VA, www.matmod.com.
26b. W.W. Goldberger. A low cost method for pressure-assisted densification of advanced materials into complex shaped parts. Mats. Tech. 10(3/4):48–51, 1995.
27. J.W. Lee, Z.A. Munir, M. Shibuya, M. Ohyanagi. Synthesis of dense TiB_2-TiN manocrystalline composites through mechanical and field activation. J. Am. Cer. Soc. 84(6):1209–1216, 2001.
28a. R.M. Spriggs, L. Atteraas. Densification of Single-Phase Systems Under Pressure.

In: R. M. Fulrath and J. A. Pask, eds. Ceramic Microstructures: Their Analysis & Production. New York: John Wiley & Sons, Inc., 1968, pp. 701–727.

28b. R.M. Spriggs , R.B. Bunk, L. Atteraas. Thermomechanical processing of ceramics, fabrication science. 2. Proc. Brit. Cer. Soc. No. 12, 3:65–82, 1969.

29. B.-N. Kim, K. Morita, Y. Sakka. A High-Strain-Rate Superplastic Ceramic. Nature 413:288–291, 2001.

30. R.W. Rice. Deformation, recrystallization, strength, and fracture of press-forged ceramic crystals. J. Am. Cer. Soc. 55(2):90–97, 1972.

31. M.H. Hodge. The Effect of Hot Working on the Texture and Magnetic Properties of The Magnetoplumbite Ferrites. PhD Thesis, Penn. State Univ., 1973.

32. J.U. Knickerbocker, D.A. Payne. Oreintation of ceramic microstructures by hot-forming methods. Ferroelectrics 37:733–736, 1981.

33. S.K. Dey, J.S. Reed. Magnetic properties of press-forged barium ferrite. Am. Cer. Soc. Bul. 64(4):571–575, 1985.

34. E. Schmid, W. Boas. Krystallplastizität" Berlin: Verlag von Julius Springer, 1935.

35. R.J. Stokes, C.H. Li. Dislocations and the strength of polycrystalline ceramics. In: H.H. Stadelmaier, W.W. Austin, eds. Materials Science Research. Vol. 1. New York: Plenum Press, 1963, pp. 133–157.

36. R.B. Day, R.J. Stokes. Mechanical behavior of polycrystalline magnesium oxide at high temperatures. J. Am. Cer. Soc. 49(7):345–354, 1966.

37. R.W. Rice, J.G. Hunt, G.I. Friedman, J.L. Sliney. Identifying Optimum Parameters of Hot Extrusions. Final Boeing Co. Report for NASA Contract NAS-7-276, 1968.

38. R.W. Rice. Hot-working of oxides. In: A. Alper, ed. High Temperature Oxides, New York: Academic Press, 1970, pp. 235–280.

39. P.F. Becher, R.W. Rice. Strengthening effects in press forged KCl. J. Appl. Phy. 44(60):2915–2916, 1973.

40. R.W. Rice, M.W. Benecke, J.L. Sliney, G.I. Friedman. Semi-Continuous Pressure Sintering. Boeing Co. Summary Report for U.S. Navy Contract N00019-67-C-0314, 1968

41. R.W. Rice, W.J. McDonough, G.Y. Richardson, J.M. Kunetz, T. Schroeter. Hot rolling of ceramics using self-propagating high-temperature synthesis. Cer. Eng. Sci. Proc. 7(7-8):751–760, 1986.

42. S.L. Wikinson. Squeezing powders into shape. Chem Week 12:1988.

43. R.V. Raman. Rapid consolidation of aluminides. Adv. Mats. Procs. 4/1990, pp. 109–110.

44. A. Sawaoka, S. Saito. Hot Isostatic compaction using a molten glass as pressure transmission media. Jpn. J. Appl. Phy. 13:751, 1974.

45. A.J. Pyzik, A. Pechenink. Rapid omnidirectional compaction of cerami-metal composites. Cer. Sci. Eng. Proc. 9(7-8):965–974, 1988.

46. T. Clark, J.S. Reed. Closed-die hot-forging of vitrified bonded abrasives. Am. Cer. Soc. Bul. 61(7):733–736, 1982.

47. J.J. Svec. Hot pressing glass-bonded mica. Cer. Ind. 1980, pp. 28–29.

48. H.C. Yeh, P.F. Sikora. Consolidation of Si_3N_4 by hot isostatic pressing. Am Cer. Soc. Bul. 58(4):444–447, 1979.

49. A. Takata, K. Ishizaki. Mechanical properties of HIPed porous copper. K. Ishizaki, L Sheppard, S. Okada, B. Huybrechts, eds. Ceramic Trans. Vol. 31. Porous Materials. Westerville, OH: Am, Cer. Soc., 1993, pp. 233–242.
50. J. Adlerborn, M. Burström, L. Hermansson, H.T. Larker. Development of high temperature high strength silicon nitride by glass encapsulated hot isostatic pressing. Matls. & Design. 8(4):229–232, 1987.
51. L.M. Sheppard. Trends in HIP Equipment Capabilities. Am. Cer. Soc. Bul. 71(3):323–327, 1992.
52. L.M. Sheppard. The Evolution of HIP Continues. Am. Cer. Soc. Bul. 71(3):313–322, 1992.
53. J.B. Huffadine, A.J. Whitehead, M.J. Latimer. A simple hot isostatic pressing technique. Proc. Brit. Cer. Soc. Fabrication Science No. 12, 2:201–209, 1969.
54. S.-L. Hwang, P.F. Becher, H.-T. Lin. Desintering process in the gas-pressure sintering of silicon nitride. J. Am. Cer. Soc. 80(2):329–335, 1997.
55. O. Yeheskel, Y. Gefen, M. Talianker. Hot isostatic pressing of Si_3N_4 with Y_2O_3 additions. J. Mat. Sci. 19:745–752, 1984.
56. K.G. Ewsuk, G.L. Messing. Densification of sintered lead zirconate titanate by hot isostatic pressing. J. Mat. Sci. 19:1530–1534, 1984.
57. W-B. Li, M.F. Ashby, K.E. Easterling. On densification and shape change during hot isostatic pressing. Acta Metall 35(12):2831–2842, 1987.
58. M.J. Andrews, A.A. Wereszczak, T.P. Kirkland, K. Breder. Strength and Fatigue of NT551 Silicon Nitride and NT551 Diesel Exhaust Valves. Oak Ridge National Lab. Report ORNL/TM-1999/332, 2/2000.
59. R.W. Rice. Ceramics from polymer pyrolysis, opportunities and needs—A materials perspective. Am. Cer. Soc. Bul. 62(8):889–892, 1983.
60. K.J. Wynn, R.W. Rice. Ceramics via Polymer Pyrolysis. Ann. Rev. Mater. Sci. 14:297–334, 1984.
61. R.W. Rice. Advanced Ceramic Materials and Processes. In: D.L. Cocke, A. Clearfield, eds. Design of New Materials. New York: Plenum Pub. Corp., 1987, pp. 169–194.
62. K.M. Taylor. Improved silicon carbide for high temperature parts. Materials and Methods 44:92–95, 1956.
63. K.M. Taylor. Dense Silicon Carbide. U.S. Patent 3,205,043, 9/7/1965.
64. G.R. Sawyer, T.F. Page. Microstructural Characterization of "REFEL" (reaction-bonded) silicon carbide. J. Mat. Sci. 13:885–904, 1978.
63. D.J. Godfrey. The effects of impurities, additions and surface preparation on the strength of silicon nitride. In: R.W. Davidge, ed. Proc. Brit. Cer. Soc. No. 25. Mechanical Properties of Ceramics (2). 1975, pp. 325–337.
66. G. Ziegler, J. Heinrich, G. Wotting. Review Relationships Between Processing. Microstructure and Properties of Dense and Reaction-Bonded Silicon Nitride. J. Mat. Sci. 22:3041–3086, 1987.
67. P.E.D. Morgan, E. Scala. The formation of fully dense oxides by pressure calsintering of hydroxides. In: G. Kuczynski, Hooton, and Gibbon, eds. Sintering and Related Phenomena. New York: Gordon and Breach, 1967, pp. 861–892.
68. T.A. Wheat, T.G. Carruthers. The hot pressing of magnesium hydroxide and magnesium carbonate. In: G.H. Stewart, ed. Science of Ceramics. Vol. 4. Stoke on Trent: Brit. Cer. Soc. 1968, pp. 33–51.

69. R.W. Rice. The effect of gaseous impurities on the hot pressing and behavior of MgO, CaO, and Al_2O_3. Proc. Brit. Cer. Soc. 3:99–123, 1969.
70. M.S. Newkirk, A.W. Urquhart, H.R. Zwicker, E. Breval. Formation of Lanxide™ ceramic composite materials. J. Mater. Res. 1:81–89, 1986.
71. M.K. Aghajanian, N.H. Macmillan, C.R. Kennedy , S.J. Luszcz, R. Roy. Properties and microstructures of Lanxide® Al_2O_3-Al ceramic composite materials. J. Mat. Sci. 24:658–670, 1989.
72. S. Antolin, A.S. Nagelberg, D.K. Creber. Formation of Al_2O_3/metal composites by the directed oxidation of molten aluminum-magnesium-silicon alloys: part I, microstructural development. J. Am. Cer. Soc. 75(2):447–454, 1992.
73. S. Antolin, A.S. Nagelberg, D.K. Creber. Formation of Al_2O_3/metal composites by the directed oxidation of molten aluminum-magnesium-silicon alloys: part II, growth kinetics. J. Am. Cer. Soc. 75(2):455–462, 1992.
74. D.K. Creber, S.D. Poste, M.K. Aghajanian, T.D. Claar. AlN composite growth by nitridation of aluminum alloys. Cer. Eng. Sci. Proc. 9(7-8):975–982, 1988.
75. P. Barron-Antolin, G.H. Schiroky, C.A. Andersson. Properties of fiber-reinforced alumina matrix composites. Cer. Eng. Sci. Proc. 9(7-8):759–766, 1988.
76. A.W. Urquhart. Molten metals sire MMC's, CMC's. Adv. Mats. & Proc. 7:25–29, 1991.
77. T.D. Claar, W.B. Johnson, C.A. Anderson, G.H. Schiroky. Microstructure and properties of platelet-reinforced ceramics formed by the directed reaction of zirconium with boron carbide. Cer. Eng. Sci. Proc. 10(7-8):599–609, 1989.
78. M.B. Dickerson, R.L. Snyder, K.H. Sandhage. Low-temperature fabrication of dense, near net-shaped tungsten/zirconium carbide composites with tailored phase contents by the PRIMA-DCP process. Cer. Eng. Sci. Proc. 22(4):97–107, 2001.
79. W.G. Fahrenholz, D.T. Ellerby, K.G. Ewsuk, and R.E. Loehman. Forming Al_2O_3-Al composites with controlled compositions by reactive metal penetration of dense aluminosilicate Preforms. J. Am. Cer. Soc. 83(5):1293–1295, 2000.
80. N. Claussen, T. Le, S. Wu. Low-Shrinkage Reaction-Bonded Alumina. J. Eur. Cer. Soc. 5:29–35, 1989.
81. S.P. Gaus, P.M. Sheedy, H.S. Caram, H.M. Chan, M.P. Harmer. Controlled firing of reaction-bonded aluminum oxide (RBAO) ceramics: Part II, experimental results. J. Am. Cer. Soc. 82(4):909–915, 1999.
82. E. Suvaci, G. Simkovich, G.L. Messing. The reaction-bonded aluminum oxide process: I, the effect of attrition milling on the solid-state oxidation of aluminum powder. J. Am. Cer. Soc. 83(2):299–305, 2000.
83. S. Lathabai, D.G. Hay, F. Wagner, N. Claussen. Reaction-bonded mullite/zirconia composites. J. Am. Cer. Soc. 79(1):248–256, 1996.
84. S. Wu, N. Claussen. Reaction bonding and mechanical properties of mullite/silicon carbide composites. J. Am. Cer. Soc. 77(11):2898–2904, 1994.
85. S. Wu, A.J. Gesing, N.A. Travitzky, N. Claussen. Fabrication and properties of Al-infiltrated RBAO-based composites. J. Eur. Cer. Soc. 7, 277–281, 1991.
86. K.H. Sandhage, S.M. Allameh, P. Kumar, H.J. Schmutzler, D. Viers, X.-D. Zhang. Near net-shaped, alkaline-earth-bearing ceramics for electronic and refractory applications via the oxidation of solid, metal-bearing precursors (the VIMOX process). Mats. Manu. Proc. 15(1):1–28, 2000.

87. M.M. Antony, K.H. Sandhage. Barium titanium/noble metal laminates prepared by the oxidation of solid metallic precursors. J. Mater. Res. 8(11):2968–2977, 1993.
88. H.J. Schmutzler, M.M. Antony, K.H. Sandhage. A novel reaction path to $BaTiO_3$ by the oxidation of a solid metal precursor. J. Am. Cer. Soc. 77(3):1994.
89. R. Citak, K.A. Rodgers, K.H. Sandhage. Low-temperature synthesis of $BaAl_2O_4$/aluminum-bearing composites by the oxidation of solid metal-bearing precursors. J. Am. Cer. Soc. 82(1):237–240, 1999.
90. K.A. Rogers, P. Kumar, R. Citak, K.H. Sandhage. Dense, shaped ceramic/metal composites at ≤1000°C by the displacive compensation of porosity (DPC) method. J. Am. Cer. Soc. 82(3):757–760, 1999.
91. K.H. Sandhage, N. Claussen. Innovative processing: reaction casting of ceramic/metal and ceramic/intermetallic composites. In: M.N. Rahaman, ed. Vol. 1. Handbook of Advanced Ceramic Science and Technology. New York: Marcel Dekker, Inc.
92. R. Rice. Processing of ceramic composites. In: J.G.P. Binner, ed. Advanced Ceramic Processing and Technology. Vol. 1. Park Ridge, NJ: Noyes Pubs, 1990, pp. 123–213.
93a. D. Lewis, R.P. Ingel, W.J. McDonough, R.W. Rice. Microstructure and thermomechanical properties in alumina- and mullite-boron-nitride particulate composites. Cer. Eng. & Sci. Proc. 2(7-8):719–727, 1981.
93b. R.W. Rice, W.J. McDonough, S.W. Freiman, J.J. Mecholsky, Jr. Ablative-Resistant Dielectric Ceramic Articles. U.S. Patent 4,304,870, 12/8/1981.
94. W.S. Coblenz, D. Lewis, III. In situ reaction of B_2O_3 with AlN and/or Si_3N_4 to form BN-toughened composites. J. Am. Cer. Soc. 71(12):1080–1085, 1988.
95. Z.A. Munir, J.B. Holt, eds. Combustion and Plasme Synthesis of High-Temperature Materials. New York: VCH Publishers, 1990.
96. R.W. Rice, W.J. McDonough. Intrinsic volume changes of self-propagating synthesis. J. Am. Cer. Soc. 68(5):C-122–123, 1985.
97. R.W. Rice. Review microstructural aspects of fabricating bodies by self-propagating synthesis. J. Mat. Sci. 26:6533–6541, 1991.
98. R.W. Rice. Assessment of the application of SPS and related reaction processing to produce dense ceramics. Cer. Eng. Sci. Proc. 11(9-10):1226–1250, 1990.
99. G.Y. Richardson, R.W. Rice, W.J. McDonough, J.M. Kunetz, T. Schroeter. Hot pressing of ceramics using self-propagating synthesis. Cer. Eng. Sci. Proc. 7(7-8):761–770, 1986.
100. C.P. Cameron, J.H. Enloe, L.E. Dolhert, R.W. Rice. A comparison of reaction vs conventionally hot-pressed ceramic composites. Cer. Eng. Sci. Proc. 11(9-10):1190–1202, 1990.
101. R.W. Rice, C.P. Cameron, P L. Berneburg. Lower Cost Ceramic Armor Materials. In: Proceedings Second Ballistics Symposium on Classified Topics, Johns Hopkins Applied Physics Lab. American Defense Preparedness Assn., 1992.
102. I. Zhitomirsky. New developments in electrolytic deposition of ceramic films. Am. Cer. Soc. Bul. 57–63, 2000.
103. R.W. Haskell, J.G. Byrne. Studies in chemical vapor deposition. In: H. Herman, ed. Treatise on Materials Science and Technology. Vol. 1. New York: Academic Press,1972, pp. 293–340.
104. H.O. Pierson, ed. Chemically Vapor Deposited Coatings. Westerville, OH: Am. Cer. Soc., 1981.

105. D.P. Stinton, T.M. Besmann, R.A. Lowden. Advanced ceramics by chemical vapor deposition techniques. Am. Cer. Soc. Bul. 67(2):350–355, 1988.
106. J.S. Goela, R.L. Taylor. Monolithic material fabrication by chemical vapor deposition. J. Mat. Sci. 23:4331–4339, 1988.
107. P.D. Dapkus. Metalorganic chemical vapor deposition. A. Rev. Mater. Sci. 12:243–269, 1982.
108. P.A. Dowben, J.T. Spencer, G.T. Stauf. Deposition of thin metal and metal silicide films from the decomposition of organometallic compounds. Mats. Sci. Eng. B2:297–323, 1989.
109. M.D. Rigterink. Material systems, fabrication and characteristics of glass fiber optical waveguides. Am. Cer. Soc. Bul. 55(9):775–780, 1976.
110. Various Authors. Glass Fibers for Optical Communications. Phy. Chem. Glasses 21(1):66, 1980.
111. L.A. Ketron. Fiber optics: the ultimate communications media. Am. Cer. Soc. Bul. 66(11):1571–1578, 1987.
112. C. Racault, F. Langlais, R. Naslain, Y Kihn. On the chemical vapor deposition of Ti_3SiC_2 from $TiCl_4$-$SiCl_4$-CH_4-H_2 gas mixtures. Part II An experimental approach. J. Mat. Sci. 29:5041–5048, 1994.
113. T. Goto, T. Hirai. Chemical vapor deposited Ti_3SiC_2. Mat. Res. Bull. 22:1195–1201, 1987.
114. M. Touanen, F. Teyssandier, M. Ducarroir, M. Maline, R. Hillel. Microcomposites and nanocomposites structures from chemical vapor deposition in the silicon-titanium-carbon system. J. Am. Cer. Soc. 76(6):1475–1481, 1993.
115. M. Moriyama, K. Kamata, I. Tanabe. Mechanical properties of SiN_xC_y ceramic films prepared by plasma CVD. J. Mat. Sci. 26:1287–1294, 1991.
116. M.I. Aivazov, T.A. Stenashkina. Synthesis of complex carbide phases in the systems Ti-B-C and Ti-Si-C bt crystallization from the gas phase. Izvestiya Akademii Nauk SSSR, Neorganicheskie Materialy 11(7):1223–1226, 1975.
117. R.W. Rice, R. Engdahl, unpublished results.
118. T. Hirai. CVD of Si_3N_4 and its composites. In: R.F. Davis, H. Palmour III, R.L. Porter, eds. Mat. Sci. Res. Vol.17. Emergent Process Methods for High Tech Ceramics. NY: Plenum Press, 1984, pp. 329–345.
119. K. Hiraga, M. Hirabayashi, S. Hayashi, T. Hirai. High-resolution electron microscopy of chemically vapor-deposited β-Si_3N_4-TiN composites. J. Am. Cer. Soc. 66(8):539–542, 1983.
120. T.Hirai, S. Hayashi. Density and deposition rate of chemically vapour-deposited Si_3N_4-TiN composites. J. Mat. Sci.18:2401–2406, 1983.
121. R.W. Rice. Porosity of Ceramics. New York: Marcel Dekker, Inc., 1998.
122. R.W. Rice. Effects of amount, location, and character of porosity on stiffness and strength of ceramic fiber composites via different processing. J. Mat. Sci. 34:2769–2772, 1999.
123. R. Hozl, R. Benander, R. Davis. Tungsten Alloys Containing A-15 Structure and Method for Making Same. U.S. Patent 4,427,445, 9/24/1984.
124. R. Hozl, J. Stiglich. Wear Performance of CM 500 Alloy as Compared to Conventional Hard Metals. 10th Plansee Proc, 973–980, 1981.
125. J.J. Stiglich, Jr., D.G. Bhat, R.A. Hozl. High temperature structural ceramic materials manufactured by the CNTD process. Cer. Intl. 6(1):3–10, 1980.

126. R.W. Rice, K.R. McKinney. Residual stresses and scaling CNTD SiC to larger sizes. J. Mat. Sci. Let. 1:159–162, 1982
127. S. Dutta, R.W. Rice, H.C. Graham, M.C. Mendiratta. Characterization and properties of controlled nucleation thermochemical deposition (CNTD)-silicon carbide. J. Mat. Sci. 15:2183–2191, 1980.
128. A.R. Cooper. A critical compilation of ceramic forming methods IV. Melt forming processes. Am. Cer. Soc. Bul. 44(1):1–8, 1965.
129. J.J. Rasmussen. Surface tension, density, and volume change on melting of Al_2O_3 Systems, Cr_2O_3, Sm_2O_3. J. Am Cer. Soc. 55(6):326, 1972.
130. B. Granier, S. Heurtault. Density of liquid rare-earth sesquioxides. J. Am. Cer. Soc. 71(11):C-466–468, 1988.
131. Newberg, J. Pappis. Fabrication of fluoride laser windows by fusion casting. In: C.R. Andrews, C.L. Strecker, eds. Proc. Fifth Conf. on Laser Window Materials, Air Force Materials Lab., Wright-patterson Air Force Base, 1066–78, 1976.
132. R.L. Gentilman. Fusion-casting of transparent spinel. Am. Cer. Soc. Bul. 906–907, 1980.
133. J.B. Huffadine, A.G. Thomas. Flame spraying as a method of fabricating dense bodies of alumina. Pwd. Metal. 7(14):290–299, 1964
134. K.T. Scott, A.G. Cross. Near-net shape fabrication by thermal spraying. In: R.W. Davidge, ed. Novel Ceramic Fabrication Processes and Applications. Proc. Brit. Cer. Soc. No. 38, 203–211, 12/1986.
135. R.W. Babbitt. Arc plasma fabrication of ferrite-dielectric composites. Am. Cer. Soc. Bul. 55(6):566–571, 1976.
136. L.-W. Tai, P.A. Lessing. Plasma spraying of porous electrodes for a planar solid oxide fuel cell. J. Am. Cer. Soc. 74(3):501–504, 1991.
137. Y. Ishiwata, Y. Ito, H. Kashiwaya. Densification of plasma-sprayed ceramic coatings by HIP treatment and their cracking behavior. J. Cer. Soc. Jpn. IntlEd., 99:665, 1991.
138. T. Mäntylä. P. Vuoristo, P. Kettunen. Chemical vapor deposition of plasma-sprayed oxide coatings. Thin Solid Films 118:437–444, 1984.
139. K. LaPierre, H. Herman, A.G. Tobin. The microstructure and properties of plasma-sprayed ceramic composites. Cer. Eng. Sci. Proc. 12(7-8):1201–1221, 1991.
140. V.I. Aleksandrov, V.V. Osiko, A.M. Prokhorov, V.M. Tatarintsev. Synthesis and crystal growth of refractory materials by RF melting in a cold container. In: Current Topics in Materials Science, Vol. 1. New York: North-Holland Pub. Co.,1978, p. 421.
141. K. Nassau. Cubic zirconia: an update. Lapidary J. 35(6):1194–1200, 1210–1214, 1981.
142. R.P. Ingel, D. Lewis, B.A. Bender, R.W. Rice. Temperature dependance of strength and fracture toughness of ZrO_2 single crystals. J. Am. Cer. Soc. 65(9):C-150–152, 1982.
143. R.D. Olt, R.G. Rudness. Single crystals as engineering materials. Mats. Design Eng. 84–91, 1963.
144. S. Simov. Review morphology of hollow crystals of II-VI compounds. J. Mat. Sci. 11:2319–2332, 1976.
145. H.E. LaBelle, Jr., A.I. Mlavsky. Growth of controlled profile crystals from the melt: part 1—sapphire filaments. Mat. Res. Bul. 6:571–580, 1971.
146. H.E. LaBelle, Jr. Growth of controlled profile crystals from the melt: part II—edge-defined film-fed growth (EFG). Mat. Res. Bul. 6:581–590, 1971.

147. B. Chalmers, H.E. LaBelle, Jr., A.I. Mlavsky. Growth of controlled profile crystals from the melt: part III—Theory. Mat. Res. Bul. 6:681–690, 1971.
148. J.T.A. Pollock. Filamentary sapphire, part 1, growth and microstructural characterization. J. Mat. Sci. 7:631–648, 1972.
149. J.T.A. Pollock. Filamentary sapphire, part 3, the growth of void-free sapphire filaments at rates up to 3.0 cm/min. J. Mat. Sci. 7:787–92, 1972.
150. F. Théodore, T. Duffar, J.L. Santailler, J. Pesenti, M. Keller, P. Dusserre, F. Louchet, V. Kurlov. Crack generation and avoidance during the growth of sapphire domes from an element of shape. J. Crystal Growth, 2000.
151. Private communication, J. Locher. Engineering manager, Saphikon, 2000.
152. F. Schmid, C.P. Khattak, D.M. Felt. Producing large sapphire for optical applications. Am. Cer. Soc. Bul. 73(2):39–44, 1994.
153. C.P. Khattak, F. Schmid. Growth of near-net-shaped sapphire domes using the heat exchanger method. Mats, Let. 7(9-10):318, 1989.
154. F. Schmid, C.P. Khattak, M.B. Smith, D.M. Felt. Current status of sapphire for optical. SPIE Proc. CR67, 1997.
155. C.P. Khattak, F. Schmid. Growth of 240 KG Multicrystalline HEM™ Silicon Ingots. Proc. 2nd World Conf. & Exhib. of Photovoltaic Solar Energy Conversions, Vienna, Austria, II:1870–1872, 1998.
156. N. Claussen. Hot-Pressed Eutectics of Oxides and Metal Fibers. J. Am. Cer. Soc. 56(8):442, 1973.
157. U. Krohn, H. Olapinski, U. Dworak. U.S. Patent 4,595,663, 6/17/1986.
158. T.-I. Mah, T.A. Parthasarathy, R.J. Kerans. Processing, microstructure, and strength of alumina-YAG eutectic polycrystals. J. Am. Cer. Soc. 83(8):2088–2090, 2000.
159. J. Homeny, J.J. Nick. Microstructure-property relations of alumina-zirconia eutectic ceramics. Mat. Sci. Eng. A127:123–133, 1990.
160. Y. Abe, T. Kasuga, H. Hosono, K. de Groot. Preparation of high-strength calcium phosphate glass-ceramics by unidirectional crystallization. J. Am. Cer. Soc. 67(7):C–142–144, 1984.

7

Special Fabrication Methods

7.1 INTRODUCTION

This chapter addresses speciality fabrication methods, i.e., those yielding special bodies or constituents for them, and thus of narrower application. Specifically, this chapter addresses fabrication of filaments, fibers, and related reinforcements, and bodies with designed porosity, followed by rapid prototyping/solid free form fabrication (SFF), then fabrication of ceramic fiber composites and ceramic coating technology are summarized.

7.2 FABRICATION OF FILAMENTS, FIBERS, AND RELATED ENTITIES FOR REINFORCEMENT AND OTHER APPLICATIONS

7.2.1 Introduction to Miscellaneous and Polymer-Derived Ceramic Fibers

Fabrication of filaments, ribbons, fibers, and related entities, such as whiskers and platelets, is a major step in producing ceramic composites, whether for their wide usage as "reinforcement" or as toughening for structural or other applications. Other applications include the large area of low-weight thermal insulators, as well as important use as filters, and as separators and other applications in batteries and fuel cells, and use as electrical conductive "wires," e.g., of superconductors. Fabrication of whiskers and related (single-crystal) platelets typically involves vapor phase processes, typically CVD, by themselves or in conjunction with local liquid droplets, ideally of selected composition (usually with additives), size, and placement. This is addressed in Section 3.4 and is not further addressed here.

This section addresses fabrication of filaments and fibers, which are not explicitly defined, but are distinguished by the former typically being larger in diameter (e.g., > 100 μm, and fibers commonly having diameters of well less than 100 μm). There are also often some differences in structure; for example, filaments may have compositional or mocrostructural gradients, which reflect differences in fabrication processes from those for fibers, which are typically more uniform. Another difference is that filaments are almost always produced and used as continuous entities, that is, they are very long in length; while fibers are often similarly produced and used, they may be produced directly as, or subsequently by chopping to, short fibers or made into cloth or fiber structures, by weaving, braiding, or knitting.

Fabrication methods include some miscellaneous ones, but are primarily via extrusion/drawing, conversion of other fibers, CVD (a major source of ceramic filaments), and melt-based processes, which are addressed in the order listed, with most having various subprocesses under them. Each of these fabrication methods and their major subprocesses—e.g., the increasing methods for fiber (filament) coatings for structural composites—are summarized.

Miscellaneous fiber and filament fabrication includes various methods, some of which are briefly summarized for perspective, e.g., growth of discontinuous fibers and those that may have other than structural uses. Thus, for example, some very porous shorter (e.g., 1–15 cm long) fibers with high surface areas (e.g., 500–1000 m^2/gm) of nominally polygonal cross section (~ 50 μm dia.) of SiO_2 [1] and of Al_2O_3 [2] have been grown by the directional freezing of a container of the appropriate precursor sol. The modest strengths obtained (e.g., of ~ 200 MPa) should be sufficient for many applications, such as for catalysis, because of their surface area or other attributes. Though such strengths are not sufficient for normal reinforcement applications, they can be increased by sintering to reduce porosity, e.g., to 300–700 MPa. Additions of salts such as NH_4Cl can increase the ~200-nm pore sizes and change the resultant shapes, for example from fibers to flakes. Another process produces discontinuous carbon fibers that appear to be a hybrid of a whisker seed and a more conventional carbon fiber (except for its CVD character) [3]. These carbon fibers, which grow like blades of grass on carbon plates can be 1 cm or more in length, have good properties, but are projected to cost of the order of $100/lb, tenfold higher than substantially smaller fibers produced by the same basic process, but with lower properties that may be useful as reinforcement for plastics for high volume, e.g., automotive, applications. There are also in situ methods of growing shorter or possibly some longer fibers or filaments, of which eutectic systems (Sec. 6.7.3) are an important one. Though costs and technical constraints are a factor, in some cases it may be possible to use this as a means of producing fibers or ribbons themselves by chemically removing the eutectic matrix. For example, while eutectic composites may have useful mechanical properties, they generally fall far short of those of many artificial fiber composites where matrix and fiber choices are made based on performance rather than process chemistry. However, there is some in situ development of fibers that can yield promising toughness (Secs. 7.6 and 8.2.3).

Extrusion (often referred to as spinning) of filaments and especially fibers is one of the most diverse and widely used method of making fibers, primarily based on one of three types of feedstock. The first type of feedstock is an inorganic glass, of which silicate glasses are particularly important examples, a key one being the current commercial production of optical fibers from various preform processing routes as discussed in Section 6.6. While fiber forming is done by extrusion, fibers are often drawn down to finer diameters by controlling the tensile stresses in the take-up system to spool the fiber. Such extrusion/drawing can be at substantial speeds—e.g., 125-μm optical fibers produced at rates of a meter per second.

The second and largest, most diverse type of fiber feedstock is that of preceramic polymers. The dominant and guiding technology here is forming the various experimental and commercial carbon, mainly graphite, continuous fibers that are produced [4,5]. Growth of the carbon fiber market and the technology to a substantial scale have lead to the availability of a diverse array of products. These range from lower cost fibers with reasonable Young's moduli (e.g., ~ 200 GPa) and strength to high-cost fibers with high Young's moduli (e.g., > 700 GPa) and strengths (i.e., costs, strengths, and elastic moduli roughly scale with one another). However, despite the market scope and size, it still depends on other uses of raw materials as shown by the switch from rayon (cellulose) precursors to other precursors, mainly PAN or pitch. This occurred since rayon fiber production greatly diminished after use of rayon fibers in tires ceased, wiping out the major market and volume for rayon fibers; continued production of rayon fibers of the quality for producing carbon fibers became unattractive.

The above use of polymeric precursors to yield carbon fibers (as well as glassy carbon bodies) by forming them in the polymeric state then pyrolyzing them to carbon served as the model for extending such processing to a variety of compound ceramic fibers (and bodies). While this extension probably occurred to various individuals, it obviously occurred to Yajima and colleagues [6] who first demonstrated a compound ceramic, SiC-based fiber by this route (which was commercialized as Nicalon fiber by Nippon Carbon Tokyo, Japan). It also occurred to Rice [7], who anticipated such, as well as much broader, use of preceramic polymers prior to Yajima's development. Since the original demonstration and discussion of such pyrolysis of preceramic polymers to compound ceramics, there has been substantial development of precursors and their processing, much of it for fibers. A number of basic compositions have been made, including AlN, BN, Si_3N_4, and B_4C, as well as a number with three or four atomic constituents, especially of Si containing ceramic compositions [4,6–10], most of which are not stoichiometric compounds. Because of the low temperatures for pyrolysis, (1200-1600°C) and the general multiconstituent character, such fibers generally have nano-scale grains, but some constituents or additives can accelerate grain development and growth, mainly at higher temperatures. Fiber diameters are typically 5–10 μm for graphite fibers and commonly 10–20 μm for other fibers.

The above work has shown the overall requirements for making fibers by this route, the first and most basic is that the polymer and processing used to yield the approximate composition needed for the properties desired. While this requirement is obvious, it is important to state it since, contrary to making carbon fibers or bodies by pyrolysis, where high purity carbon is typically the result, this is uncommon in making compound ceramics by polymer pyrolysis. Compositional variations commonly occur in making compound ceramics by polymer pyrolysis, since some constituents of the desired ceramic compounds often do not exist in the preceramic polymer in stoichiometric quantities. Further, there are often stoichiometric variations in the decomposition during pyrolysis, the nature of which can depend substantially on pyrolysis conditions, especially atmosphere, which is often used as a method of influencing the process outcome. Thus, some polymers may pyrolyze to a Si-C based composition in a neutral atmosphere, e.g., Ar, but to a Si-N based composition in a N_2 atmosphere.

Another basic requirement for successfully making fibers of ceramic compounds by polymer pyrolysis is that the pyrolysis gives both a high enough ceramic yield (typically measured as the percent of ceramic mass produced per unit mass of starting polymer) and does so as a coherent fiber. Clearly too low a ceramic yield means too high a polymer to ceramic shrinkage to avoid fiber distortions, and especially cracking or crumbling on a global scale. However, the nature of the decomposition can also be a factor since two different polymers having the same ceramic yield may decompose in different fashions; one maintaining basic solid coherency on both a local and global scale giving a sound fiber, and the other decomposing to more of a granular character, thus destroying the solid coherency. (The latter is better for making ceramic powder via pyrolysis, and especially for use of preceramic polymers as binders in forming and sintering ceramic bodies.)

The final two requirements for making ceramic compound fibers via polymer pyrolysis is that the polymer first have suitable plasticity, e.g., that it is not highly cross linked, but can be suitably rigidified after extrusion, e.g., by polymerization (and some possible drawing, while also accommodating pyrolysis shrinkage). Note that the latter can be an important practical factor as illustrated by the original development and production of the Nicalon fiber. Curing of the drawn fiber presented some difficulties and related costs, which were initially overcome in production by allowing some limited oxidation during curing at the expense of greater oxygen content in the final pyrolyzed fiber, which resulted in some performance limitations. Such oxidation cured fibers are 58% Si, 31% C, and 11% O and have a density of 2.55 g/cc, a Young's modulus of 194 GPa, and tensile strength of 3 GPa, while more expensive electron-beam-cured fiber have less oxygen and higher density (2.74 g/cc) and Young's modulus (257 GPa), but slightly lower strength (2.8 GPa, possibly reflecting larger grain size with less second phases). The above lower Young's moduli than that for theoretically dense SiC (416–451 GPa) is due to the extra phases diluting the SiC. To put such technology into production requires a great deal of development and refinement of the process, to reduce the size and number of fiber defects to acceptable levels (Fig. 7.1).

In close analogy with the above polymer pyrolysis to produce fibers is their production via sol-gel processing, much of this by Sowman and colleagues [11–13], which was developed relatively independently of polymer pyrolysis. This independence in part arises since polymer pyrolysis has been used mostly for

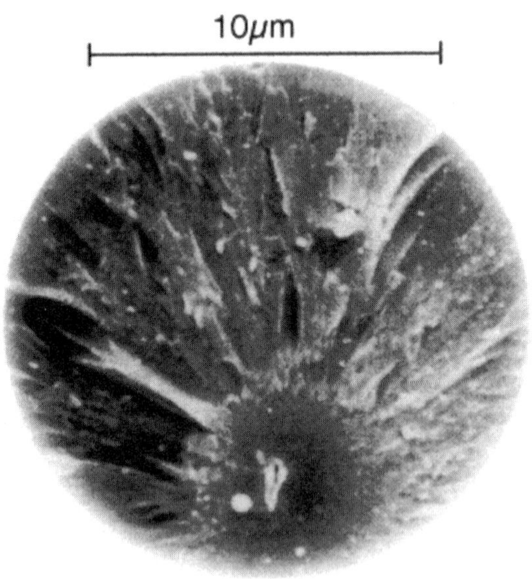

FIGURE 7.1 Fracture origin in polymer-derived SiC-based fiber failing from an internal pore illustrating the type of defect that must be reduced in size and occurrence. (Photo courtesy of J. Lipowitz, Dow Corning).

nonoxide fibers for which it is most suited and sol-gel mostly for oxide fibers for which it is readily applicable. Most extensively developed are fibers based mainly on alumina, with differing amounts of silica or boria to provide easier processing versus some property/performance limitations (e.g., high temperatures). Gelling is typically an important factor in rigidizing fibers following extrusion (with limited or no drawing). Sol-derived fibers are similar to polymer-derived fibers in diameter, lower firing temperatures, and fine, (nanometer), grain sizes. Different alumina-based fiber compositions have been commercialized by a few major corporations, with costs and properties reflecting the amount of alumina and its phase (delta, gamma, or alpha) and the amount and type of second phase. Thus, densities vary from 2.7–3.9 g/cc while Young's moduli vary from 150–373 GPa (theoretical alpha alumina ~ 420 GPa). Tensile strengths follow a similar, but more variable, correlation with density, with broader variations at higher density, again possibly reflecting larger grain sizes in purer fibers.

7.2.2 Preparation of Ceramic Fibers from Ceramic Powders and by Conversion of Other Fibers

The final precursor category for fibers formed by extrusion is that consisting of the diverse array of fibers that can be produced by various means from mixes of ceramic powders with suitable plastic binders. With finer ceramic particles (1 μm or less) and suitable binders with a good ceramic powder loading, green ceramic fibers can be extruded at modest temperatures, with use of melting polymer binders [14,15] or at room temperature with water-based binders [16], e.g., following earlier work [17–19]. Such methods have been used for both oxide and nonoxide fibers, spun and spooled as individual fibers as with the above extrusion of fibers from preforms, preceramic polymers, or sols (e.g., producing green fibers ~ 300 μm dia). There is substantial literature [20–23] on development and production of such fibers, primarily of alumina. Such fiber processing has obvious close parallels to sintering of bulk ceramics, e.g., use of ZrO_2 to limit alumina grain growth and hence obtain higher strength [21,24].

Individually handled fibers are more costly, as those above, so for lower cost fibers for less demanding application, other fiber processes have been demonstrated, For example, Lessing [25,26], used slurries with very volatile solvents such as acetone. Such slurries are placed in a rotating chamber with fine orifices in its walls so the slurry, slung outward from the rotation, forms fibers that are rigidified by flash evaporization of the solvent, forming green fiber mats that may be useful for various applications, e.g., for batteries or fuel cells. Such ceramic fiber forming is analogous to making cotton candy (equipment for which has been used for making some ceramic fibers), except for the need to subsequently fire the ceramic fibers (in mat form). Individual fibers are commonly 1–20 μm in diameter, and have substantial aspect ratios.

Other manifestations of such centrifugal spinning of fibers to make fiber mats exist, such as using sols that are spun to form fibers that are rigidified by rapid drying, as shown for stabilized zirconia fibers(27). Such fibers can be finer than those made using powders in binders and are commonly fired to high density and much finer (nm), grain size at lower temperatures.

Consider now making ceramic fibers by converting some other fiber to a ceramic fiber, which also has several variations depending on the nature of the starting fiber and the conversion process. One of the largest sources of starting fibers are organic fibers, often based on cellulose—especially rayon. Thus, Hamling and coworkers and others [28-30] reported making several oxide as well as some nonoxide fibers by impregnating rayon fibers with either an aqueous or organic solution of an inorganic salt. After impregnation of the salt(s) in the fibers in a bath, the fibers are removed, excess surface salt removed, and the organic fiber material slowly pyrolyzed, the salt decomposed to the oxide, and the relic oxide sintered to replicate the starting organic fiber. There are several variations of this versatile method, one of which is reaction processing to convert oxides from salt decomposition to nonoxide ceramics, such as carbides or nitrides. Another attractive aspect of this process is that it can often be successfully performed on various fiber forms, such as woven, knitted, and felted. preforms of the organic fiber, while maintaining much of the flexibility of the original organic fiber body in the resultant ceramic fiber replica. This process is apparently the basis of the Zircar® process for making zirconia-based felt boards for high temperature insulation. Other natural fibers have been used in laboratory trials; for example, jute fibers were found to be better than some other natural fibers, with the process of ceramic replication of the organic structure seen as similar to that observed in production of SiC from rice hulls [31]. Some earlier trials by others [32] [colleagues at the Boeing Co., 1960] with cotton gauze pads gave good fiber/gauze replication, but the resultant ceramic "pads" were brittle, apparently due to sintering of many fiber contacts with one another.

The other major manifestation of fiber conversion processes is the direct chemical conversion of a ceramic or other inorganic fiber to one of desired chemical character, which can again follow different routes, mainly whether ingredients are being removed from the starting fiber or added to it. An example of removal of an initial fiber constituent is an approach to more versatile, and possibly lower cost, production of silica fibers, which in pure form require extrusion (spinning) at ~ 1800°C using graphite tooling [2]. A demonstrated alternative is similar to the Vycor® process of forming a phase-separated glass from which almost all of the nonsilica phase can be leached out and the remaining silica readily sintered to full density. In the fiber case, a glass of silica with 25 w/o Na_2O was made into fibers at 1100°C, which had the soda leached from it. Another example is Simpson's preparation of alpha alumina fibers by drawing of glass fibers with 60% alumina, then heat treating the fibers to thermally remove most of the non-alumina constituents, mainly the B_2O_3 and P_2O_5 [33]. However, such conversion often leaves porous areas near the fiber center and larger grains and irregular surfaces, especially as the fiber diameters and conversion temperatures increase (Fig. 7.2).

More common and extensive manifestation of chemical conversion of fibers is addition of constituents by reaction of a fiber to form a chemical compound different from the initial fiber composition. This is typically done by atomic components of the desired fiber composition being deposited on the fiber surface and allowing them to difuse into the precursor fiber and react with it to convert much or all of the fiber to the desired compound. The first of two key examples is the making of BN fibers by first making borea fibers by conventional extrusion (spinning) and drawing in the glassy state, then exposing the borea fibers to NH_3 to convert them to BN fibers, usually in two stages: one at > 350°C and the other at > 1500°C [34,35]. The other common case is conversion of carbon fibers to carbides (and in come cases to mixed carbide/nitrides) by addition of the metal to the carbon fiber surface via CVD, e.g., from halide precursors. In principle, such fibers could often be similarly made by conversion of at least some more refractory metal wires, to carbides or nitrides. However, this would tend to accentuate the disadvantages of the process since the wires would often be more expensive, and they are larger (diameters of 25 μm or more) thus exac-

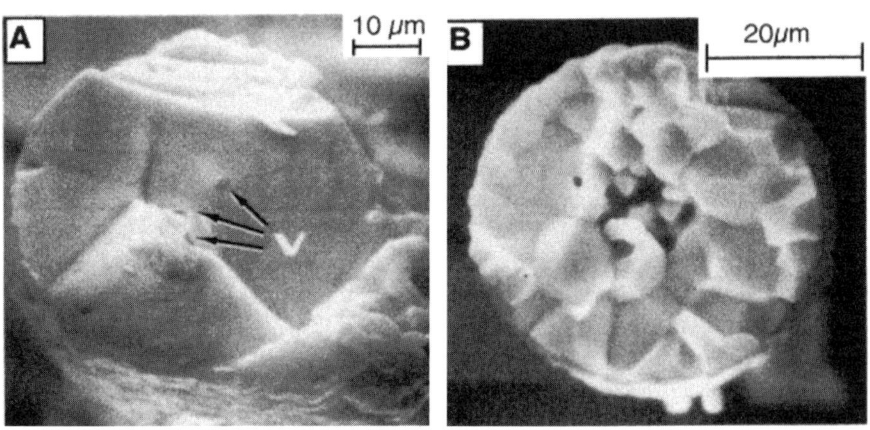

FIGURE 7.2 Examples of fracture cross sections of alpha-alumina fibers made by drawing of alumina-boria glass fibers, then thermally removing the boria to yield the alumina fiber. Note the typical occurrence of both pores (usually at or near the center of the fiber and of larger grains from growth during the thermal removal of most nonalumina ingredients and resultant rougher surface. (Original photos courtesy of F. Simpson, The Boeing Co. From Ref. 24.)

erbating the porosity and grain size issues generally common to such reaction processing of fibers (Fig. 7.2). Such problems are very serious for making fibers for structural applications, but may be tolerated more in other applications, such as for superconductivity, which motivated making some of these fibers, though their costs may be a problem.

7.2.3 CVD of Ceramic Filaments and Melt-Derived Fibers and Filaments

Consider now production of filaments by CVD, where several ceramics have been demonstrated by such processing. The focus here is on B [67–38] and SiC [36–40] filaments which are commercially produced by this process. Though there are variations for both materials, both use a core on which the fiber material is deposited. Originally, both apparently used W wires as cores, but the W resulted in high-temperature performance limitations for SiC fibers. Generally a 25-μm-diameter W wire is used for B and a 33-μm pitch-derived carbon fiber (apparently PG coated) is used for SiC. Boron halides, e.g., BCl_3, are used for B filaments, and various chlorosilanes for SiC, with the CVD conducted in long, > 1m, glass tubes with one filament being formed along the axis of each tube. The tubes are sealed with mercury, which also acts as the electrical contact to resistively heat the W or C core on which deposition occurs, and allows feeding the core and filament through the reactor. Both filaments, commonly with diameters of 75–150 μm, have high strength and Young's moduli, very fine, nanometer grain structures, but are complex. Both have high internal stresses (e.g., that can allow the B filaments to be longitudinally split into three equal segments). The SiC filaments have substantial, mostly radial, grain elongation over much of the deposit giving varying preferred orientation, as well as some radial compositional variations. SiC filaments, which were developed later than the B filaments, were developed in part since costs of the B filaments could not be reduced sufficiently to further expand the market for them, while SiC had more potential for lower cost. SiC filaments are available with different surface compositions to enhance compatibility in different composite matrices.

Preparation of ceramic fibers or filaments from the melt, while posing a basic problem of liquid column instability, is a large source of ceramic fibers with substantial further potential. The problem is that any liquid column intrinsically is driven by surface tension effects to break up into droplets, called the Rayleigh instability. How fast and the extent to which a liquid column distorts its surface toward breaking into droplets is a function of factors resisting such changes in shape, with higher viscosity being an important intrinsic resistance. (This instability, though an important limitation in making fibers from a liquid column, is an important factor in making ceramic beads and balloons, see Sec. 7.3.2.)

Glass fibers can thus generally be readily pulled (extruded/drawn) because of their much higher viscosities combined with their excellent plasticity and its uniformity, i.e., superplasticity. This fiber producing capability of glasses is utilized not only in extrusion/drawing and spooling of individual glass fibers as discussed above, but also in large volume production of lower cost glass fibers produced in mass, not individually [41]. Such fibers are commercially made from various raw materials and mixes, with an important one being alumina silicates (often with other limited additives), with 45–60 w/o alumina to be in practical melting and viscosity ranges. In this case, the first of two forming methods is blowing of low-cost fibers from pouring a molten sheet of glass across a high velocity blast of steam or compressed air such that the glass stream is blown into droplets—many of which are blown into fibers before the glass becomes too viscous. The alternative method is to spin fibers by pouring a molten stream of glass onto a vertically oriented disk rotating at high speed, which spins off glass droplets at high speed so they more effectively form fibers. Both processes leave a fair amount of glass particles, called shot (e.g., 5–50 w/o), with blowing producing more and spinning fibers less shot. Such fibers cover a range from < 1 to > 10 μm in diameter and 1 to several centimeters long, with spun fibers being smaller diameter and longer in length. Resulting fibers are used in large volumes for thermal insulation, but also for more sophisticated applications (with preforms for metal and other composites being an important growing one). Thus, with lower shot content (e.g., by washing) these fibers are formed into papers and other preforms for making glass fiber reinforced components, such as aluminum engine components [42].

Crystalline ceramics (and intermetallics) often melt to low viscosity liquids which, unless countered by other factors, readily results in droplet formation from a liquid (molten) stream. However, physical constraint can be effective in preventing the distortion and breakup of liquid streams. This is demonstrated by the earlier Taylor wire method of preparing fibers of some metals, intermetallics, and ceramics of moderate melting temperatures by placing them in glass tubes in which they can be melted [43]. The tube prevents the melt from breaking into droplets, and can subsequently be removed or used, for example as an insulator for electrically conducting fibers formed in them.

More recently, a significant extension of this type of constraint of the liquid has been developed, called the *invisid melt spinning* (IMS) process [44-46]. This process is based on observations that rapid coating of an extruded fiber/filament of molten ceramic with a thin layer of compatible solid can retain the liquid fiber/filament shape until it solidifies and thus fully stabilizes its fiber shape. A practical low-cost method of making such a fiber coating is to extrude the molten fiber into a closed chamber with a gas that is readily pyrolyzed or cracked on the hot surface of the liquid fiber to produce a solid coating on the fiber surface. Use of propane gas to produce a carbon coating (e.g., a few hundred nanometers in thickness has been used), which can be subsequently readily removed by oxidation but may also be useful for protection from surface damage in some subsequent fiber handling/processing. This process has been demonstrated on CaO-Al_2O_3 compositions, from which filaments could not be made by pulling from the melt nor by melt spinning (extrusion) alone, but the filament (e.g., 100–500 μm dia.) shape in which its melt had been extruded could be retained by the thin carbon coating. Compositions that gave continuous amorphous filaments in the range of 50–80% alumina with melt temperatures of 1340–1840°C had promising strengths, for example, to 1 GPa, while compositions with ≥ 81.5% alumina with melting points of 1850–2070°C were crystalline and broken into pieces that had low strengths of < 165 MPa. Low strengths were attributed to heterogeneous crystallization and formation of voids and cracks.

A promising extension of the IMS process is redrawing of the resultant amorphous fibers, thus identified as *RIMS*. Thus, amorphous fibers of 46.5 w/o CaO and 53.5 w/o Al_2O_3 were redrawn at 1200–1300°C , i.e., 100–200°C below the liquidus. Resultant strength was reasonable, 750 MPa, while Young's modulus more than doubled to 164 GPa. Initial drawing rates from the melt approach those of drawing glass fibers of meters per second and those of redrawing can also approach such levels. Thus, the technology, though mainly or exclusively for compositions that yield amorphous fibers, appears promising, but, as for any partially developed technology, is uncertain in its true potential.

There are some materials and conditions under which ceramic fibers or filaments can be successfully drawn from the melt via at least one of two techniques. Both rest on keeping the length of the melt forming the fiber short and it, and the fiber's diameter large—usually a filament since short lengths and larger diameters of melt increase the stability of the liquid against breaking into droplets [45,46]. The first is the EFG shaped crystal growth process (Sec. 6.7.2). There, a die of a material compatible with the ceramic melt is held on the melt surface such that molten liquid is fed via capillary action through an orifice in the die to form a local molten pool from which a single crystal is grown with the shape of the die orifice. The EFG process was, in fact, originally discovered and developed in a successful effort to grow sapphire filaments, For example, 250 μm diameter, which showed good strengths of ~3 GPa, despite some limitation due to some small, mainly isolated pores [47–50]. Crystal filament pull rates of ~ 2.4 m/hr are substantially slower than for many methods but are useful and can be multiplied many fold by having multiple orifices in a die for one crucible of molten alumina. Such sapphire filaments have been of interest for structural composites, and are apparently used for some optical purposes such for medical lasers and high temperature sensors. The EFG filament growth method is also applicable to some other, mainly oxide, ceramics, including eutectics. Thus, YAG/alumina eutectic filaments (125 μm in dia.) have been grown with a mixed-axial-oriented-lamellar and random structure and strengths of 1.9 GPa. Growth rates were 2.5 cm/min, i.e., 1.5 m/hr, similar to sapphire [51].

The other major, and potentially more diverse, method of growing single-crystal filaments is the floating zone process using laser heating. This uses laser beams to melt a narrow, moving zone of a green or bisque-fired polycrystalline feed rod to convert it to a single-crystal filament. This has been successfully demonstrated with several materials and different experimental conditions. Thus, sapphire filaments 200–250 μm in dia. have been grown to give strengths of up to 9-10 GPa at growth rates of up to 250 cm/hr [52)\], i.e., similar to EFG sapphire. Similarly, stabilized zirconia filaments 400–600 μm in dia. have been grown at 15 cm/hr giving room temperature strengths of ~ 1 GPa and ~ 0.5 GPa at 1400°C [53], and other materials have been grown, including ternary compounds, such as, barium titanate [54]. This method in its various manifestations has the advantage of material versatility since the melt is self-contained [55]. It also allows different directions of filament pulling, e.g., downward, which reduces bubble entrapment that is still a factor for EFG. Growth rates are similar for the two processes, but feed rod preparation is an added cost for the floating zone method and multiplying the number of filaments grown simultaneously is probably more costly for this method than for the EFG method. However, it is not necessarily an either/or choice for the two methods, since they each may have their role to play. What is certain is the growing importance of single -rystal fibers and filaments [55]. For other possibilities of making single-crystal fibers and filaments, see conversion of polycrystalline to single-crystal bodies (Sec. 3.5).

7.2.4 Fiber and Filament Behavior, Uses in Composites, and Future Directions

Since much of the above fabrication, especially of continuous fibers and filaments, is for structural composites, consider their mechanical properties, especially strength and Young's modulus, which are critical to such applications. Young's modulus (E) depends on the material, that is, composition (and for single-crystal fibers or filaments, their axial crystal orientation), and porosity (its amount and character determine its reduction of E [56]. While strength generally scales with E, it also decreases with increasing fiber/filament diameter and gauge length, and with both the amount and size of pores present. Fiber/filament strength also generally increases as the grain size of polycrystalline fibers/filaments decreases, i.e. similar to monolithic ceramics [24], especially once processing defects have been sufficiently reduced. However, while in monolithic ceramics such grain size effects generally result from the size on machining flaws relative to the grain size, in fibers/filaments the grain size dependence arises mainly from grain size effects on surface roughness. Thus, as with monolithic ceramics, fiber strengths have been increased by use of grain growth inhibitors, e.g., of zirconia in alumina [21]. Over the finer grain regime, strength increases with decreasing grain size are attributed mainly to grain boundary grooving when grosser defects such as frequent CVD colony structure and resultant botryoidal (i.e., knobby) surface are minimized. At larger grain sizes, more severe strength limitations result from more knobby, faceted surfaces (e.g., Fig. 7.2), where surface coatings such as silica on alumina fibers may limit strength reductions [21,24]. At high temperatures creep becomes a limiting factor, which is often more limited in ternary than binary compounds, favoring the former for high temperatures, and may also be constrained in some eutectic systems.

While fabrication of ceramic fiber composites is addressed in Section 7.6 (Sec. 8.2.3), it should be briefly noted that while handling of individual filaments is reasonably practical, it is not practical to handle individual fibers, especially finer ones. Instead, in most cases where it is feasible, fibers are formed into bundles of hundreds to thousands of fibers called *tows* with an organic coating (called *sizing*) for handling tows, that is, for keeping them from "fuzzing up". Such handling may be for fabricating uniaxial composites or more commonly for making fabrics for composites by weaving, braiding, knitting, etc., where fiber diameters and Young's moduli are fine enough to allow sufficient flexibility to do so (this commonly cannot be done with filaments). Fabrication of ceramic fibers by replication of organic precursors in cloth or multifiber forms such as felts, braided forms, and so forth is thus desirable since fabrication of such fiber forms as organic rather than ceramic fibers has cost and versatility advantages.

Regardless of the fiber form, structural composites of ceramic matrices with ceramic fibers generally preform best when there is very limited bonding between the fiber and matrix. Thin ceramic coatings are often used on the fibers to assure limiting such fiber-matrix bonding. For mainly nonoxide fiber composites for non- or limited high temperature oxidation condition use, the primary coating is BN applied by CVD [57–59], which is available commercially. BN is applied to individual fibers or filaments and especially to fibers in tow form (after removal of the sizing, by solvent or by burning it off in a flame) and can also be applied to cloth forms, with bolt to bolt cloth coating becoming available [R. Engdahl, President, Synterials, Reston, VA circa 2000). For composites of oxide fibers in oxide matrices for use at high temperature use in oxidizing conditions, considerable work to identify suitable bond limiting coatings is underway with rare earth phosphates the leading candidates [61].

Finally, consider briefly two other areas of fiber development concerning varying fiber composition and geometry. Thus, the above summary of fiber fabrication focused on the bulk of work on fibers with the same composition throughout the fiber. However, glass fibers made from two glass compositions, apparently by fusing two randomly twisted fibers together, such that the axial plane along which the two halves of the resultant fiber are joined twists randomly [62]. Such fibers are reported to be highly flexible and resilient. Coextrusion of one preceramic polymer over another has also been reported [63], giving a coating on a core fiber or a thicker surface layer, i.e., a composite core-shell fiber. Much more extensively, optical fibers are often made with various radial gradients or steps of refractive indexes via corresponding changes in dopants. Further compositional tailoring of such bi- or multicompositional fibers seems feasible depending on imagination and needs.

The second area of fiber development is their shape, mainly their cross sectional shape, especially for hollow fibers of various shapes. The simplest manifestation is fibers in the form of a cylindrical shell, i.e., having a circular internal and external boundary of the solid forming the fiber, as developed with glass fibers [64,65]. Such fibers of a few oxides and of Ni metal have been made by Card [64] by the conversion method. Thus, a porous surface layer in graphite fibers was prepared by controlled oxidation, followed by impregnation of ceramic salt precursor in the pores, then pyrolyzed in air to yield the ceramic and remove the remainder of the graphite fiber.

More versatile and extensive is the demonstration of fabricating hollow fibers of various sizes and especially shapes (Fig. 7.3) from many materials by various processes, e.g., as demonstrated and reviewed by Hoffman and coworkers [66,67]. While quartz tubes can be drawn down to 2 μm ID, a variety of replication techniques are being used to prepare a diverse array of hollow fibers, with fugitive tube processing being an important practical example, with possible costs of the order of 1 cent per cm. Such technology ranges from the nanometer to the micrometer scale and from the microelectromechanical systems (MEMS) with diverse applications from composites to catalysis, filters, miniature heat exchanges, and so forth.

7.3 FABRICATION OF POROUS BODIES

7.3.1 Introduction

Fabrication of porous ceramic bodies is an important and growing area of both research and application, as recently reviewed, since, while porosity decreases may properties, it can provide important functions that are best done with a variety of controlled porosity [56,68–71]. These include large and diverse applications in catalysis and thermal insulation, as well as a diversity of less extensive applications for lightweight materials for mechanical functions, and various existing and developing filtering needs and burner applications.

The various applications require a diversity of needs for combinations of the amount of porosity and its character, e.g., size, shape, location, and degree of its interconnection and orientation. Thus, substantial quantities of highly interconnected, very fine pores are needed to give both high surface area and its ready

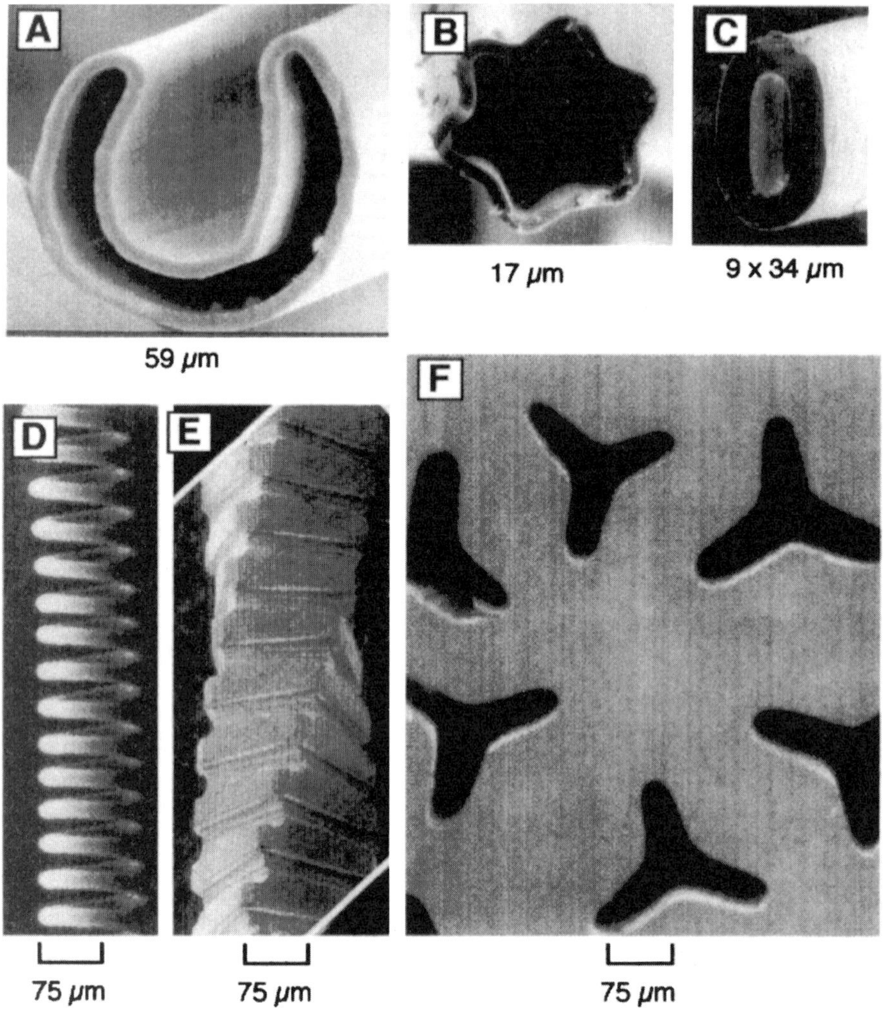

FIGURE 7.3 Examples of microtube fabrication versatility showing tubes with (A) thin, (B) thick, and (C) porous walls; (D) with a liner, (E) a spiral tube on a straight tube, (F) a hollow bellows of noncircular and rotating cross section, and (G) a multichannel tube. (Photos courtesy of Dr. W. Hoffman, Air Force Philips Lab.)

accessibility for catalytic applications. Gradients or anisotropy of the porosity may be needed—for example, in filters to allow good filtration in one fluid flow direction, but limited back pressure for cleaning via reversing the flow direction. In both of these, and many others, these porosity needs must be met while limiting the compromise of other (e.g., mechanical) properties. This must be done in a diversity of component sizes and shapes, for a variety of environments.

The simplest, and most common, method of fabricating porous bodies is to partially sinter compacts of powder of varying character. The particle size and its distribution and the method and extent of powder consolidation along with the extent of sintering are the key controls for the type and amount of porosity obtained. The first of three key limitations of this method of fabricating porous bodies is the amount porosity, e.g., to 50% or less; the second is the limited range of pore character, that is, of pores between partially sintered particles; and the third is its limitations on other properties. Thus, pores between partially sintered particles typically give the greatest reduction in key physical, e.g., mechanical, properties per volume fraction porosity (and give substantial constraint to fluid flow) versus other basic types of pores.

Another simple and widely used method of fabricating porous bodies is by mixing ceramic powder with particles of some fugitive material, for example a burnable material such as plastic or carbon particles of fibers (e.g., chopped), then sintering. Other burnable materials used in industry for their low costs are saw dust or crushed nut shells, (used in making refractory bricks). Pore size and shape of such bodies is thus mainly controlled by the size and shape of the fugitive material, with both of these, especially shape, playing an important role in physical properties. Shapes range from quite angular to polyhedral/spherical to cylindrical, with reductions in key properties such as strength and elastic moduli decreasing in the order listed, i.e., greatest decreases with sharp angular pores (but generally less so than for pores between partially sintered grains), less reduction for polyhedral pores and generally least for tubular pores (especially when their axes are aligned parallel with the stress). Such porosity is generally not open until higher porosity levels, e.g., > 50%, but can theoretically approach a limit of 100% porosity, so there are significant opportunities for bodies with similar (somewhat constrained) fluid flow as with pores between partially sintered grains, but better properties such as strength and elastic moduli.

Clearly, both of the above methods can be combined with one another, as is often done to varying extent. Another important combination is of either or both of the above with extrusion or other fabrication of bodies with highly oriented, fine, uniform tubular channels, e.g., of a honeycomb monolith, which is done to produce automotive exhaust, as well as various industrial, catalyst supports (Fig. 4.10). Thus, the tubular channels add some to the body surface area and a great deal to its permeability, such that pores in the cell walls are highly accessible. Other methods of forming such tubular channel honeycomb bodies are by rolling up calendared tape and possibly via electrophoretic deposition onto many axially oriented, consumable electrodes.

There are, however, other important possibilities for fabricating bodies of designed porosity, including various methods of making foams, making bodies from preformed subunits, and even some melt processing possibilities. Consider first making ceramic foams, mainly open cell foams of oxides, by replication of polymeric, especially polyurethane, foams, which can be prepared in a variety of shapes and microstructures [71]. This entails infiltrating an organic foam with a ceramic slurry to coat the foam struts, drying, calcining, and firing the ceramic, with accompanying burnout of the original organic foam, to leave a ceramic foam with hollow struts (due to removal of the organic foam struts). A key issue with this method is cracking of the ceramic struts due to various combinations of drying shrinkage of the ceramic slip, high thermal expansion of the organic struts prior to their burnout, and sintering shrinkage of the ceramic struts. However, ceramic foams suitable for various applications, especially filtering out impurity particles from molten metals prior to casting, have been industrially produced for a number of years (Fig. 7.4A). Another, more recent replication method uses open-cell, carbon-based (mainly glassy carbon) foams and CVD/CVI to deposit

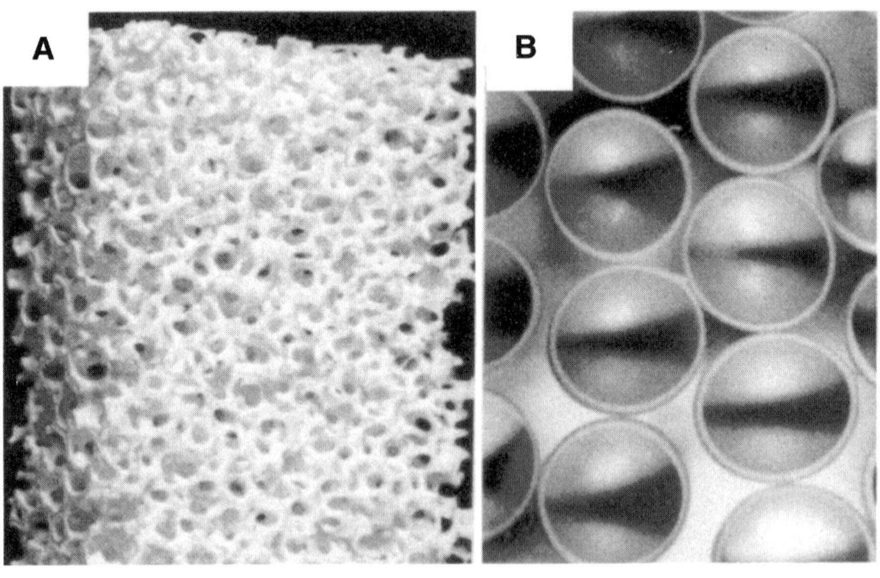

FIGURE 7.4 Examples of highly porous ceramic bodies. (A) Reticulated foam made by replication of polyurethans foam. (B) Closed cell foam made by bonding ceramic balloons together. (Photo courtesy of Prof. J. Cochran of Georgia Tech. From Ref. 56.)

metal, carbon, nonoxide ceramics, as well as some oxide ceramics onto the carbon struts to form a composite foam [72]. Both oxidative removal of the carbon for oxide foams, and reaction between metals and the carbon as an alternative to directly depositing carbides, should be feasible.

Direct fabrication of foams requires first a source of gas to do the "blowing" and sufficient plasticity of the material to be blown to form the foam structure, which must then be rigidized. It is often necessary to follow this with thermal treatment to obtain the final ceramic structure desired. Air or other gasses can be beaten or otherwise mechanically introduced into very fluid ceramic precursors, e.g., ceramic slips; or a material that releases suitable amounts of a gas due to chemical reaction, heating, or both may be used. In either case, the liquid precursor may form a foam by themselves or with surfactant additives, e.g., soaps, which leaves a key issue, namely rigidifing the foam. In some cases this can be done with benign additives, e.g., using plaster of paris (calcium sulfate, calcined gypsum) with ceramics in which the resultant CaO is benign or useful, as is the case with many zirconia bodies.

There are other more general methods of rigidizing foams, with polymerization of an ingredient being an important one. Thus, many ceramic producing sols can be foamed then rigidified by gelling(73), and many preceramic polymers can often similarly be made into foams by frequently using their polymerization for rigidizing (and their decomposition gasses as part or all of the blowing) [74]. Commercial foaming of glassy carbon, which has been in production for years, is a good example of this, and the more recent offering of POCO® graphite foam may be another. Binder constituents may also be another source of rigidifyng foams, with gelcasting of foams being an important example of this [75]. There are also variations of this, for example, very fine foam structures have been produced using a polyethylene-mineral oil binder system, which phase separates on cooling such that frequently most fine ceramic powder remains in the polyethylene and little in the mineral oil, which is readily solvent or thermally removed, leaving a green ceramic foam structure [76)] as in the above cases. A ceramic foam has also been demonstrated via sol precursors that appears to entail phase separation of the sol from some other fluid ingredients [77]. Ceramic foams have also been demonstrated via infiltration of a body of partially sintered, readily soluble, particles, e.g., of NaCl, into which a ceramic precursor is infiltrated and rigidified, then the soluble phase is dissolved out.

There are also opportunities for foaming some ceramics via processing in the melt state or with some melt present. The substantial commercial production of foamed silicate glasses is a prime example, where the key is having a "blowing agent," commonly $CaCO_3$, that decomposes at high temperature where glass foaming is feasible and limited cooling can freeze in the foam structure. Self propagating high-temperature reactions often have transient liquid phases, e.g., of metal constituents, and frequently considerable gas release, e.g., of adsorbed gases on the powder particles, due to the high temperatures rapidly reached. Such reactions frequently show some foaming, but a key issue is likely to be reproducibly controlling this process. Another process based on both melting and leaching is that of Saphikon (Milford, NH) to produce an open cell porous sapphire made by first leaching Cu from W-Cu composite used for high thermal conductivity in electronic systems. Then the porous W body is infiltrated with molten alumina, which is directionally solidified, with the W subsequently leached out to leave the sapphire crystal with the W induced pore structure.

7.3.2 Porous Bodies via Ceramic Bead and Balloon and Other Fabrication Methods

A very promising, method of fabricating bodies of designed porosity, especially of closed cell foams, is via fabrication of ceramic beads that may be dense, porous, foamed, or hollow (balloons), for which there are a variety of fabrication methods [56]. Glass beads are manufactured by atomization of glass melt streams. Glass balloons are fabricated commercially by injecting either: (1) spray-dried agglomerates of glass frit or glass producing materials or (2) slurry droplets of these constituents, both with suitable binder, into a vertical gas flame. Successful balloon forming occurs when much of the outgassing of volatile species is complete as the glass formation and melting is about to seal off the surface such that the remaining gas released will form balloons from forming glass beads, which are light enough to be carried up by the flame, and captured above, with shot and other debris falling down and being collected below. (Sound balloons are separated from most remaining defective ones by putting them in water, where they are "floaters" and "sinkers, "respectively.)A somewhat similar process for some polycrystalline ceramics is by passing appropriate suitable agglomerate particles through a plasma torch, as used commercially to produce larger, coarser balloons of alumina or zirconia, primarily used for thermal insulation.

There are other important processes for making glass or other ceramic beads or balloons [56,78–80]. Basically, both require forming a droplet of a ceramic precursor, rigidizing it, then converting it to a ceramic. One important commercial method is by dripping a sol precursor into a fluid that will cause gelling of each very uniform drop before they are collected and fired. Another, more versatile method is to make an emulsion of a ceramic precursor, e.g., a ceramic slurry, sol, or preceramic polymer, by vigorously mixing it with an immiscible liquid and surfactants, such that the spherical precursor droplets formed become rigidified and hence can be readily sieved out of the remaining liquid, then converted to ceramic beads. Rigidization can again be done via several routes to polymerize the precursor, or part of the binder content for slurries, which can be done by release, e.g., thermally, of a rigidizing agent, e.g., of H_2O

for sols, by catalyzing a polymerization reaction, or basic thermal polymerization. Such emulsion methods produce a range of particle diameters, the mean of which can be shifted some by processing parameters, such as surfactants, and mixing. Another important method of making ceramic beads is to form a downward stream of uniform diameter of a liquid precursor of ceramic—for example, a slurry, sol, or preceramic polymer—and take advantage of the Rayleigh instability that inherently turns a cylindrical stream into one of uniform droplets, which can be rigidified in their downward flight. This is commonly done by having the stream contained in a larger pipe with an upward draft of air (heated) to remove slurry solvent, or with moisture or other polymerization agent for sols or preceramic polymers, or to do so by heating them. Droplets, which are usually very uniform in size and shape, commonly can be rigidized while falling one to two stories in a building, collected, and fired to beads a few millimeter in diameter. The basic simplicity of the process, its potential efficiency, e.g., up to a few ten thousands of beads per minute from one nozzle, make it a promising process demonstrated on some oxides [79,80].

There are two general methods of making balloons of a variety of ceramics. The first is the above downward oriented fluid stream of ceramic slurry, sol, or polymeric precursor, since a nozzle to produce a hollow liquid stream breaks up into uniform liquid balloons that are converted to solid balloons of fairly uniform thin walls and uniform size and sphericity (Fig. 7.4B). Some commercial production indicates further potential for this process. The second method consists of coating plastic, typically polystyrene, spheres with a ceramic slurry that rigidizes by drying, or possibly with a sol or preceramic polymer that are regidized by polymerization, all followed by firing to produce the ceramic balloon via pyrolysis of the plastic kernel bead. Again some commercial use indicates good future potential for producing thin wall balloons a few mm in diameter of oxides, SiC, and a some metals.

There are also ways of producing beads that are at various points along the continuum from dense bead to balloon. The first three are to (1) simply not fully sinter the bead, leaving various amounts of porosity; (2) form beads with various amounts of a fugitive pore forming agent; or (3) use combinations of these with each other or with other techniques outlined as follows. The above emulsion process for making beads can be used to produce a broad range of porous beads, by making two or more sequential emulsions. Thus, a first emulsion of a ceramic precursor with some nonceramic containing liquids can in turn form a second emulsion with another nonceramic containing liquid such that the ceramic containing droplets from the first emulsion can be rigidized and filtered out. Then, chemically or thermally removing the nonceramic containing liquid from the first emulsion leaves a green ceramic bead with a pore structure from the first emulsion, in addition to any porosity subsequently left from firing the ceramic bead. Thus, such a double emulsion fabrication can produce beads that essentially have an internal foam structure, which usually does not involve opening of emulsion formed pores to the bead surface(Fig. 7.5A). However, some bead compositions and fabricating methods can yield such surface opening of larger internal pores to the bead surface by some foaming of the beads (Fig. 7.5B).

A key aspect of fabricating porous bodies from such beads or balloons is forming them into desired shapes and bonding them to form the desired body. In any of the bead or balloon processes bulk bodies can be made from them in either the green or fired state, each route with its advantages and issues. In either case, large and complex shapes can very practically be made using simple molds made of plastic sheets that form the part since beads or balloons only need to be poured into the mold (possibly with some vibratory compaction, especially for beads or balloons with considerable variation in diameter). Then a bonding slip, often of the same composition as used to make the beads or balloons, is poured into the mold, possibly with surfactants to cause wetting primarily at bead or balloon contacts, leaving pores at the interstices between the beads or balloons. (Such pores at the interstices are very effective for reducing weight and fairly benign for other properties since fully filling the interstices between particles is the least effective use of mass to enhance properties.) Upon drying of the slip, the part is removed from the mold and fired. Sealing of body surfaces can also be done via slips, either in conjunction with bead or balloon bonding or in a separate operation.

Using green beads results in the same or similar shrinkage of the beads or balloons and the bonding material, which may give a stronger bond between

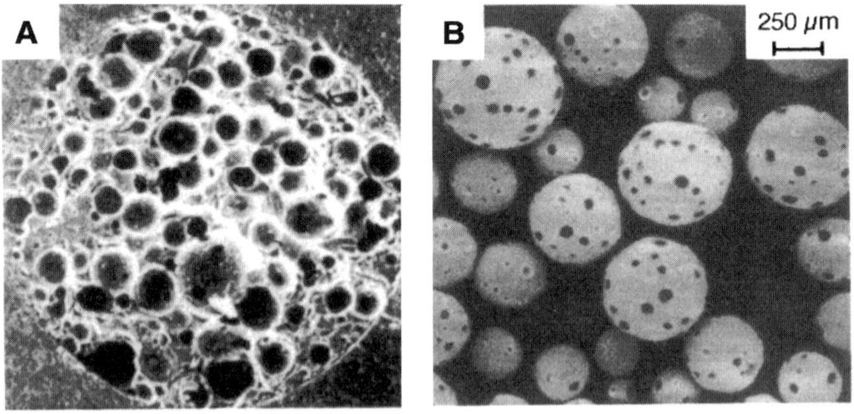

FIGURE 7.5 Examples of porous ceramic beads. (A) Cross section of bead made by the double emulsion process. (B) Ceramic beads foamed in the green state, then fired. Note foam pores intersecting bead surfaces. From Ref. 56.)

them, but presents possible problems of substantial shrinkage as well as damage, e.g., distortion, to the green and sintering beads and especially balloons. On the other hand, using previously sintered beads or balloons to make a body results in a zero-shrinkage body, which is a large advantage, especially for making large and complex shaped bodies. Such use of fired beads or balloons results in some strains between them and the sintering slip bond. However, even thin bonds, where the strain differences are small, result in good strength [56,80]. In some cases other bonding materials/methods, e.g., glasses or cements result in good strength, with CVI being a potentially very important one since it deposits a strong bond with no shrinkage, but again forms generally favorable interstitial pores, and is also potentially superior for sealing surfaces.

There are also some more specialized methods of fabricating some novel porous bodies, in particular to prepare bodies with tubular pores [25,81]. Thus, it has been shown that electrophoretic deposition from aqueous suspensions at higher deposition rates where bubble formation at each electrode, which is normally a serious problem, can yield important pore structures (Fig. 7.6). Thus gas bubbles from electrolysis of the water nucleate at or near the electrode surface and grow with the deposit forming nominally parallel channel pores that are normal to the electrode surface and taper to larger cross sections as deposit thickness increases, i.e., with a cross section like a combination of the letters U and V. Thus, the resultant pores are extremely anisotropic in the shape, having opening of the order of a micron at their bottom of the U/V and orders of magnitude more at their open end that should be very favorable for a very anisotropic filter with very low back pressure for cleaning. Such structures are also very strong for their porosity levels, since there is very little porosity in the ceramic walls around the pores. Further, ceramic tubes with such U/V shaped pores can be grown in groups to possibly form a major element of a filter module. Another,

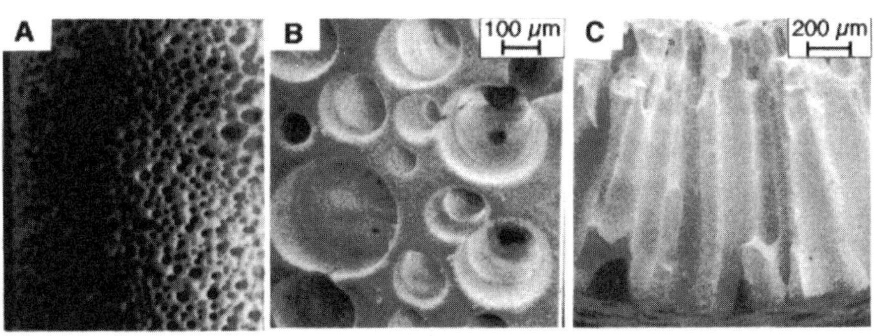

FIGURE 7.6 Radial U/V-shaped pores made by electrophoretic deposition (EPD) of a tube. From Ref. 24.)

lesser developed method is making such through pores of reasonably controlled, adjustable sizes normal to a thin sheet of alumina [56].

Finally an important source of making an important range of porous bodies is via use of fibers. This can be via use of shorter fibers, commonly in the form of felts, or in bodies of continuous fibers. The latter may be used in various architectures, e.g., in various weaves and resultant cloth layup, or various filament winding. In all of these cases various extents and ways of introducing matrices can be used, as discussed in Section 8.2.3.

7.4 RAPID PROTOTYPING/SOLID FREE-FORM FABRICATION (SFF)

7.4.1 Introduction and Methods

A recent area of ceramic fabrication development that has become very active and diverse was initially focused on rapid prototyping of components as an important step in their evaluation and design [82]. Prototyping of ceramic components has existed for a long time, e.g., via green machining them from isopressed logs of the desired composition, or using temporary pressing or injection molding dies, and especially slip-casting molds. However, the newer focus has not just been on making prototype components in shorter turnaround times and with lower overall costs than many fabrication methods that often have expensive tooling requirements, e.g., for die pressing and especially injection molding; instead the interest has widened to developing the technology to produce components directly from a computer design, such as via a rapid prototyping machine controlled by a computer, e.g., via CAD (computer-aided design) files. This shift to a broader range of interests is reflected in the change of the process name or designation from rapid prototyping to solid free-form fabrication (SFF).

Two further general extensions of this have also become of interest. The first is "electronic warehousing", that is, not storing spare components in inventory, but instead storing their design in a computer that could produce components on demand via the developing SFF technology. The second is for manufacturing of customized components on a limited scale where such manufacturing could be cost-effective. It should be noted that much of the focus has been on structural components, but there is substantial potential for electrical and especially electronic components, particularly multilayer ceramic electronic packages, as discussed below. Also, besides the above overall interests in an effective way of rapidly producing parts, e.g., prototype ones, there are interests in developing some of these processes for designing microstructures on a research or limited production basis. The more extensively developed fabrication processes are summarized followed by some discussion of other methods that have been considered, followed by some discussion of extension of such processing to produce designed microstructures. Then the processes are compared from several aspects, and future directions are discussed.

The concept behind the various methods is to dissect the component to be built into various layers in a computer, which then drives the fabrication of the part layer by layer. One of the logical early manifestation of this was to use tape casting and automate cutting, stacking, and laminating of designed tape sections to build up a green component to be fired, which is still in active development [83]. This method, as some others, has considerable tape technology on which to draw.

Another early and still expanding and developing collection of technical approaches is based on photolithography, which was the basis of developing stereolithography of plastics via use of photocuring. This entails building up of polymeric components by doctor blading a layer of polymer with photoactive additives, e.g., photoinitators, so that the pattern of that section of the component can be photocured, i.e., polymerized, and the uncurred excess polymer removed. The formed portion of the component, which is on an elevator system controlled by the computer, is then lowered to allow the next layer to be formed. It has been shown that either ceramic or metal powder particles can be mixed into the photopolymer system that also acts as the binder for resultant metal or ceramic photopolymer slips that can still be photocured [84-88]. This also generally requires use of some additives to limit viscosities with suitable powder loading, and limiting using too fine powders, e.g., lower limits are often $^1/_2$ μm to keep viscosities below 2000 centipoise. Also, use of metal and some ceramic powders limit the thickness of curable layers of such slips to 50–100 μm versus thicknesses of the order of 250–500 μm with other ceramics, e.g., alumina. Such procedures allow green metal or ceramic components to be photoformed layer by layer, with suitable metal or ceramic powder loadings in the photopolymer system, , e.g., > 50 v/o, to obtain reasonable to good sintering of the parts.

Two variations of the photolithography approach exist. The first is using a laser beam to photocure the photopolymer-based slip by moving the laser beam over the layer to be patterened by computer control of mirrors directing the laser beam. Most commonly, photocuring uses UV wavelengths, but photopolymers curing at other wavelengths, e.g., in the visible spectrum, are becoming more available. Completely automated systems for such laser curing and layer by layer development of a component have been developed. Again, these consist of a "build platform" on an "elevator" that moves in the z direction to accurately lower the portion of the green sample already formed so the next layer can be formed by spreading the photopolymer-based slip, curing it, and removing uncured slip, then proceeding to the next layer (Fig. 7.7). The other photocuring method uses a lamp, i.e., "flood" illumination, system with masks to determine the areas of curing each layer. Both systems have extensive technical infrastructure, especially flood illumination, since it is the basic method by which much

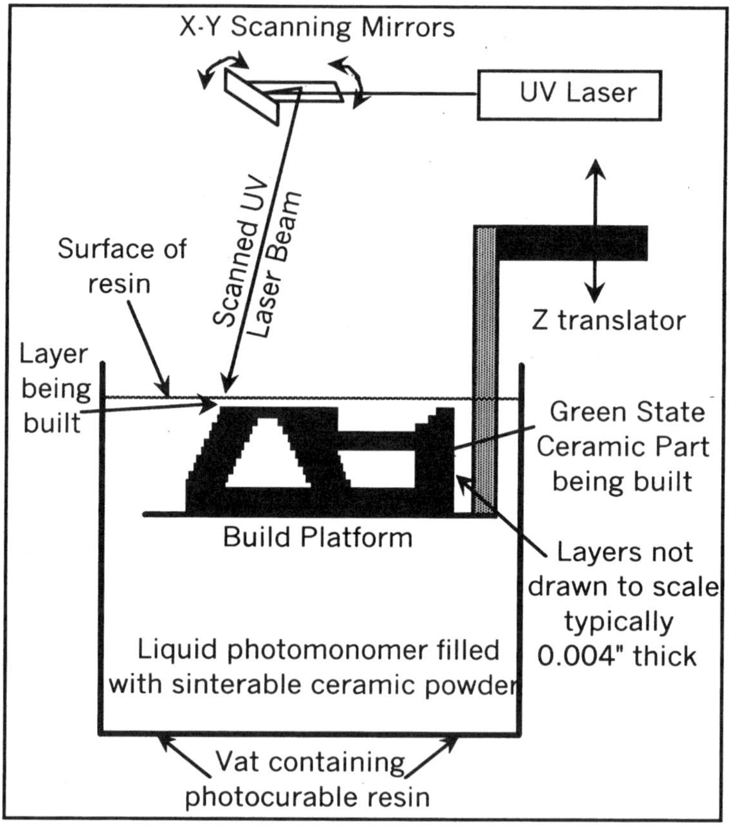

FIGURE 7.7 Schematic of the photolithographic approach to green body formation. (Schematic courtesy of W. Zimbeck of Ceramic Composites, Inc., Millersville, MD.)

printing is done, e.g., many newspapers and many food packages in supermarkets are printed from plates made from photocured photopolymers. Such systems, for example, for making plates for printing two full newspaper pages at a time, allow 1 to 2 orders of magnitude larger areas to be simultaneously photocured at a time than in systems using robotized laser beams for curing. Either illumination system can produce a diversity of shapes and structures with among the best surface quality (Fig. 7.8). Further, electronic photomask systems, e.g., based on liquid crystal technology, are being developed to replace the more cumbersome separately produced plastic photomasks. Another promising development is formulation of preceramic polymers that are also photosensitive to serve as photocuring binders that also contribute to the end ceramic product, hence effectively increasing the ceramic green density.

Different SFF processes are being developed base on inkjet printing technology. One uses printing to pattern a preform of ceramic (or metal) from suc-

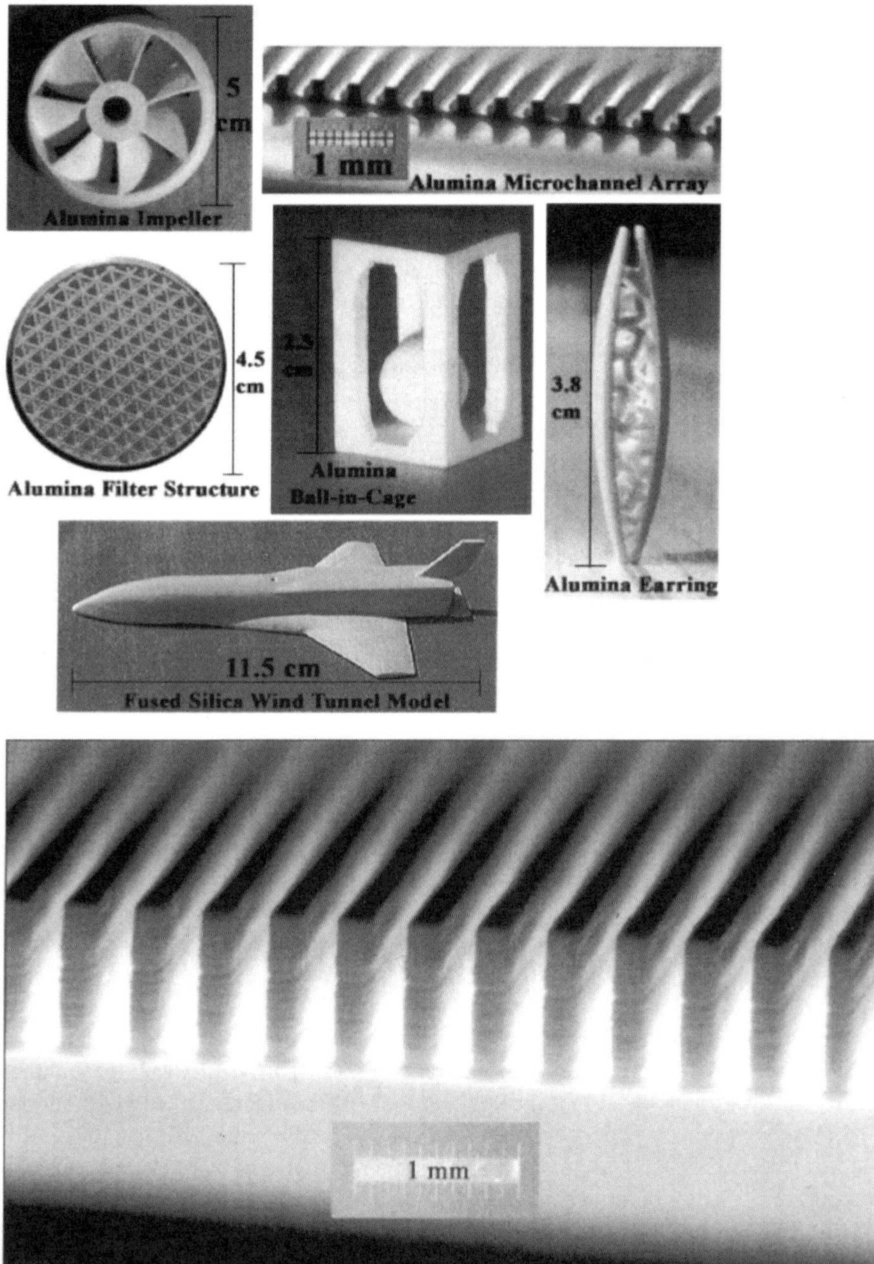

FIGURE 7.8 Examples of bodies fabricated by SFF using photolithography. Top figure portion shows five alumina and one silica body, and the bottom shows the superior surface finish and detail obtained via photolithography. (Figures courtesy of W. Zimbeck, Ceramic Composites, Inc., Millersville, MD.)

cessive layers of powder bed (e.g., of spray-dried granules 50–100 μm in dia. in layers 180 μm thick) by printing a liquid-based binder system in the pattern needed for that layer, then another layer of powder is spread and the binder again printed in the pattern for that layer [91]. The pattern is frozen in by evaporation of liquid constituents of the fluid binder system. A critical issue with this method is the low density of the powder preform produced as discussed further below. Binder drying time may also be a factor. Other methods print droplets of a molten organic, e.g., a wax, mixed with ceramic or metal powder in the desired pattern for each layer that is rigidified on cooling and freezing of the droplets. Limited volume fraction of solids in such binders, e.g., 30 v/o, have been a limitation, but printing speeds are high, e.g., 165 cm/s.

Processes are also being developed based on the use of melts. The first, called *fused deposition modeling* (FDM), uses thermal polymers as binders made into filaments, e.g., 1.8 mm in diameter, by extrusion [92]. Such filaments are the feedstock for the process that consists of reextruding the filaments through a heated, articulated nozzle forming an extrudate of 250 to 1300 mm in diameter depending on nozzle size selected. The pattern is formed for each planar section of the part by articulation of the nozzle over the *x-y* coordinates of the build platform, allowing for some expansion of the extrudate for the build, e.g. to 1.2–1.5 times the as-extruded diameter, as well as some slumping of the extrudate. (Note: Other extrusion processes can be used to produce some finer structures, e.g., to 10 μm sizes, by multiple reextrusions that may compete with some SFF processes as discussed by Halloran [93].)

Another process based on melting is a process called *selective laser sintering* (SLS), which was originally intended to actually melt selected areas of layers of ceramic or metal powder in order to build a part layer by layer [94]. However it was soon found that such laser melting was not feasible in terms of amount of energy or time to melt unless the liquefying temperatures were quite low. Some processing using low melting constituents, e.g., B_2O_3 or phosphate precursors, to react with other constituents, such as alumina, to at least partially form refractory materials, such as aluminum borates or phosphates (often requiring posttreatment to complete the reaction), was demonstrated. Thus, much of the focus of development included using a laser beam to selectively melt organic binder constituents, e.g., of thermal polymers to form a green part. However, reasonable success has been achieved in melting thin layers of metals in a pattern with much higher power lasers, and even more melting versatility may be feasible with electron beams as long as materials do not have high vapor pressures at high temperatures, as for ceramics, such as SiC, BN, SiN, and to some extent MgO.

Other approaches to SFF have been tried to varying extents and others may be feasible. Thus, systems for SFF have been used to apply layers of a ceramic slurry with a fluid binder system that can be gelled by selective addition of a fluid gelling agent to form the desired pattern on a layer and thus build a part layer by layer. While this has been demonstrated with alginate binders [95], it should be applicable to other gelling systems [96], possibly including sols. A significantly different SFF process that may be very desirable, but very challenging is laser stimulated CVD. Some demonstration of this has been made by using either one or two focused laser beams, the latter having a common focus, with formation of a solid ceramic product at the focal point, which can be controlled to form a desired planar, as well as apparently a three-dimensional pattern [97]. There are also further possibilities for SFF based on buildup of melt layers.

While not successful using normal power lasers to melt most ceramic and many metal powders, as noted above, some success has apparently been in SFF of refractory metal parts by instead using an electron beam, which can provide far more power than a laser beam. Though there are added challenges with ceramics due to greater sensitivity to thermal cracking, outgassing, greater solidification shrinkages (Sec. 6.7), and some not melting, electron beam methods might have some useful application for some SFF of ceramics. This and further possibilities for SFF using ceramic melts is shown by recent reporting of making a sapphire irdome by making the EFG crystal growth-method into an SFF process [98,99]. This was done by using the EFG method to provide a small source of liquid alumina and the arm holding the seed crystal being robotized such that it progressively builds the dome spherical shell form by spiraling the seed and the arm holding it and the growing dome to progressively build up a thin layer of melt and have it solidify to build up the dome shell in a spiral fashion (Fig. 7.9). The rate of build, hence the parameters of the spiral (e.g., its angle, speed, and thickness) were computer controlled to allow directional solidification of the melt toward the liquid or free surface in a fashion to avoid thermal stress cracking.

7.4.2 SFF Applications, Comparisons, and Trends

The primary focus of most SFF development has been on structural (e.g., engine), components that have sufficient complexity so rapid prototyping is of value to refine the component design by making test components; the higher costs of doing this are both limited and reduced by SFF. However, though not widely recognized, it is increasingly clear that there are important applications of SFF for nonstructural components, especially electronic components, and particularly ceramic multilayer packages. These reflect markets that dwarf those for structural components and generally have high value added as well as significant technical needs and opportunities for SFF. Consider the primary opportunity of ceramic electronic packages, many of which are complex and can be significantly aided by SFF to produce prototypes for refining design, as well as possible limited production of packages customized for particular applications. Such packages are made by casting ceramic, mainly alumina-based and some AlN,

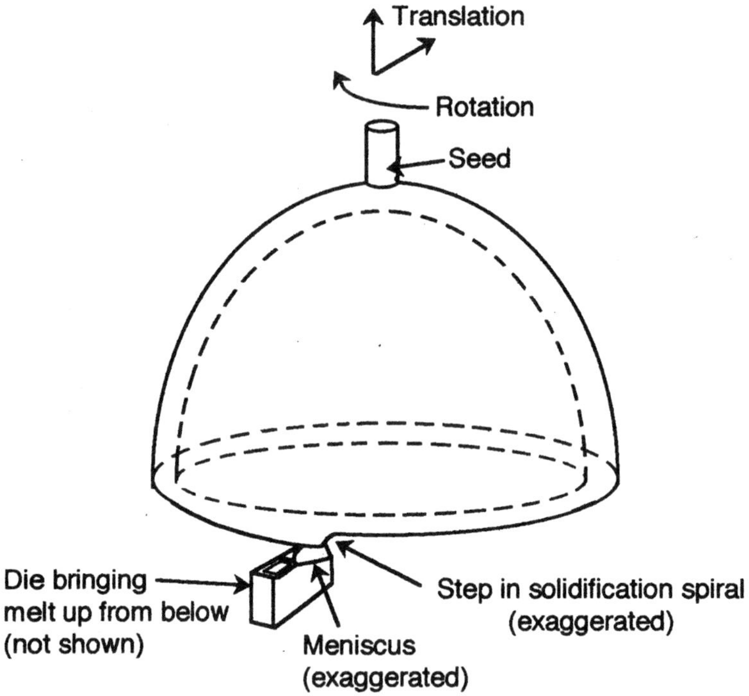

FIGURE 7.9 Schematic of growing sapphire domes, i.e., spherical shell sections via a modified EFG process, as used by Théodore [98].

tapes which have metallic, e.g., Mo or W (for cofiring ceramic and metal) patterns for lateral (x,y) conduction screen printed on them. For conduction in the z direction, holes are punched in the tape, then filled with conductor paste. This can be very costly since production tooling, typically WC to resist wear, is very expensive and very large numbers of holes ("vias") must be punched, e.g., many tens of thousands per layer and often several tens of tape layers to be punched and laminated make sequential punching of via holes impractical.

Photolithography has long been a potential alternative to the above conventional tape fabrication of ceramic electronic packages (and some related products), since it not only offered a lower cost alternative to vias, but also offers important size reductions and accuracy improvements. Thus, key limitations of conductor widths and spacings by screen printing are in the range of 50–125 μm versus 10–20 μm for photolithography. However, photolithography application to ceramic electronic packaging was initially limited to forming a single layer, then firing, before the next green layer was applied. This severe constraint was removed by demonstrating the technology for forming a multilayer stack of green alumina layers with green cofireable metallization that could be densified in a single firing [100]. However, solid loadings in metal and ceramic slips resulted in too high a viscosity, requiring too much solvent dilution to be practical, a limitation that was removed in developing ceramic and metal slip compositions that gave good solids loadings with reasonable viscosities (< 2000 centipoies) [84]. Some work is underway to develop SFF of ceramic electronic packages as a result of these developments [W. Zimbeck, Ceramic Composites, Inc., Millersville, MD, personal communications, 2000–2001.]

Another development leading to added SFF investigation is the possibility of using it to make bodies of designed microstructure, especially of porous or composite bodies. Thus, current technologies allow pores or second phases to be placed in a body with both their shape, orientation, and spacing determined within the resolution of the particular process being used, which for some better systems can be in the range of 10–50 μm. Such resolution may be further increased by the potential of handling ceramic particles one at a time [102]. Thus, as noted earlier, placing spherical pores at the center of the material filling the interstices between spherical beads or balloons cuts weight with little reduction of most physical properties. SFF appears to offer an alternative method of forming such compound pore structures. Also note that Lakes has proposed that hierarchal pore or composite reinforcement structures should give better properties than bodies with random pores or reinforcing particles [103]. Such hierarchal structures are those in which the pores or reinforcing particles in one part of the body are identical in their location, shape, and orientation, but not size, in one area are the same as in other corresponding areas. Thus, in a foam, the cell walls or struts would have the same pores as those forming the foam, except being proportionally smaller, and in turn, there could another one or two levels of smaller hierarchal pores in the cell walls or struts.

SFF methods that have been developed have demonstrated the ability of making test bars of ceramics, metals, or both that have generally achieved room temperature test bar strengths comparable with conventionally made sintered test bars. As far as mechanical quality is concerned, this leaves issues such as the scaling to larger, more complex parts, and of the laminar character of parts. There is some inherent laminar character that can often arise due to some preferred orientation of some powder particles in extrusion or doctor blading, however, residual porosity between layers [24] is often more important. This can have some lamellar character, which clearly leads to anisotropic properties, but considerable anisotropy can occur due to laminar arrays of even equiaxially shaped pores. Such arrayed pores are generally inherent in laminating cast layers and to some extent extruded rods due to particle gradients, e.g., from more settling of larger particles in casting. Thus, interfaces between layers consist of a surface of coarser particles being bonded to a surface of finer particles, which inherently generates a change in porosity at the interface that is often as well removed as porosity within the layers on sintering. Such residual laminar pore arrays result in definition of the laminations on cross sections normal to the lamination, especially fractures (Fig. 7.10).

Most current strength and elastic moduli tests do not reflect anisotropy or other limitations of such residual laminar character, but tests of 3-D printed alumina test bars in various orientations showed substantial strength anisotropy [91]. Tests of bars with the tensile axis normal to the plane of the powder bed had only 30% the strengths of conventionally prepared sintered alumina. However, there was also substantial anisotropy for specimens with the tensile axis in the plane of the powder bed relative to the fast or slow axis of printing—the latter giving 75–95% of strengths of conventionally prepared aluminas, and the former only 50–60% of normal strengths. However, while this is an issue that needs more attention, 3-D printing appears to be an extreme case, and there are potential solutions. Thus, isopressing the 3-D printed alumina specimens eliminated the anisotropy, giving slightly higher strengths with warm versus cold isopressing. Though isopressing adds an extra step and may distort shapes some, especially when starting from such low densities as the above noted 3-D printed part, it may be an allowable operation. More generally some parts may be hot pressed or HIPed to reduce residual porosity and any anisotropy from it as well as its reductions in properties, since costs of this may often be more tolerable for SFF.

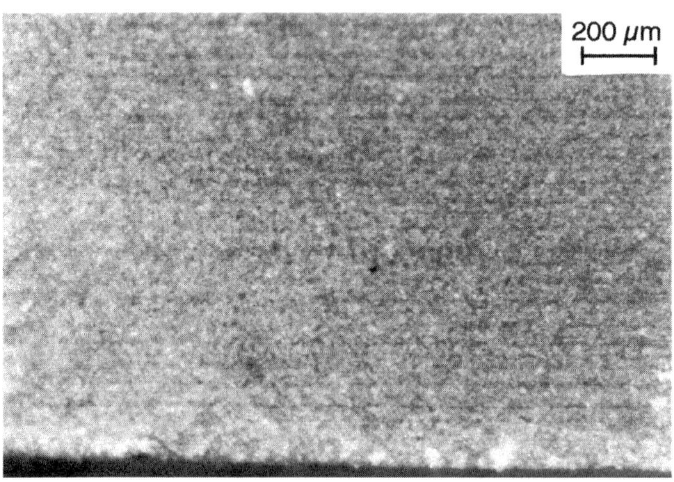

FIGURE 7.10 Fracture cross section of a stainless steel bar made by photolithography showing diffuse, but definite, remnants of the laminar processing of such metal and ceramic parts. (Original photo courtesy of W. Zimbeck of Ceramic Composites, Inc., Millersville, MD.)

Thus, while mechanical properties achieved from various SFF processes may be a factor in selection of specific processes, other factors as summarized below will generally be primary ones in the choice of SFF processes.

Trends in the factors impacting the selection of an SFF method are summarized in Table 7.1, with observations on some possible deciding factors as follows. For forming larger components or larger numbers of components at a time (hence also high average build speed), flood illumination curing of photopolymers has significant potential, while tape lamination also has good size potential for larger components, though binder removal could be a limitation for both methods in thicker parts. For working with two or more materials in the same component, as needed for some composites, and many electronic components, especially multilayer packages, photolithography is a leading candidate. Tape lamination is also a possibility, and fused deposition modeling (FDM) may also be possible, but coarseness of structure, especially with FDM, is a limitation. For the more specialized area of forming internal cavities (e.g., for designed pore

TABLE 7.1 Summary Evaluation of More Developed SFF Methods for Ceramics

Method	E. C.[a]	Advantages	Disadvantages
Tape lam., cutting	L–M	Fast build, especially for large parts, can do 2 or more materials, can cover cavities	High binder content, coarser resolution, surface finish and structure
Photocure binder	M–H	Finer resolution, surface finish, and structure, do two or more materials at a time, large area/ multiple parts via flood illumination	High binder content, covered cavities more difficult, limited area/number of parts with laser beam cure
Inkprinting	L	High build speed, low cost system	Low green density, limited resolution/surface finish, one material at a time
Laser melting binder (SLS)	M	More limited binder content, potentially easier for covering cavities	More materials limited, one material at a time
Extruded thermalplastic binder (FDM)	L–M	Good build speed, may be able for two or more materials at a time	High binder content, coarse layers/surface finish/ limited resolution, some support needed to cover internal cavities

[a]E. C. = equipment costs, with low (L), medium (M), and high (H) being respectively 50–100, 200–500, and 700–1000 thousands of dollars.

structures), tape lamination is probably the most effective for caping internal cavities, and photolithography more limited, but not without means of doing a considerable amount. Thus, for example, potential internal pores can be left filled with uncurred photopolymer during the build phase if the uncured photopolymer can be drained out later, commonly during part removal from the build platform.

The issue of what, if any of the, SFF methods may become commercially established for ceramics is still uncertain since much more needs to be demonstrated, but some factors and trends are noted. Major factors could be (1) the importance of doing larger components, larger numbers of them, or both; (2) the importance of internal structure in components, as needed for most electronic, some composite, and some porous materials applications, and; (3) the opportunities for metal components and the applicability of their methods to fabrication of ceramics. A factor that may bear on both metal and ceramic components with regard to electronic warehousing is the extent to which the large number of materials from which components are conventionally made can be narrowed down so a more reasonable number of raw materials for SFF could be chosen. Another factor for components in which producing the external shape is the goal, e.g., as for many structural components, is that there is often a basic choice of SFF methods, namely to directly produce (metal or) ceramic components by SFF, or instead to produce a temporary die, e.g., of plaster (via a rubber or plastic model) mold for slip casting, or a composite die for die pressing, extrusion, or injection molding. This is not a feasible route for any component that has internal structure formed in the SFF process, e.g., for some composite or porous bodies and for most electronic applications. However, the depth and diversity of capabilities demonstrated and the potential for further developments indicate that some forms of SFF will be established in industry and continue to grow and evolve.

7.5 CERAMIC FIBER COMPOSITES

Ceramic fiber composite fabrication is a large, diverse, and growing topic that could easily fill a separate volume, and hence can only be summarized in outline form here, mainly from other reviews [58, 104, 105]. The topic encompasses ceramic or glass matrices with short (e.g., ceramic whiskers, other as-grown discontinuous, or chopped) fibers, continuous fibers (typically 5–30 microns in dia., usually in bundles, i.e., tows of hundreds to thousands of fibers), or continuous ceramic or metallic filaments (typically 100 or more microns in dia.). The type of fibers, their chemical nature and that of the matrix, as well as the size and shape and use of the composite, all impact the fabrication route used for making the component.

Whisker composites are made primarily by mixing them with the matrix powder, e.g., alumina, commonly by milling, then hot pressing (Sec. 6.2) or occasionally by HIPing, for more shape versatility. Important limitations are the health hazards posed by some whiskers in their processing, machining, or use. Other short, e.g., chopped, fiber composites can be similarly fabricated, but have received limited attention since properties achieved (especially toughness) are far less desirable than with most other ceramic fiber composites. Three other possibilities that may hold promise have received little attention. The first is making thin paper-like or thicker felt preforms of whiskers or especially short fibers by paper- or felt-making techniques and infiltrating these with matrix precursors and laminating infiltrated layers. The second possibility for whiskers or other, e.g., carbon [3], fibers that grow like blades of grass is to use their growth mode to form tapes of highly aligned whiskers or short fibers. Then with the whiskers or fibers coated with a matrix precursor before or after making tapes, the tapes can be laminated in various orientations to give greater planar isotropy and subsequently appropriately densified, usually by hot pressing (Sec. 6.2). Third, alignment of whiskers or fibers with some matrix precursor may allow forming a high density, e.g., 60 v/o, of them into pseudofilaments that are then laid up with the same or different matrix in the pseudofilaments, with or without some whiskers or fibers.

Even more versatile fabrication alternatives are available for continuous fiber composites, where much of the versatility arises from the fiber architecture obtained by operations such as weaving into various cloth weaves or braiding or knitting into various shape, e.g., cylinders, by applying or adapting methods used for polymeric matrix composites (Sec. 7.2). Thus, tows, typically after chemical or thermal removal of organic sizing agents can be infiltrated with liquid matrix precursors such as sols, preceramic polymers, or especially generally lower cost slips (usually by drawing the tow through a bath of the liquid matrix precursor), then laid up as tapes to be stacked in various configurations of filament wound into various shapes. (Note: Good ceramic composite toughness usually requires limited chemical bonding between the fibers and the matrix, which often requires a fiber coating, e.g., BN for SiC and related fibers, with some phosphate coatings showing promise for oxide fibers in oxide matrices, as discussed in Section 7.2. BN coatings can now be applied on a bolt to bolt manufacturing scale [60]. Such coated fibers already have sizing removed.) Cloth sheets may be similarly individually filled with liquid matrix precursor and stacked into a preform, or in some cases stacked up, then infiltrated, e.g., to the extent that slips may penetrate uniformly over the thicknesses required.

There are two basic routes to densifying the above preforms. The first, most diverse in shape and size capability, but by far the most limited in terms of densities and properties achievable, is matrix pyrolysis followed by various numbers of further matrix precursor infiltrations and pyrolysis steps, especially for use with preceramic polymers for the matrix. The other basic alternative (since sintering and HIPing are generally very limited by fibers not axially shrinking with the matrix and often being damaged by stresses along the axis of fibers) is hot pressing, which is only applicable to fairly basic laminar composites (with the uniaxial pressing pressure mainly or exclusively normal to the lamination plane, i.e., not parallel with the fibers). Hot pressing of composites infiltrated with slips to produce glass matrices is also readily done, aided by the high temperature plasticity of the glass matrix, and is extensively used for them (Fig. 7.11). However, more versatile derivatives of hot pressing such as simple versions of transfer and injection molding (which thus eliminate the slip infiltration step) are also feasible for glass matrix composites (Fig. 7.11).

Ceramic (or refractory metal) filaments are much more limited in preforms feasible due to filament stiffness and resistance to bending, generally ruling out any fabric formation and limiting them to mainly uniaxial tape formation and lay up, and matrix infiltration and densification as above. Some three (or higher "dimension" composites due to fibers at addition angles other simple orthogonality) dimensional composites, e.g., the important case of carbon-carbon composites are made by multiple matrix precursor polymer impregnation and pyrolysis (which is an important factor in their higher costs).

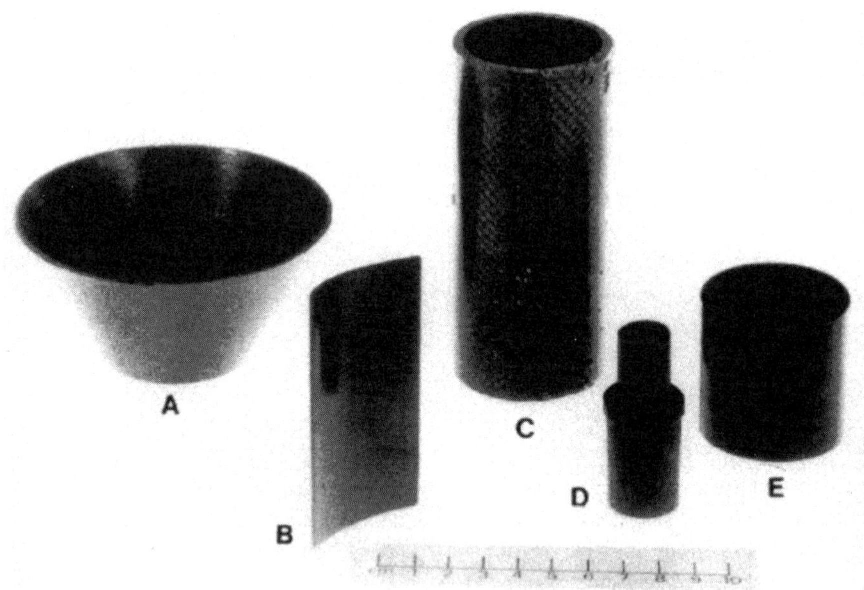

FIGURE 7.11 Ceramic composites with glass matrices and SiC fibers densified by hot pressing and various derivatives giving considerable shape versatility. A and B, hot pressed; C, matrix transfer molded; and D and E, injection molded. (Photo courtesy of K. Prewo, United Technologies Res. Center.)

Another major method of fabricating ceramic composites, including carbon-carbon, is chemical vapor infiltration (CVI) (Sec. 6.6). This is basically CVD with gas flow through the preform to deposit matrix material in the interstices between fibers. Though there are depths of penetration of matrix deposition depending on factors such as preform character, this is a very useful method. Low deposition temperatures, often from < 1000°C to 1200°C, reasonable process times, sizes, times, and reasonable deposition times and chemistries make this an important method (Fig. 7.12).

Some trials of melt forming of ceramic fiber composites have been made with varying results. Forming SiC composites by infiltrating carbon fiber preforms or pyrolyzed pieces of wood with molten Si have had some success. Some attempts to form matrices in fiber preforms by melt spraying have not proved successful. Eutectic composites, which are clearly a type of fiber composite, can clearly be made, but generally have too strong a bond between the

FIGURE 7.12 Example of large ceramic fiber composite for a flame tube made from a braided tube of oxide fibers infiltrated with a SiC matrix by CVI. (From Ref. 58, published with permission of Technomic Pub. Co. Inc.)

"fibers" (rods or lamella) and the matrix to yield the high toughnesses, i.e., noncatastrophic, fiberous fracture that can be achieved with composites made by incorporating (often coated) fibers in a matrix rather than by eutectic phase separation. However, directional solidification of calcium phosphate bodies has resulted in considerable fiberous fracture at very respectable strength levels, indicating promise [106].

7.6 COATINGS

Application of ceramic and some metallic surface coatings on ceramic and metal components is briefly summarized here. Traditional use of ceramic glass-based coatings has been for decorative, surface sealing, or both purposes, such as glaze coatings on ceramic dinnerware. Some of these coatings (and sometimes just quenching surfaces from high temperatures) also may contribute to mechanical reliability by providing useful levels of surface compressive stresses [107]. Use of ceramic coatings on ceramics to extend surface compressive stressing is briefly discussed below. Coating of ceramics for electrical, dielectric purposes, or optical purposes, are also important for specialized applications. Metals are extensively coated with ceramic or intermetallic coatings for friction and wear as well as corrosion protection, as well as for thermal or electrical insulation, increased emittance for better heat radiation, or combinations of these. Metal coatings are also applied to some ceramics for conductive purposes and sometimes as a step in joining ceramics with other ceramics, and metals (Sec. 8.3.3). Major processing technologies for such coatings, especially ceramic, are outlined below.

Major and simple approaches to applying ceramic coatings on metals or ceramics that are widely used in industry is application via slips or powders. In the latter case, a major advance was the development of electrostatic spraying of dry frit powder for application of porcelain enamel coatings on metal surfaces, e.g., for kitchen appliances [108–112]. This approach replaced much application by spraying slips and has advantages over electrophoretic application, and is widely used in industry. However, coatings of many components is done by dipping metal or ceramic parts in the selected slip, or spraying the parts, or both, then drying, followed by firing [112] Many of these ceramic coatings contain glass constituents to limit firing temperatures, allow better thermal expansion matching between coating and substrate, give an impermeable coating, or combinations of these. An example of this is the coating of electrical resistance heating elements with chromia containing glasses for refractoriness and higher emittance.

Metals can also be electrophoretically coated with thin layers, as well as via electrolytic coating of oxides or hydroxides from solutions [113], and some nonoxide coatings may be applied electrolytically [114]. Some metal parts have hard boride coatings formed on them commercially by exposing the part to B powder, e.g., as a coating or more commonly in a powder bed, and heating to allow the B to diffuse into and react with the surface and near surface metal [115,116]. Sol-gel coating is of increased interest and use for coating, e.g., of ceramic surfaces for optical purposes, particularly with advances in techniques to control shrinkage stresses and possible crazing and peeling [117,118]. Application of coatings via other liquid precursors, e.g., preceramic polymers has also been demonstrated and should have good commercial potential if such polymers become more available and at lower prices.

Thicker coatings and coatings for surface compressive effects are frequently applied to the green body and cofired with it. Such coatings may be applied to parts of very simple shapes by laminating tapes to the surface, but more versatile methods are needed for most real components. Where conductivity issues do not preclude it, electrophoretic deposition, though limited in thickness, may be useful. While coatings can be designed for surface compressive stressing, the coarse stepping of resultant steps, poor bonding along the interface, or both may be serious problems. Electrophoretic deposition can give finer grading, and CVD can be an even better method since it allows grading composition and expansion differences, and does not result in discrete interfaces at tape or other green surface layers, as discussed below.

A large area of extensively commercialized coating of mainly, but not exclusively, metal components is melt spraying (mainly plasma spraying) of ceramic coatings, especially for engine and other high temperature applications [119–122].This basically entails passing powders of coating materials through a plasma torch, which melts much of the particles and accelerates them so many splat onto the surface to be coated where they are rapidly (unidirectionally) solidified. Modern systems give a variety of types of torches, their capabilities in terms of melting and acceleration of particles, as well as protective environments for the component to be coated, including in situ surface cleaning, e.g., by sputtering. Metal parts are commonly first coated with a bond coat that provides some grading from the metal to the final ceramic coating, of which zirconia coatings are common. While many of the parts coated are modest in size, there have been significant increases in the size of components plasma coated, e.g., Fig. 7.13.

Consider vapor phase coating processes, which also have a fair amount of commercialization, first addressing physical vapor deposition (PVD), that is where all or a key ingredient are vaporized by heating, for example, via an arc or electron beam [122,123]. This is often used for metal coatings, but is also used for some ceramics, e.g., ZrO_2-based coatings which compete with some plasma sprayed coatings, but with different behavior reflecting basic differences in microstructure. (Note: Fabrication of ZrO_2 billets for the electron beam evaporization required specific design and fabrication parameters to give the necessary

FIGURE 7.13 Piston of a marine diesel engine being experimentally hand coated with ceramic via plasma spraying. (From Ref. 119, published with permission of Plenum Pub. Corp.)

open porous microstructure for the needed thermal shock resistance and avoiding trapped gases and their release during evaporization.) As noted above, melt-sprayed coatings have a splat (generally microcracked) structure parallel to the surface, while PVD produces a substantially elongated grain structure normal to the surface, e.g., grains like grass blades of fibers in a deep carpet that give a more compliant coating [122,123]. Another method of PVD is based on sputtering which has many variations and uses, especially for thin coatings (e.g., for wear and corrosion uses).

An important extension of PVD is where the vaporized species reacts with ingredients in the atmosphere of the PVD. An important example of this is the arc vaporization of Ti or Zr metals and reaction of their vapors with methane, nitrogen, or ammonia to produce the carbides or nitrides of the vaporized metal, respectively, or combinations of these. This process, developed in Russia, was first commercialized for coating industrial metal cutting tools with Ti nitride or carbonitride coatings, then was introduced to the consumer tool market. Recently, this process has been introduced for consumer plumbing fixtures and door fixtures, to replace conventional brass-plated fixtures [124].

The other major vapor-based coating process is CVD [122,125]. Conventional CVD, that which generally uses halide precursors, is used to coat ceramic bodies and for applying debond coatings on ceramic fibers for ceramic composites. An important extension of materials that can be deposited by chemical vapor routes is that of diamond and diamond-like materials [126]. These, particularly the former, significantly extend the ranges of hardness-wear resistance, as well as other properties achievable in coating materials—e.g., IR transmission and band gap. For CVD coating of metals, halide precursors often require too high a temperature and too aggressive reaction products, e.g., HCl, so lower temperature reactants such as organometallic compounds are used, though they often present problems of toxicity and cost.

Finally, another source of coating technology is reaction processing. While boride coatings noted earlier are one example of this, there are many variations of this despite frequent limitations on temperature capabilities of the substrate, especially for most metals. An extension via reduction in thermal exposure is use of SHS reactions (Sec. 6.5) which can make some coatings feasible via the very transient heating. Similarly, application of a reactant to form a desired reaction coating or the coating material itself as a powder then reacting it with, or bonding it to, the surface via a scanned laser beam expand possibilities [127].

7.7 DISCUSSION AND SUMMARY

Fabrication technologies used for more specialized uses, e.g., fabrication of ceramic fibers (or filaments), fiber composites, bodies of designed porosity, rapid prototyping/solid free form fabrication (SFF), and coatings have been reviewed. This was done since, though of narrower use than the generally broader fabrication methods addressed in Chapters 3–6, the methods of this chapter play critical roles for many important and growing applications. Further, the technologies for these more specialized fabrication methods also provide insight into the broader picture of fabrication technology. Thus note, for example, the extensions of use of not only powder-based processing to these more specialized areas, but also the important, and often growing, role of secondary processing methods, e.g., of CVD and melt processing.

Overall development and use of the technologies of this chapter are expected to increase significantly. Particular examples of this are likely to be for more versatile fiber production in terms of methods, as well as production of more complex and single-crystal fibers, fabrication of ceramic fiber composites, especially with continuous fibers, and porous bodies with more designed porosity. Some aspects of SFF are expected to become established, e.g., for rapid prototyping, with further extensions for melt processes, e.g., single crystal, components, and microstructural design, possibly becoming established. Development and application of coating technology is expected to continue substantial further development and use.

REFERENCES

1. W. Mahler, U. Chowdhry. Morphological consequences of freezing gels, L.L. Hench, D.R. Ulrich, eds. Ultrastructure Processing of Ceramics. New York: John Wiley, 1984, p 207.
2. T.F. Cooke. Inorganic fibers—a literature review. J. Am. Cer. Soc. 74(12): 2959–2978, 1991.
3a. Max L. Lake, Jyh-Ming Ting. Vapor grown carbon fiber composites. In: T.D. Burchell, ed. Carbon Materials for Advanced Technologies. New York: Elsevier, 1999.
3b. M. Lake. Large scale production of VGCF. In: L.P. Biro et al., eds. Carbon Filaments and Nanotubes: Common Origins, Differing Applications. Kluwer Academic Publishers, The Netherlands, 2001, pp. 187–196.
3c. M. Lake. Novel Applications of VGCF. IN: L.P. Biro et al., eds. Carbon Filaments and Nanotubes: Common Origins, Differing Applications. Kluwer Academic Publishers, The Netherlands, 2001, pp. 331-341.
4. P. Bracke, H. Schurmans, J. Verhoest. Inorganic Fibers and Composite Materials. New York: Pergamon Press, 1984.
5. J.-B. Donnet, R.C. Bansal. Carbon Fibers. New York: Marcel Dekker, Inc., 1984.
6. Y. Hasegawa, M. Iimura, S. Yajima. Synthesis of continuous silicon carbide fibre. part 2. Conversion of polycarbosilane fibre into silicon carbide fibres. .J. Mat. Sci. 15:720-728, 1980.
7. R.W. Rice. Ceramics from polymer pyrolysis, opportunities and needs—a materials perspective. Am. Cer. Soc. Bul. 62(8):889–892, 1983.
8. K.J. Wynn, R.W. Rice. Ceramics via polymer pyrolysis. Ann. Rev. Mater. Sci. 14:297–334, 1984.
9. J. Lipowitz, J.A. Rabe, K.T. Nguyen, L.D. Orr, R.R. Androl. Structure and properties of polymer-derived stoichiometric SiC fiber. Cer. Eng. Sci. Proc. 16(4):55–62, 1995.
10. J. Lipowitz, J.A. Rabe, R.M. Salinger. Ceramic fibers derived from organosilicon polymers. In: M. Lewin, J. Preston. Handbook of Fiber Science and Technology: V. III High Technology Fibers. Part C. New York: Marcel Dekker, Inc., 1993.
11. H.G. Sowman. A new era in ceramic fibers via sol-gel technology. Am. Cer. Soc. Bul. 67(12):1911–1916, 1988.
12. H.G. Sowman, D.D. Johnson. Ceramic oxide fibers. Cer. Eng. Sci. Proc. 6(9):1221–1230, 1985.
13. T.L. Tompkins. Ceramic oxide fibers: building blocks for new applications. Cer. Ind. pp. 45–50, 1985.
14. G.V. Srinivasan, V. Venkateswaran. Tensile strength evaluation of polycrystalline SiC fibers. Cer. Eng. Sci. Proc. 14(7–8):563–572, 1993.

15. F. Frechette, B. Dover, V. Venkateswaran, J. Kim. High temperature continuous sintered SiC fiber for composite applications. Cer. Eng. Sci. Proc. 12(7–8):992–1006, 1991.
16. S.D. Nunn, D. Popovic, S. Baskaran, J.W. Halloran, G. Subramanian, S.G. Bike. Suspension dry spinning and rheological behavior of ceramic-powder-loaded polymer solutions. J. Am. Cer. Soc. 76(10):2460–2464, 1993.
17. J.E. Bailey, H.A. Barker. Ceramic fibers for the reinforcement of gas turbine blades. In: W.W. Kriegel, H. Palmour III eds. Ceramics in Severe Environments, Materials Science Research. New York: Plenum Press, 5:1971, pp. 341–359.
18. H.D. Blakelock, N.A. Hill, S.A. Lee, C. Goatcher. The production and properties of polycrystalline alumina rods and fibers. Proc. Brit. Cer. Soc. No. 15:69–83, 1970.
19. N.D. Nazarenko, V.F. Nechitailo, N.I. Vlasko. The manufacture and properties of oxide fibers. Soviet Pwd. Metall. 4:265–267, 1969.
20. J.D. Birchall. The Preparation and Properties of Polycrystalline Aluminum Oxide Fibers. Trans. J. Br. Cer. Soc. 82:143–145, 1983.
21. A.K. Dhingra. Advances in inorganic fiber developments. In: E.J. Vandenberg, ed. Contemporary Topics in Polymer Science, Vol. 5. New York: Plenum Press, 1984, pp. 227–260.
22. J.C. Romine. New high-temperature ceramic fiber. Cer. Eng. Sci. Proc. 8(7–8):755–765, 1987.
23. M.H. Stacey. Developments in continuous alumina-based fibers. Trans. J. Br. Cer. 87:168–172, 1977.
24. R.W. Rice. Mechanical properties of ceramics and composites, grain and particle effects. New York: Marcel Dekker, Inc., 2000.
25. P. Lessing. High Temperature Fuel Cell Research and Development. Final Technical Status Report 4FC-DOE-F/80, Montana Energy and MHD, 1980.
26. P. Lessing, M. Johnston. High Temperature Fuel Cell Research and Development. Final Technical Status Report, Montana Energy and MHD, for contract DE-AC03-77ET 11320, Mod. A004, 1981.
27. S.S. Jada, J.F. Bauer. Discontinuous ZrO_2 fiber: precursor solution chemistry-morphology relations. Cer. Eng. Sci. Proc. 11(9–10):1480–1499, 1990.
28. B.H. Hamling, A.W. Naumann, WH. Dresher. Ceramic fibers and textiles from organic precursors. Appl. Polymer Symposia. No. 9. 1969, pp. 387–394.
29. P.A. Vityaz, I.L. Fyodorova, I.N. Yermolenko, T.M. Ulyanova. Synthesis of alumina and zirconia fibers. Cer. Intl. 9(2):46, 1983.
30. R.J. Card, M.P. O'Toole. Solid ceramic fibers via impregnation of activated carbon fibers. J. Am. Cer. Soc. 73(3):665–668, 1990.
31. M. Patel, B.K. Padhi. Titania fibers through jute fiber substrates. J. Mat. Sci. Let.12:1234–1235, 1993.
32. Reference deleted.
33. F.H. Simpson. Continuous Oxide Filament Synthesis (Devitrification). Boeing Co. Final report for contract AFML-TR-71-135, 1971.
34. J. Economy. Now that's an interesting way to make a fiber. Chemtech 10:240–247, 1980.

35. J. Economy. Present status and future for high strength fibers. SAMPE J. 1976, pp. 5–9.
36. J.-O. Carlsson. Review techniques for the preparation of boron fibers. J. Mat. Sci. 14:255–264, 1979.
37. J. Bhardwaj, A.D. Krawitz. The structure of boron fibers. J. Mat. Sci. 18:2639–2649, 1983.
38. F.E. Wawner et al. Various papers in Proc. of the 2nd and 3rd Annual Confs. on Composites and Adv. Mtls., American Ceramic Society, Westerville, OH, 1980.
39. R.L. Crane, V.J. Krukonis. Strength and fracture properties of silicon carbide filament. Am. Cer. Soc. Bul. 54(2):184–188, 1975.
40. S.R. Nutt, F.E. Wawner. Silicon carbide filaments: microstructure. J. Mat. Sci. 20:1953–1960, 1985.
41. J.J. Svec. Glass fibers produced with platinum alloys. Cer. Ind. Mag. 1978, pp. 20–21.
42. L. Caspersen, T. Bradley. Advanced ceramic insulating materials. Adv. Mtls. Procs. 2000, pp. 57–59.
43. I.W. Donald, B.L. Metcalfe. The preparation, properties and applications of some glass-coated metal filaments prepared by the taylor-wire process. J. Mat. Sci. 31:1139–1149, 1996.
44. S.A. Dunn, E.G. Paquette. Redrawn inviscid melt-spun fibers–potential low cost composite reinforcement. Adv. Cer. Mats. 12(4):804–808, 1987.
45. F.T. Wallenberger, N.E. Weston, K. Motzfeldt. Invisid melt spinning of alumina fibers: jet stabilization mechanism. Cer. Eng. Sci. Proc. 2(7–8):1039–1047, 1991.
46. F.T. Wallenberger, N.E. Weston, K. Motzfeldt, D.G. Swartzfager. Invisid melt spinning of alumina fibers: chemical jet stabilization mechanism. J. Am. Cer. Soc. 75(3):629–636, 1992.
47. S.A. Newcomb. Temperature and Time dependence of the Strength of C-Axis Sapphire from 800-1500° C. MS Thesis, Penn State Univ., May 1992.
48. S.A. Newcomb, R.E. Tressler. Slow crack growth of sapphire at 800 to 1500° C. J. Am. Cer. Soc. 76(10):2505–2512, 1993.
49. S.A. Newcomb, R.E. Tressler. High temperature fracture toughness of sapphire. J. Am. Cer. Soc. 77(11):3030–3032, 1994.
50. R.W. Rice. Corroboration and extension of analysis of C-axis sapphire filament fractures from pores. J. Mat. Sci. Let. 16:202–206, 1997.
51. J.-M. Yang, S.M. Jeng, S. Chang. Fracture behavior of directionally solidified $Y_3Al_5O_{12}/Al_2O_3$ eutectic fiber. J. Am. Cer. Soc. 79(5):1218–1222, 1996.
52. J.S. Haggerty, K.C. Wills, J.E. Sheehan. Growth and properties of single crystal oxide fibers. Cer. Eng. Sci. Proc. 12(9–10):1785–1801, 1991.
53. K.J. McClallan, H. Sayir, A.H. Heuer, A. Sayir, J.S. Haggerty, J. Sigalovsky. High-strength, creep-resistant Y_2O_3-stabilized cubic ZrO_2 single crystal fibers. Cer. Eng. Sci. Proc. 14(7–8):651–659, 1993.
54. M. Saifi, B. Dubois, E.M. Vogel. Growth of tetragonal $BaTiO_3$ single crystal fibers. J. Mater. Res. 1(3):452, 1986.
55. R.S. Feigelson. Opportunities for research on single-crystal cibers. Mats. Sci. Eng. B1:67–75, 1988.

56. R.W. Rice. Porosity of Ceramics. New York: Marcel Dekker, Inc., 1998.
57. R.W. Rice. BN Coating of Ceramic Fibers for Ceramic Fiber Composites. U.S. Patent 4,642,271, 1987.
58. R.W. Rice, D. Lewis III. Ceramic fiber composites based upon refractory polycrystalline ceramic matrices. In: S.M. Lee, ed. Reference Book for Composites Technology. Vol. 1. Lancaster, PA: Technomic Pub. Co., Inc., 1989, pp. 117–142.
59. B. Bender, D. Shadwell, C. Bulik, L. Incorvati, D. Lewis III. Effect of fiber coatings and composite processing on properties of zirconia-based matrix SiC fiber composites. Am. Cer. Soc. Bul. 65(2):363–369, 1986.
60. Reference deleted.
61. T.A. Parthasarathy, E. Boakye, M.K. Cinibulk, M.D. Petry. Fabrication and testing of oxide/oxide microcomposites with monazite and hibonite as interlayers. J. Am. Cer. Soc. 82(12):3575–3583, 1999.
62. Announcement. Two glass fibers fused to form stronger filaments. Adv. Mats. Processes 13:1995.
63. G. Curran. Bicomponent extrusion of ceramic fibers. Adv. Mats. Processes 25–37:1995.
64. R.J. Card. Preparation of hollow ceramic fibers. Adv, Cer. Mats. 3(1):29–31, 1988.
65. Announcement. Glass gets tough. Chem Week 1964.
66. W.P. Hoffman, H.T. Phan, P.G. Wapner. The far-reaching nature of microtube technology. Mat. Res. Innovat. 2:87–96, 1998.
67. Phillip G. Wapner, Wesley P. Hoffman. Utilization of surface tension and wettability in the design and operation of microsensors. Sensors and Acxtuators B 71:60–67, 2000.
68. R.W. Rice. Evaluation and extension of physical property-porosity models based on minimum solid area. J. Mat. Sci. 31:102–108, 1996.
69. R.W. Rice. Comparison of physical property-porosity behavior with minimum solid area models. J. Mat. Sci. 31:1509–1528, 1996.
70. R.W. Rice. Microstructural dependance of fracture energy and toughness of ceramics and ceramic composites versus that of their tensile strengths at 22 C. J. Mat. Sci. 31:4503–4519, 1996.
71a. J. S.-Woyansky, C.E. Scott, W.P. Minnear. Processing of porous ceramics. Am. Cer. Soc. Bul. 71(11):1674–1882, 1992.
71b. R.W. Rice. Fabrication of ceramics with designed porosity. Cer. Eng. Sci. Proc. (in press) 2002.
72. A.J. Sherman, R.H. Tuffias, R.B. Kaplan. Refractory ceramic foams: a novel, new high-temperature structure. Am. Cer. Soc. Bul. 70(6):1025–1029, 1991.
73. T. Fujiu, G.L. Messing. Analysis and modeling of sol foaming for the preparation of low-density ceramics. J. Mats. Synthesis Proc. 1(1):33–42, 1993.
74. T.J. Fitzgerald, V.J. Michaud, A. Mortensen. Processing of microcellular SiC foams, part II. Ceramic foam production. J. Mat. Sci. 30:1037–1045, 1995.
75. P. Sepulveda. Gelcasting foams for porous ceramics. Am. Cer. Soc. Bul. 76(10):61–65, 1997.
76. R. Brezny, R.M. Spotnitz. Method of Making Microcellular Ceramic Bodies. U.S. Patent No. 5,427,721, 1995.

77. J. Block. Inorganic Membrane. U.S. Patent No. 4,980,062, 1990.
78. R. Brezny. Porous Ceramic Beads. U.S. Patent No. 5,322,821, 1994.
79a. Pohrong R. Chu, "A Model for Coaxial Nozzle Formation of Hollow Spheres", PhD Dissertation, Materials Science and Engineering, Georgia Institute of Technology, Atlanta, GA, June, 1991.
79b. J.K. Cochran. Georgia Institute of Technology, Atlanta, GA, private communication 2001.
80. J.H. Chung, J.K. Cochran, K.J. Lee. Compressive mechanical behavior of hollow ceramic spheres. In: D.L. Wilcox, M. Berg,T. Bernat, J.K. Cochran, D. Kellerman, eds. Spheres and Microspheres: Synthesis and Applications. Pittsburg, PA: MRS Proc., 372:179–186, 1995.
81. A.V. Kerkar. Manufacture of Conical Pore Ceramics by Electrophoretic Deposition. U.S. Patent No. 5,340,779, 1994.
82. F.B. Prinz et al. Rapid Prototyping in Europe and Japan. JTEC/WTEC Panel Report Pub., SME, 1997.
83. C. Griffin, J. Daufenbach, S. McMIllin. Desktop manufacturing: LOM vs pressing. Am Cer. Soc. Bul. 73(8):109–113, 1994.
84. T. Quadir, S.K. Mirle, J.S. Hallock. Three Dimensional Sintered Inorganic Structures Using Photopolymerization. U.S. Patent No. 5,496,682, 1996.
85. W. Zimbeck, J. Jang, W. Schulze, R.W. Rice. Automated fabrication of ceramic electronic packages by stereo-photolithography. In: S.C. Danforth, D.B. Dimos F. Prinz, eds. Proc. Solid Freeform and Additive Fabrication 2000 Symposium, Materials Research Society. Vol. 625. 2000, pp. 173–178.
86. W.R. Zimbeck, R.W. Rice. Freeform fabrication of components with designed cellular structure. In: D. Dimos, S.C. Danforth, M.J. Cima, eds. Proc. Solid Freeform and Additive Fabrication Symp. Materials Research Soc. Vol. 542. 1998.
87. W. Zimbeck, M. Pope, R. Rice. Microstructures and strengths of metals and ceramics made by photopolymer-based rapid prototyping. Austin, TX: Solid Freeform Fabrication Symposium Proceedings, 9/1996, pp. 411–418.
88. W. Zimbeck, R. Rice. Stereolithography of Ceramics and Metals. Proceedings of the 50th Annual Conf. Soc. for Imaging Science & Technology, Cambridge, MA, IS&T Society, 1997, pp. 649–655.
89. M. Griffith, J.W. Halloran. Freeform fabrication of ceramics via stereolithography. J. Am. Cer. Soc. 79(10):2601–2608, 1996.
90. J.W. Halloran, M. Griffith, T.-M. Chu. Stereolithography resin for rapid prototyping of ceramics and metals. U.S. Patent 6,112,612, 2000.
91. B. Giritlioglu, M.J. Cima. Anisotropy on rectangular bars fabricated via three-dimensional printing. Cer. Eng. Sci. Proc. 16(5):763–772, 1995.
92. M.K. Agarwala, A. Bandyopadhyay, R. van Weeren, A. Safari, S.D. Danforth, N.A. Langrana, V.R. Jamalabad, P.J. Whalen. FDC, rapid fabrication of structural components. Am. Cer. Soc. Bul. 75(11):60–65, 1996.
93. C.V. Hoy, A. Barda, M. Griffith, J.W. Halloran. Microfabrication of ceramics by coextrusion. J. Am. Cer. Soc. 81(11):152–158, 1998.
94. D.L. Bourell, H.L. Marcus, J.W. Barlow, J.J. Beaman. Selective laser sintering of metals and ceramics. Intl. J. Pwd. Met. 28(4):369–381, 1992.

95. S. Baskaran, G.D. Maupin, G.L. Graff. Freeform fabrication of ceramics. Am Cer. Soc. Bul. 77(7):5358, 1998.
96. A. Matsuda, Y. Matsuno, M. Tatsumisago, T. Minami. Fine patterening and characterization of gel films derived from methyltriethoxysilane and tertaethoxysilane. J. Am. Cer. Soc. 81(11):2849–2852, 1998.
97. F.T. Wallenberger. Rapid prototyping directly from the vapor phase. Science 267(3):1274–1275, 3/1995.
98. F.Théodore, T. Duffar, J.L. Santailler, J. Pesenti, M. Keller, P. Dusserre, F. Louchet, V. Kurlov. Crack generation and avoidance during the growth of sapphire domes from an element of shape. J. Crystal Growth 204:317–324, 2000.
99. V.N. Kurlov, F. Théodore. Growth of sapphire crystals of complicated shape. Cryst. Res. Tech. 34(3):293–300, 1999.
100. J.W. Lau, K.E. Bennet. Imaging Process for Forming Ceramic Electronic Circuits. U.S. Patent 4,828,961, 1989.
101. Reference deleted.
102. W.D. Teng, Z.A. Huneiti, W. Machowski, J.R.G. Evans, M.J. Edirisinghe, W. Balachandran. Towards particle-by-particle deposition of ceramics using electroststic atomization. J. Mat. Sci. Let. 16:1017–1019, 1997.
103. R. Lakes. Materials with structural hierarchy. Nature 361:511–515, 1993.
104. R. Rice. Processing of ceramic composites. In: J.G.P. Binner, ed. Advanced Ceramic Processing and Technology. Vol. 1. Park Ridge, NJ: Noyes Pubs, 1990, pp. 123–213.
105. K.M. Prewo, J.J. Brennan. Fiber reinforced glasses and glass ceramics for high performance applications. In: S.M. Lee, ed. Reference Book for Composites Technology. Vol. 1. Lancaster, PA: Technomic Pub. Co., Inc., 1989, pp. 97–116.
106. Y. Abe, T. Kasuga, H. Hosono, K. Groot. Preparation of high-strength calcium phosphate glass-ceramics by unidirectional crystallization. J. Am. Cer. Soc. 67(7):C-142–144, 1984.
107. H.P. Kirchner. Strengthening of Ceramics: Treatments, Tests, and Design Applications. New York: Marcel Dekker, Inc., 1979.
108. W.D. Faust. Electrostatic enamel application theory and practice. Am. Cer. Soc. Bul. 59(2):220–222, 1980.
109. H. Hoffmann. Theory and practice of electrocoating porcelain enamel. Am. Cer. Soc. Bul. 57(6):605–608, 1978.
110. G. Hein. Electrostatic deposition of powdered frit. Cer. Ind. 1973, pp. 20–21.
111. A. Lacchia. Electrostatic spraying of powdered enamels. Cer. Ind. 1977, pp. 36–38.
112. S.B. Lasday. Nature of ceramic coatings and their benefits in thermal processes. Ind. Heating 1982, pp. 49–52.
113. I. Zhitomirsky. New developments in electrolytic deposition of ceramic films", Am. Cer. Soc. Bul. 2000. pp. 57–63.
114. D. Elwell. Electrolytic growth from high-temperature solutions. In: E. Kaldis, H.J. Scheel, eds. Current Topics in Materials Science, 1976. Crystal Growth and Materials. 2. New York: North-Holland Pub. Co., 605–637, 1977.
115. J. Cataldo, F. Galligani, D. Harraden. Boride surface treatments. Adv. Mats. Processes 2000, pp. 35–38.

116. K. Stewart. Boronizing protects metals against wear. Adv. Mats. Processes 1997, pp. 23–25.
117. H.G. Floch, J.-J. Priotton,"Colloidal sol-gel optical coatings. Am. Cer. Soc. Bul. 69(7):1141–1143, 1990.
118. H.G. Floch, P.F. Belleville, J.-J. Priotton. Colloidal sol-gel optical coatings-coatings for lasers, II. Am. Cer. Soc. Bul. 74(11):84–89, 1995.
119. R.W. Rice. Advanced Ceramic Materials and Processes. In: D.L. Cocke, A. Clearfield, eds. Design of New Materials. New York: Plenum Publishing Corp. 1987, pp. 169–194.
120. G. Fisher. Ceramic coatings enhance performance engineering. Am. Cer. Soc. Bul. 65(2):283–287, 1986.
121. G. Geiger. Ceramic Coatings. Am. Cer. Soc. Bul. 71(10):1470–1481, 1992.
122a. D.R. Biswas. Review deposition processes for films and coatings. J. Mat. Sci. 21:2217–2223, 1986.
122b. R.F. Bunshah, Handbook of Deposition Technology for Films and Coatings," Noyes Publication, Inc., Park Ridge, NJ, 1994.
123. H. Lämmermann, G. Kienel. PVD coatings for aircraft turbine blades. Adv, Mats. Proc. 1991, pp. 18–32.
124. T. O'brien. LifeShine non-tarnish coating. Adv, Mats. Proc. 2000, pp. 39–40.
125a. H.O. Pierson, ed. Chemically Vapor Deposited Coatings. Westerville, OH: Am. Cer. Soc. 1981.
125b. H.O. Pierson, "Handbook of CVD" Noyes Publication, Inc., Park Ridge, NJ, 1992.
125c. F.S. Glasso, "Chemical Vapor Deposited Materials," C.R.C. Press, Inc., Boca Raton, FL, 1991.
126a. C.B. Collins, F. Davanloo, T.J. Lee, J.H. You, H. Park. The production and use of amorphic diamond. Bul. Am. Cer. Soc. 71(10):1535–1542, 1992.
126b. W.A. Yarbrough. Vapor-phase-deposited diamond-problems and potential. J. Am. Cer. Soc. 75(12):3179–3200, 1992.
127. A. Agarwal, N.B. Dahotre. Laser surface engineering of titanium diboride coatings. Adv, Mats. Proc. 2000, pp. 43–45.

8

Crosscutting, Manufacturing Factors, and Fabrication

8.1 INTRODUCTION

This chapter provides an overview of ceramic fabrication technology in three ways. The first is by addressing three factors that can have significant impact across several fabrication technologies, namely anion/gaseous impurities, heating methods (e.g., use of microwave and other newer methods), and fabrication of ceramic composites. Next, three manufacturing factors are addressed, namely surface finishing, inspection and nondestructive evaluation (NDE), and joining/attachment. Finally, a summary comparison of ceramic fabrication technologies used on a substantial scale is given along with some observations on manufacturing control.

8.2 IMPORTANT CROSSCUTTING FACTORS

8.2.1 Anion/Gaseous Impurities and Outgassing Prior to or During Densification

Adequate outgassing is a basic need for suitably densifying powder-derived components before measurable closed porosity occurs. Adequate outgassing of atmospheric gasses between powder particles is a basic necessity to reach, or closely approach, theoretical density, i.e., transparency of dielectrics. It was noted that hot pressing without a vacuum generally allows near theoretical density to be achieved (Sec. 6.2). This is consistent with theoretical evaluation showing that < 0.5% porosity can be achieved by conventional sintering of fine powders (but substantially higher porosity can remain with coarser particles and higher sintering gas pressure) [1]. However, there are issues of retaining larger pore clusters and possible reduced density near the center of larger bodies due to the central areas taking longer to get to temperature while the surface is being sealed off by earlier densification there. There are also cases where residual gases in pores may be a factor in some special applications. An example of the latter is failure of small microwave tubes due to internal cracking of the 96% alumina tube housing allowing gases in the closed pores connected by cracks to raise the tube vacuum pressures above failure level [2].

The major source of outgassing problems in densification of powders is formation and release of gaseous species that are not constituents of the ceramic powder nor the environment remaining in the powder. A major source of these gases is impurities themselves or their interactions with other nonconstituent, e.g., adsorbed species discussed below. Thus, for example HF and SO_x gases are given off from firing of bodies containing clays or a broader, but still limited, range of raw materials respectively, [3,4]. Though SO_x can arise from combustion of fuel in firing it also arises from impurities in the powders used, which are the exclusive source of HF emissions. Though these outgassings may have some effects on the resultant fired bodies, the basic concerns are environmental, which, as noted in Section 2.2 is also an increasing factor in preparation of high purity ceramic raw materials, e.g., avoiding use of nitrate oxide precursors because of NO_x emissions.

A major source of outgassing problems are volatile species adsorbed on powder particle surfaces or entrapped in particles or agglomerates of them that occur to varying extents depending on the powder material, its preparation, the fabrication/processing conditions, as well as the end use of the component [5–7]. Adsorbed species from the atmosphere are a problem that varies with the powder history and external conditions, e.g., seasonal variations of humidity and temperature. However, a major source of such outgassing problems are species left from the precursor to the powder, especially for oxides. The problem can be serious for two reasons. The first is that complete decomposition may not occur, some released gas may be trapped by grain growth during calcining, or adsorption of released gas species onto powder surfaces (possibly with some rereaction with the oxide surface), all giving strongly bonded, hard to remove species. Again, there are interactions with impurities that may enhance retention of adsorbed species. The second reason for the seriousness of the problem is that the species, e.g., adsorbed or rereacted, are effectively in the solid state, but upon decomposition yielded gases with of the order of 10^4 volume expansion that can cause serious blistering, bloating, or both.

There are three key factors regarding the above anion impurity problem, namely the amount of gas producing species, their nature, and that of the powder they are with (which impacts both the amount of impurity as well as its character and thus the ease of its decomposition/volatilization), and the opportunity for them to be removed before densification proceeds to inhibit and preclude their release from the body. Thus larger bodies present more problems because of longer times required to diffuse resulting gases out of the body and greater temperature gradients that may inhibit gas exiting the surface due to more sintering there and lower temperature in the interior to drive gases out. Finer particle sizes also increase the problem, both by providing finer pores through which the gases must diffuse (often with pore dimensions being less than the gas mean free path) as well as by lowering the sintering temperature, which causes more sintering at lower temperatures and hence greater gas entrapment potential with less thermal driving force to remove the gas. This issue of opportunity for gas removal is also very important in the method of densification, with pressureless sintering giving greater opportunity, and pressure sintering progressively less opportunity as the pressure increases and as one goes from hot pressing to HIPing.

Though not reported extensively (in part since publication is generally focused on successes, not on problems) there is clear evidence of a problem that is often serious and a reasonable outline of its patterns and causes, as outlined above and as follows. MgO derived from hydroxides, bicarbonates, or carbonates by calcining or direct reactive hot pressing is quite susceptible, e.g., frequently yielding small hot-pressed specimens that are transparent, but have IR absorption bands in their optical transmission (Fig. 8.1) showing the retention of mainly hydroxide or carbonate impurities. Such bodies blister, become opaque, and bloat upon subsequent high temperature exposure as the problem increases, e.g., in larger bodies or higher pressure densification (Fig. 8.2). Thus, HIPing of hydroxide-derived high purity, very fine MgO powder (previously CIPed at 630 MPa to give 55% of theoretical density) with 100 MPa pressure at 800°C resulted in retention of 5% hydroxide (Brucite) despite prior vacuum outgassing of the canned compact at 250°C for 4 hrs [8]. Some residual gaseous species were still present on subsequent thermal exposure to 1750°C.

While MgO is probably more susceptible to this problem, it is a fairly general one, e.g., occurring to varying extents in other oxides from other precursors, e.g., in hot pressed Al_2O_3 and $MgAl_2O_4$ from sulphate precursors where similar, but probably somewhat less severe effects and not occurring till higher temperatures (Fig. 8.3). Similarly, use of hydroxide and other precursors for Y_2O_3 have resulted in outgassing of powders calcined at 1000°C and green bodies presintered to 1400°C and hydroxide absorption bands in transparent sintered bodies [9,10]. Use of phosphate precursors in sintering transparent ZrO_2 has also resulted in some porosity generation and clouding upon exposure to higher sinter-

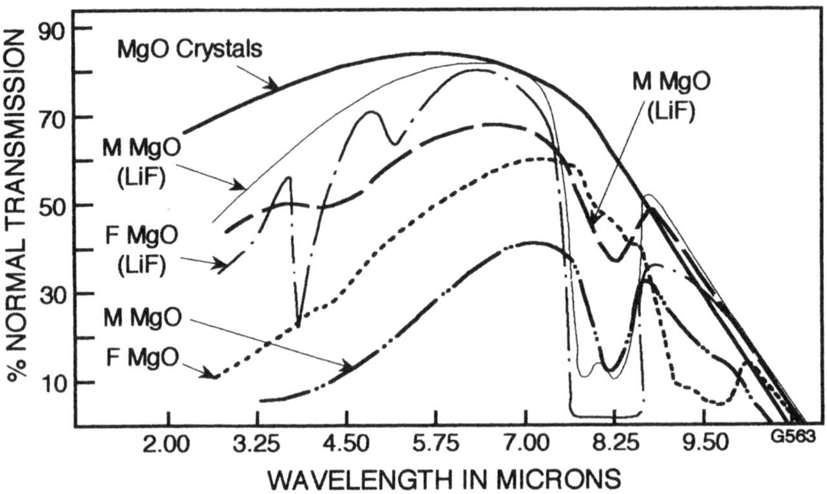

FIGURE 8.1 IR absorption band in transparent hot-pressed MgO showing retention of measurable anion impurities, i.e., OH. (From Refs. 5 and 6. Reproduced with permission of the American Institute of Chemical Engineers. Copyright© 1990 AIChE. All rights reserved.)

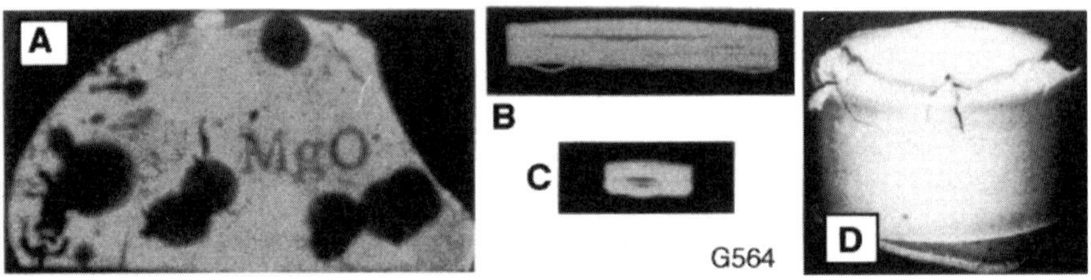

FIGURE 8.2 Examples of clouding, blistering, and bloating due to entrapped anion species on post densification heating of MgO. (From Refs. 5 and 6. Reproduced with permission of the American Institute of Chemical Engineers. Copyright© 1990 AIChE. All rights reserved.)

ing temperatures. Other problems can occur, e.g., reduced densification in Al_2O_3 powder derived by CVD from oxidation of $AlCl_3$ due to residual content of Cl (Sec. 2.4) and similar retention in similar production of TiO_2 [11], apparently due to the residual Cl enhancing grain growth in the powder compacts.

The persistence of such anion impurities and the extremes under which they can be a problem is shown by their causing problems in melt-derived ce-

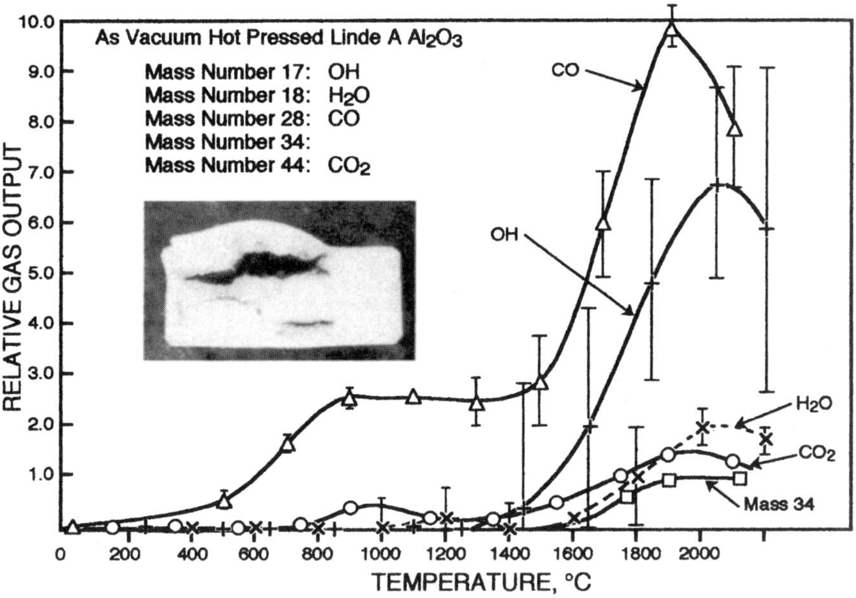

FIGURE 8.3 Plots of outgassing of dense, high purity, hot-pressed, fine-grain alumina versus temperature on heating in a Knudsen cell in a mass spectrometer. (From Refs. 5 and 6. Published with permission of AIChE.)

ramics. Again, MgO is an important, though possibly somewhat more extreme example. As noted in Section 6.7, MgO is fused by arc skull melting for various refractory and other uses, the latter including as the electrical insulator between a central heating wire and an (often flattened) external metal tube in heating elements used for kitchen stoves and many industrial uses. The fused MgO grain meeting this application for many years was produced from MgO that was a byproduct of phosphate fertilizer processing. When changes of the latter manufacturing eliminated this MgO source, another one was needed, with seawater-derived MgO being a source. However, fusion of such MgO presented problems, which with substantial research were traced to retention of hydroxyl impurities that remained despite experiencing temperatures of > 2800°C (the melting point of MgO). Much of this hydroxyl was associated with lattice defects and impurities [12]. Upon subsequent heating such hydroxyl impurities decomposed at moderate temperatures, resulting in small pores filled with resultant hydrogen at high pressures, resulting in degradation of the high thermal conductivity of the MgO. It took several years to solve the problem to again produce suitable fused MgO grain for heating elements. Another example is efforts to use the MgO-BeO eutectic as a high temperature heat storage media by using its good heat of fusion for a stable heat source for temperature cycling about the melt temperature. However, all efforts to eliminate outgassing from the compacted powder before sealing the material in tungsten containers by electron beam welding, resulted in failure due to resultant outgassing distorting and rupturing the relatively thick W container walls.

While MgO may again be a more extreme case, effects of anion impurities present problems in other melting operations, especially single-crystal growth. Thus, commercial growth of sapphire and other oxide crystals presents challenges in selecting raw materials since powders commonly retain enough adsorbed species to present bubble entrapment problems in the resultant crystals [13] (Sec. 6.7.2). One solution to this is to melt material more than once, by using fused grain and recycled crystal scrap left from machining components out of single crystals, which greatly reduces the problem, but adds to costs. Another approach is to use coarser powders and bake them, probably under vacuum at elevated temperatures, but this again adds to costs.

Much less is known of anion species and their effects in nonoxide ceramics other than effects of oxygen or oxygen-containing species, in part since most nonoxide powders are not calcined from precursors, e.g., often being made by carbothermal reduction and reaction (Sect. 2.5). However, increasing use of decomposable precursors, e.g., for silicon nitride, may reveal some analogous problems to those with calcined oxide powders. However, significant hydrogen evolution has been reported in silicon nitride bodies, possibly due to reduction of adsorbed water [14]. Clearly, reaction processing, which commonly involves some nonoxides, can be complicated by interaction of anion species from different constituents.

Other, nonpowder based fabrication may also present problems; for example, there is substantial possibility of residual species that may later volatilize on subsequent heating, but there is little or no data on this. Though limited, there is also some data for the important area of CVD. For ceramics this is mainly the demonstration of residual anions, usually Cl, in powders from CVD via oxidation of metal halides. There is some more specific data for W from deposition of the reduced halides, where Cl-derived W had less grain size stability on subsequent high temperature exposure with resultant increased ductile-brittle transition, while F-derived W had more stable grain size and properties [15].

8.2.2 Effects of Alternate Heating Methods

Alternate methods of heating ceramics such as various types of plasma and especially microwave heating have received substantial attention in recent years [16–23]. There are a variety of uses of such heating, especially microwave heating, ranging from drying (long used in industry), calcining, joining, ignition of reactions (briefly noted below), and especially for heating for pressureless sintering. There is also some application to hot pressing as discussed in Section 6.2.2 and below.

The following is a brief summary of this field, focusing mostly on microwave heating, which is the dominant area of investigation, starting with the basic driving forces for such work. One motivation has been direct reduction of the energy needed based on the mechanism of heating often being operative only in the specimens to be sintered so excess energy is not expended in heating large furnace masses as in conventional sintering using resistance or gas fired furnaces. While this is true, there are also costs of conversion of normal alternating current to microwave (or plasma) energy that counters much or all of the benefits of using less microwave energy. Compared to gas firing, the disadvantage of microwave heating is increased by the extra costs of converting gas heat to electricity [19,20]. However, there are also possible energy and other (e.g., higher throughput) cost savings from the very rapid heating achievable with microwave or plasma heating and potentially much shorter firing cycles. Another important motivation is the improved microstructures often obtained, e.g., finer, more uniform, less porous bodies. However, there are a variety of issues impacting future possible production uses that remain to be fully resolved. These issues along with a further outline of aspects of such heating processes follows.

One basic set of issues is the uniformity of heating both in the furnace itself and within individual samples, especially as a function of individual component size and shape, as well as with a mixture of different components. Recall that home microwave ovens (used for much of the initial work on microwave sintering) commonly heat nonuniformly, which is the reason for the "lazy Susan" (rotating) base in them. Much better uniformity has apparently been achieved in the larger industrial microwave furnaces that are becoming available [22]. However, computer simulations indicate considerable effects of component shape on thermal uniformity, which also varies with both rate of heating and microwave frequency [24]. Another important issue is the compatibility of microwave heating and binder removal since the latter can be much more sensitive to the nature of heating other than normal electrical or gas heated furnaces. Thus, much microwave sintering appears to have been with parts die pressed without binders, or with small amounts of binder, generally removed before microwave sintering, and has been noted as an issue [18], but has apparently become a more active area of evaluation.

The issue of both binder burnout and specimen outgassing are basic ones for all heating methods. Thus, conventional electric- or gas-fired furnaces heat specimens from the outside inward, the former via radiation and convection to the surface and the latter via thermal diffusion inward from the surface. This presents some problem of normal outgassing if the surface sinters too much before the gasses from the interior of the specimen are adequately diffused out of the specimen. Microwave heating being effectively from the inside out should thus aid such outgassing (though the speed of heating may be a problem as noted below). However, the opposite is true of binder burnout, since conventional heating removes it from the outside inward, which is necessary to progressively open pore channels to allow binder products to escape the body. The internal nature of microwave heating presents the potential problem of binder removal from the more central region of the sample before pore channels are open to accommodate this removal. Because of this, and other issues, systems that are hybrids of conventional and microwave heating are being investigated.

Another set of possible problems arise from the potential advantage of very fast densification rates, e.g., densification of thin rods or thin-wall tubes in seconds at temperature. For example beta-alumina (apparently thin-wall) tubes isopressed to 9.5 mm in dia. and bisque fired at 700°C to remove binder, then held at 500°C to avoid gas adsorption prior to sintering, were sintered in an rf induction coupled plasma at feed rates of 25 mm/min giving a transition from green precursor to dense product phase in < 15 sec [23]. However, one issue is the thermal stresses of such rapid heating and their variation with size and shape factors, and possible resultant cracking in both unsintered material approaching the heating zone and the sintered material receding from the heating zone. Related issues are outgassing, which can be retarded by rapid sintering [25], and binder burnout; for example, binders may give more tolerance of thermal stresses in unsintered material from fast heating, but exacerbate binder removal issues.

Another basic issue is the broad variation in material susceptibility to microwave (or plasma) heating. This means that "firing" schedules may have to be tailored to the material, as well as possibly the size and shape of the components being fired, and probably precludes cofiring of bodies which can be done to some extent in conventional firing. Another issue is that microwave sintering is more tenuous or precluded for materials of increasing electrical conductivity. While some materials that cannot be directly heated by microwaves can be heated in dielectric, e.g., alumina, crucibles [26], this may often be more cumbersome than practical. Another issue as well as possible opportunity, is microwave sintering of ceramic composites, where there is a significant difference in the microwave coupling to the composite constituents, which presents both problems and possibilities. Microwave heating has also been shown to be useful in igniting and thus effecting propagation of self-propagating high temperature synthesis (SHS) reactions (Sec. 6.5) [27].

Overall there are a variety of issues for these novel heating methods that are not fully addressed. These range from basic differences with conventional sintering, i.e., internal heating that diffuses outward versus conventional heating progressing from the component surface inward, and their ramifications for factors such as binder removal. This leads to consideration of hybrid new-conventional heating and how to best operate these. Other issues include practical limits on the speed of heating and working with materials that microwaves do not couple well with. Lacking a full evaluation of these and related issues, a clear future for such newer heating is not apparent, but some specialized applications seem most likely—e.g., fast firing of small rods or tubes and possibly of fibers or filaments. Some specialized use for joining and with reaction processing may be feasible, e.g., ignition of reactions, and selective heating of braze or welding materials, including use of SHS reactions for joining.

It should be noted that some of the effects of hot pressing with direct resistive heating of the die (and the powder compact if it is conductive), plasma heating, or combinations, with some aspects of this being in a pulse mode, shows some similarity to effects of microwave or plasma heating. Some possible aspects of plasma type heating have in fact been cited as a possible factor in apparent enhanced densification (Sec. 6.2.2). More recently Oh and coworkers [28] have reported improved properties from first-pulse resistive heating of nanocomposite Al_2O_3-SiC and the graphite dies at lower temperature in vacuum versus conventional hot pressing. Similarly, Groza and coworkers [29] reported faster, lower temperature densification of nano-TiN powder using plasma activated sintering under vacuum with a graphite die (and 66 MPa pressure) and an initial electric pulse to aid outgassing. These and other reports indicate opportunities, but again mechanisms and practicality (e.g., repeatability, scalability) need more definition.

8.2.3 Fabrication of Ceramic Composites

Fabrication of ceramic composites with a ceramic matrix and a dispersed metal or ceramic phase or with ceramic fibers, though addressed in various earlier sections on specific applicable fabrication, deserves additional attention since it reflects significant shifts in emphasis of fabrication methods from that for monolithic ceramics. The shifts in commonly used or preferred fabrication methods for ceramic composites vary with the type of composite, generally increasing with the type of composite, as summarized in Table 8.1 and below. Such evaluation is supported by other reviews of ceramic composites [7,30–32].

TABLE 8.1 Use of Major Fabrication Methods for Ceramic Composites

	Fabrication method[a]			
Composite	Sintering	Hot pressing	HIPing	CVD/CVI
Particulate	M–H	M–H	L–M	—
Platelet/whisker	L	H	L	—
Fiber	—	H	—	M

[a]H = high, M = medium, L = low, — = none or very low. Note also some limited fabrication of particulate or platelet (e.g., crystallized glasses) and fiber (e.g., cutectic) composites, as well as some fabrication of glass matrix fiber composites using hot glass flow injection (Fig. 7.10)

Particulate composites are often fabricated by pressureless sintering of green bodies formed by processes used for monolithic ceramics. However, two sets of issues, one of green body formation and the other of sintering, must be considered. Both the size, shape, and density differentials of the particles of the dispersed phase versus those of the matrix phase have impacts on the fabrication methods used. Forming the green body by slip, tape, or pressure casting will give some differences in spatial orientation and distribution of the dispersed phase as a function of these differences in particulate character on settling and orientation. Differential character of matrix versus dispersed particles may also impact powder pressing results, especially in die pressing. Thus, differential local particle packing densities may occur around larger particles, and preferred orientation of dispersed particles, especially larger platy particles, is likely to occur on a global scale with variations in the body as a function of compaction gradients. There are also basic issues in densification via pressureless sintering, since isolated particles of a different phase inhibit sintering of the surrounding matrix particles, with such effects often being exacerbated by local green density variations that may occur as noted above.

However, besides effects of both the differing characters of matrix and dispersed particles, there are effects of the volume fraction dispersed phase, and on the mechanism(s) of sintering since liquid-phase sintering, which is common in many ceramic particulate composites, can reduce differential densification problems. Further, coating of matrix material on individual dispersed particles has been shown to frequently improve densification of ceramic particulate composites as discussed in Section 2.6 (and should also improve the homogeneity of distribution of the dispersed phase). However, there are situations in which such particle coating is not practical for technical or cost reasons. Thus, fabrication of dual composites, i.e., of composite particles of one composition dispersed in a matrix that is a composite of differing composition (usually of volume fraction of dispersed phase, but could also include a different dispersed phase) are less amenable to sintering [7]. Additionally, it is important to note that reaction processing is a substantial and promising method of fabricating many ceramic particulate composites, and that pressure sintering, especially by hot pressing, is widely used for such fabrication (Sec. 6.5). Further, melt-derived eutectic particles are typically more difficult to densify by pressureless sintering, as to a lesser extent are composites of nonoxide constituents (Sec. 6.7.3). Thus, while pressureless sintering is used for production of some ceramic particulate composites, hot pressing is also substantially used, as is some HIPing. Again recall that there are also some other methods of fabricating particulate composite such as CVD and especially melt processing (e.g., via glass crystallization).

Turning to whisker and platelet composites, the shift to pressure sintering, especially via hot pressing, is substantially greater than the shift in processing of particulate composites. Problems in consolidation of green bodies of whisker and platelet composites are a factor in this, but much of this is due to the much greater inhabitation of densification in pressureless sintering. Thus, such composites are predominately hot pressed in both laboratory preparation and in industrial production (primarily or exclusively alumina-SiC whisker composites). Note that HIPing of such composites, which produces a more isotropic composite, can be more difficult. The anisotropy produced by uniaxial densification of hot pressing aids densification, and is a factor in their use, probably a benefit for many applications, (e.g., wear), but this requires more study. Also again note that some platelet composites are produced by melt processing, e.g., some crystallized glasses.

Finally, consider fiber composites, primarily those with continuous fibers (or filaments), where there are not only significant changes in densification, but even greater ones in fabrication of the green body. (There has been limited effort on chopped or other short fiber composites, whose fabrication generally falls between that for whisker and continuous fiber composites.) These shifts are summarized below, with readers referred to appropriate reviews for greater details [30,31]. For continuous fiber (or filament) composites the challenge is to get a reasonably homogeneous infiltration of matrix material in between fairly closely spaced fibers, e.g., typically 30–60 v/o fibers. This is universally done via fluid infiltration, with the fluid method depending on whether a green composite body is to be formed then densified, or whether infiltration and densification are concurrent.

Consider first the gross shifts in green body fabrication of fiber composites versus particulate composites where liquid infiltration is used for fiber composites as opposed to powder mixing, e.g., wet or dry milling is commonly used for particulate composites, but is generally unsuited for fiber composites. Thus, a liquid source of the matrix is coated on fibers or filaments, commonly by infiltration into tows, cloth, or preforms. Also, depending on the specific nature of the liquid source of the matrix, many of the methods of forming plastic composites are available for forming green ceramic fiber composites. Thus, slurries of matrix powder are a major source of matrix infiltration in fiber composites, and sols or preceramic polymer infiltration are used, the latter being the method of making carbon-carbon composites. Slurry infiltration is normally done filament by filament, tow by tow, or cloth layer by cloth layer since greater masses of fibers is likely to result in the outer layers of fibers acting as a filter to reduce and ultimately prevent slurry constituents from penetrating to the center of fiber preforms. Sol and preceramic polymers or their liquid precursors can be infiltrated into fiber preforms, provided they can be suitably rigidized in the fiber preform, which is typically done for carbon composites with carbon producing polymers. It is also possible, in principle, to coat fibers, e.g., in their production with matrix sources, with polymeric precursors probably being most practical, e.g., from an adherence standpoint. However, this would require large volumes of coated fiber and a significant change in relations between composite producers and their material suppliers.

Once green fiber composites are made, they are densified primarily by hot pressing. Many composites are mainly planar fiber architecture with easy consolidation in the direction normal to the plane of the composite fiber architecture, and thus are a natural choice for hot pressing normal to the plane of the composite fiber architecture. Such plainer composites limit added costs of hot pressing, which are often not a large factor in view of other sources of composite costs, such as those of the fibers themselves and costs of forming the composite green body. Further, there are limited alternatives to densifying green composite bodies with complex trade-offs of body quality and costs that make hot pressing a natural process for many composites. Neither sintering nor HIPing are generally feasible since these both require shrinkage parallel as well as normal to the fibers, which is generally impractical and generally avoided in hot pressing. Hot pressing typically yields a near or fully dense composite body and can handle substantial size composites. For glass matrix fiber composites some added versatility is feasible via injection of fluid glass for the matrix (Fig. 7.10).

An alternate densification route is to do multiple impregnations (often under pressure) and pyrolysis of sol or especially polymeric matrix sources, but this never reaches full density, and also becomes an expensive process. However, the normal process for making carbon fiber composites with three- or higher dimensional fiber reinforcement is such multiple impregnation and pyrolysis steps since this is generally the only way to make such very expensive composites. One other process that may be promising with lower temperature processing of matrices, e.g., some phosphate ones made by reaction, is autoclave processing, but this is not low in cost and may require multiple impregnations and autoclave reaction processing.

The remaining matrix infiltration process that entails simultaneous densification is via chemical vapor deposition (CVD), typically termed chemical vapor infiltration (CVI) for this purpose. This can, in principle, handle a wide range of composites in terms of size, shape, and fiber architecture, as well as a reasonable to good range of matrix materials. However, it tends to be a slow process, can have some limitations on the thicknesses handled and depths infiltrated. It also inherently cannot give fully dense matrices, but the pores it tends to leave appear to be intrinsically more benign in limiting physical properties, in contrast to other methods of forming matrices that leave some porosity, i.e., methods involving multiple impregnations and thermal decomposition processing [33,34].

Finally, note that there are some possibilities of forming some ceramic fiber composites from the melt. Directional solidification of eutectics is one possible method, and may be more feasible where successfully fabricated by edge-defined film-fed growth (EFG) methods (Sec. 7.3). Some experiments to introduce matrix material into fiber composites by melt spraying were unsuccessful, but there may be possibilities for this with further development.

8.3 MANUFACTURING FACTORS

8.3.1 Machining and Surface Finishing

Many ceramic components require machining to achieve the shape needed, e.g., many single-crystal components. Further, while some ceramic components are used as-fired, many require some surface finishing for meeting dimensional or surface finish or both requirements, some have coating requirements (Sec. 7.5), and some both coating and finishing requirements. Though there are some surface finish operations such as grit blasting or tumbling, and thermal or chemical polishing, most finishing entails machining. Machining for both shaping and finishing is important since it is often a major cost factor (Table 1.2) and because it is often a key factor in determining the mechanical reliability of components.

Though, much remains to be determined, the overall trends and important specifics of machining are reasonably identified [34–46], as outlined in this section. Almost all machining is done with hard abrasives, commonly diamond, used in one of two forms. One is as fixed or bonded abrasives where the abrasive particles of specific size and concentration are bonded in various matrices, e.g., polymeric, glass, or metal for grinding, and in polymeric and, especially, metal matrices for sawing. The other abrasive form used is as a powder (or in a slurry or paste) for lapping or polishing, i.e., for finer surface finishing than most or all grinding. Note that wire sawing may use wires with imbedded abrasive particles, with abrasive coated on the surface, or with free abrasive particles added as powder, slurry, or paste are also used, often to make multiple cuts simultaneously.

A key to machining effects is the direction of the abrasive particle motion, since abrasive particles moving over a ceramic surface under pressure basically generate two sets of resulting flaw populations along the grooves formed by abrasive particle that have considerable penetration of the surface being machined (Fig. 8.4). One set of flaws formed are cracks that extend into the ceramic from the base of the groove and parallel with it. The other set of flaws that form are cracks generally centered on, and approximately normal to, the groove and hence path of the abrasive particle. Both sets of flaws, which can have some interactions, but are generally separate flaws, though of similar depth, are clearly distinguished from each other by their shapes as well as their orientation relative to the machining groove. Flaws parallel with the machining groove are typically fairly planar, but elongated along the groove either as a single, more elongated crack, or as a few less elongated, but adjacent or overlapping, cracks. The flaws forming nominally normal to the machining grooves are less planar, i.e., often having some curvature(s), but are not substantially elongated, i.e., are closer to halfpenny cracks. This difference of crack elongation persists over a broad range of materials, both of composition and structure, i.e., in glasses, single crystals (where the orientation of cleavage planes can also play a role as a function of

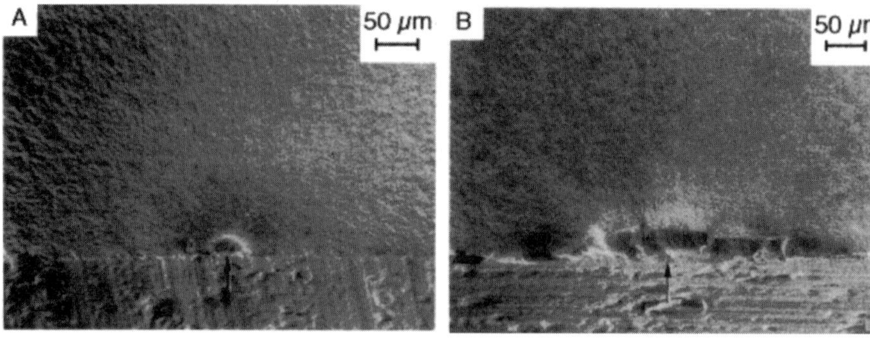

FIGURE 8.4 Examples of machining flaws from grinding at fracture origins as a result of (A) grinding parallel to the tensile axis of flexure strength testing, hence failing from a machining flaw normal with the abrasive groove and (B) grinding normal with the tensile axis and thus failing from a machining flaw formed parallel to the abrasive groove. Though in a softer ceramic, MgF_2, they are representative of other ceramics.

crystal orientation) and many porous and dense polycrystalline materials. However, the difference in flaw shape is impacted by grain size, generally with the shape/elongation difference disappearing when the flaws and grains are about the same size [7,44,45] and may be reduced by porosity in some cases [34,45].

Thus, the key difference is the general elongation of the flaws parallel with the machining grooves, which makes them more serious flaws, i.e., reducing strength by up to 50%. Besides the above noted effect of grain size on the elongation of flaws parallel with machining grooves, there are two other effects determining the impact of this flaw elongation on tensile strengths. The first is other sources of failure, which if present to control failure of most or all components will obscure or preclude a strength difference due to the two machining flaw populations. The second is the type and orientation of the tensile stress relative to the machining direction. In lapping and most polishing there is no persistent direction of the abrasive particles; the fine grooves formed and the associated parallel and perpendicular flaws are essentially randomly oriented in the component surface. Thus, when the machining flaws are the source of failure, the most severe flaws control failure, which in this case are the flaws parallel to the machining grooves. This is also true where the tensile stresses causing failure are either bi- or triaxial since the elongated flaws parallel with abrasive grooves are high stressed by such loading. However, where the tensile stresses causing failure are uniaxial, there is anisotropy of the tensile strength depending on the orientation of the stress versus any persistent machining direction, as is generally the case for most grinding (which is one of the most extensively used machining operations), and appears to also be so for much sawing. In such cases having the stress axis parallel with the machining direction results in higher strength and stressing normal to the machining direction and results in lower strengths, often by up to 50%.

While the difference in shape of the flaws parallel versus perpendicular is a major factor in the strength of machined ceramics, there are other factors that need to be considered. These include effects of machining parameters, materials and microstructures being machined, and the generation of residual stresses by machining. Consider first machining parameters, where a dominant factor is the grit size used. Except for some variations at very fine grit sizes and fine grain sizes, strengths consistently decrease with increasing grit size, which means increasing flaw sizes, depths for flaws both parallel and perpendicular to the machining direction. Similarly there is some evidence that increasing depth of cut increases flaw depths, decreasing strengths. An important ramification of both of these is that when a finer machining operation follows a coarser one, as is frequently the case, the final machining will not reflect the normal flaw population for that machining unless the net thickness of material removed by the last machining is greater than the depth of flaws from the previous machining. Otherwise, the strength after the final machining step will reflect larger flaw remnants from prior machining steps.

Machining effects vary some with material and microstructure since flaw sizes (c, measured as its depth) are

$$c \propto (E/H)^{1/3}(F/K)^{2/3} \qquad (8.1)$$

where E= Young's modulus, H=hardness, and K= fracture toughness, all being local values controlling flaw introduction, while F= the force on abrasive particles (increasing with depth of cut and increasing grit size) [35,38]. Thus there are broad, but generally limited, effects of material properties and microstructures on c via their effects on the parameters, primarily E, H, and K. This arises since the above dependences are not strong, and those of strength, which varies as $c^{-1/2}$, are further reduced. Further, while E and H vary widely for different materials, there are similar trends of each for the same material so there are not large changes of the E/H ratio between different materials. Similarly effects of porosity occur via effects on E, H, and K, but are again limited both by the low power of the dependence of c on them, and that the porosity dependences of both E and H are generally similar and thus approximately cancelling. On the other hand, H generally decreases as grain size increases, but E is generally independent of grain size, giving a general trend for c to decrease some with increasing hardness. However, the dominant effect of grain size is via its effect in limiting elongation of flaws formed parallel with the abrasive particle motion as the flaw dimensions approach those of the grain. Clearly, there are also effects of K, but local values controlling c appear to be much less that large crack values, especially as a function of microstructure.

To put the above in perspective, most machining flaw sizes for representative strengths of typical production ceramics are in the range of 20–60 μm, often the same range as for other flaws such as isolated pores. Finer machining flaws 5–20 μm in size have been identified for higher strength, e.g., > 600 MPa fine-grain bodies such as Si_3N_4. A key result of flaw sizes not changing much with grain size is that machining flaws change from being larger than the grains at fine grain sizes, and smaller than larger grains in large-grain bodies (and also often in isolated large grains) [7,44,45].

The situation for ceramic composites is both more complex and less studied, being better defined for ceramic particulate composites. These generally show *K* going through a maximum as a function of the volume fraction of dispersed particulates with *c* thus showing a modest minimum and strength a modest maximum [7,38]. These trends are shifted some, but not grossly changed by dispersed particle size and matrix grain size, i.e., finer particle size directly increases *H* some and indirectly by its trend to reduce matrix grain size some. Other composites are more complex due to significant anisotropies of properties and of the related orientations of the fiber-matrix interface that have received little or no attention.

Residual stresses are typically another factor in strengths of machined (and some other) ceramics. Though not extensively studied, there are sufficient studies to outline the trends of surface compressive machining stresses. These, are generally shallow, e.g., probably < 10 μm with significant gradients decreasing their levels from the surface inward. Data shows considerable variability in measured values, which are commonly of the order of a few hundred MPa in mainly structural ceramics studied. Thus surface compressive stresses are commonly a factor in, but do not appear to dominate, mechanical behavior. However, such stresses appear to be a factor in limiting strengths benefits of polishing over fine grinding since there appears to be less compressive stress from polishing versus grinding. Also, recall that polishing is typically done with random motion of the free abrasive particles so failure is dominated by the more elongated polishing flaws formed parallel with the local abrasive particle motion, which limits polished strengths relative to those for stressing parallel to the grinding direction of ground samples. Thus both the machining direction and residual stress effects combine to limit the benefits of polishing versus fine grinding on strengths (again polishing effects will also be limited if sufficient polishing is not done to remove more serious finishing flaws from prior finishing).

Substantial progress has been made in understanding machining interactions with, and effects on, various types of ceramic materials. This has been accompanied by substantial advances in obtaining better machined surfaces and some reductions in machining costs. Basic engineering improvements such as better abrasives and especially better (and often thinner) saw blades or wires and use of ganged blades to make multiple and thinner cuts simultaneously (thus increasing product yield from a given piece) have all helped to reduce costs. Thus, while machining costs generally remain an important factor, progress has been, and continues to be made, and clever engineering also can help, e.g., as shown for some sapphire windows [46]. Optically polishing of sapphire windows ground to dimensions is a substantial cost for many of its applications. However, it has been shown that much or all polishing can be eliminated by applying a thin, fired glass coating on ground sapphire surfaces normally needing substantial polishing. By using a glass with a close (within 1%) match to the refractive index of sapphire and a reasonable match, e.g., within 10%, of thermal expansion, both for the window orientation used, can result in suitable transparency for some applications such as armor windows with no polishing, that is, as fired. Further, if greater transparency quality is needed, the glass-coated surfaces can be polished to improve transparency quality at lower costs than the bare sapphire. While use of such glass coatings is not suitable for surfaces needed for some uses such as wear or erosion resistance, it is applicable for some uses such as armor windows.

8.3.2 Component Inspection and Nondestructive Evaluation (NDE)

An important step in the manufacturing of ceramic components is their inspection and evaluation to assess their suitability for meeting the needs for which they were manufactured. This typically entails two sets of evaluation, one to verify dimensional and surface finish requirements have been met, and one to ascertain that properties needed for the function have been achieved. Dimensional and finish verifications are generally reasonably accomplished via available technology such as optical comparitors that are simple, fast, and are thus generally cost-effective and have no effects on the components. Similarly, many properties can be checked to the extent needed with no effect on the component, e.g., its density, refractive index or dielectric constant, and optical transmission, generally with good results at moderate costs.

The greater challenge is to inspect components for their potential reliability in applications where failure from mechanical, thermal, or electrical stressing of the component must be avoided within designed values of parameters of mechanical or thermal stress or of dielectric breakdown. Proof testing—stressing components as used in service to stress levels that result in limited proof testing failures while assuring no failure in the subsequently planned use—is very effective where it is applicable. Unfortunately, proof testing is very limited in its use, primarily or exclusively to components mainly stressed in rotation, as for grinding wheels and turbine blades or rotors. Alternatively, testing to failure sets of samples can be used (and probably was in the design/selection of the component) to assess the probability of produced components can be done, but is costly, time consuming, and potentially of limited accuracy.

An alternate approach that has been the focus of much effort is nondestructive destructive evaluation (NDE), an inspection method to ascertain the capability of finished components to meet such requirements in a nondestructive fashion. There has been substantial research directed toward achieving such a general technical goal, much of it directed toward the broad and absolute goal of determining the potential failure-causing flaws and the operational property levels at which they would cause failure so components that would fail at less than acceptable stress levels are identified and rejected. There is a substantial and growing diversity of technologies for doing this [47]. However, there are still serious limitations to such approaches for highly stressed components, where resolution of fine flaws is uncertain. The first set of limitations is in the detection of flaws present that may limit components from meeting their design function. There is first the limiting sizes of flaws identifiable and the probabilities of their detection and accuracies as a function of flaw type, location in the component, and material. Sizes detectable vary with both the noted parameters and the detection method used (e.g., ultrasonic versus radiographic versus thermal methods). While the lower limits of flaw size detectability overlap with the upper end of the sizes of flaws of concern, there is a considerable range of flaw sizes that may cause failure that are generally not detectable. This inability of detecting smaller flaws again varies with several key parameters noted above, as well as the presence of different types of flaws. All of these limit both the size of flaw detection and their probability of detection, which is a critical issue that has received little or no attention.

The second major problem, that has also received very little attention, is that identifying individual flaws of small enough size is not enough for accurate selection of acceptable versus unacceptable components. This arises for three related reasons: First, detection of a flaw is not sufficient; its size, shape, orientation, and character must be determined to estimate its severity and hence probable failure stress. Second, many failures occur from two or more nearby features that may be similar, such as two nearby machining flaws or a cluster of a few pores near one another or different, such as a pore associated with a large grain, impurity particle, or machining flaw. Detection of such combinations and determining the stress at which they would cause failure are both still major limitations. Thus, failure from many individual machining flaws identified on resultant fracture surfaces agree with the known strength and fracture toughness along with the observed flaw character. However, for more irregular machining flaws, results are more variable, introducing important uncertainties, despite such flaws identified on fracture surfaces being better characterized than by NDE methods. Another indication of the challenge in identifying enough details of collective flaws is to consider differences in flaws, e.g., pores, causing dielectric breakdown versus mechanical failure. Dielectric failure is most likely from a chain of pores parallel to the electric field in the component, while mechanical failure is most likely from a closely spaced cluster of pores in a plane normal to the stress direction [2]. The accuracy of distinguishing between these two cases by NDE is very uncertain. The third complication, especially for mechanical failure, is that there may be local stresses as well as local concentration of applied or residual stresses, both of which can be difficult to detect and quantify, making quantification of expected strengths even more uncertain.

Continued research and development will create further advances in NDE methods. Whether these will be sufficient for much wider use of NDE in selecting good versus bad components remains uncertain, as does the cost-effectiveness of much NDE. Thus, note that the primary NDE methods in much of the ceramics industry are visual examination, backed up by die penetrant tests as needed. However, current NDE techniques are useful for rejecting some components, e.g., refractories, where larger, more detectable flaws are dominant sources of failure. Further, there is growing recognition that NDE is valuable for guiding improvements in fabrication methods to make better components and may be effective as a production control at various stages of fabrication, especially of green bodies, rather than in addition to a final inspection (Sec. 8.4) [48]. There are also possibilities of developing NDE techniques to reject finished components that are not based on identifying specific potential failure-causing flaws, but instead depend on correlation of behavior with use parameters (e.g., the ringing of a tea cup when struck indicating a sound cup), which deserve more attention.

8.3.3 Attachment and Joining

Attachment or joining of ceramic components to support structures or other ceramic or metal components is often needed for the ceramic component to function (some of which may also be aided by NDE). Thus, for example, insulating ceramics, e.g., bricks and fiber mats, must often be joined to each other and to furnace structures, while many structural components, e.g., turbine blades and vanes as well as exhaust port liners or valves for internal combustion engines, also must be attached to or incorporated into corresponding parts of the engines. Many wear components must also be mounted to or in devices in which they are used, as are extensive arrays of ceramic components in papermaking machines and ceramic plungers and liners for slurry pumps. Many ceramic optical and other electromagnetic windows and all IR-domes and radomes must be mounted to provide seals and mechanical and aerodynamic integrity, respectively. Finally, one of the most extensive areas of joining metals and ceramics is for electrical and electronic applications, ranging from varistors to capacitors to ceramic substrates and multilayer packages.

To meet the above extensive and diverse needs for joining and attachment of ceramics to like or other ceramics or metals, an extensive and diverse array of technology has been developed, with much of it in wide industrial use. Such technology has been revicwed elsewhere [49–53] and is summarized here. These technologies can roughly be placed in various categories ranging from mechanical, (organic) adhesive, metal (soldering and brazing), and glass to metal sealing and ceramic brazing, cementatious bonding, bonding via preceramic polymers, diffusion welding, and fusion welding. These are listed roughly in order of increasing potential temperature capability. While all can be used to join ceramics to themselves or other materials, the first three are more often used for joining ceramics to other materials, while the latter four are often used to join ceramics to themselves. These techniques are briefly outlined below and in Table 8.2.

Mechanical attachment covers a broad range of often simple and low-cost techniques ranging from joining or attachment via typical mechanical fasteners such as metal screws, nuts and bolts, clamps, and hangers. A key factor is minimizing stress concentrations from holes and localized contacts as well as generation of attachment stresses due to differences of elastic or thermal strains between the ceramic components, their attachment, and the structure to which they are attached. A key way of limiting these problems is via use of compliant materials, e.g., rubber, plastic, or soft metal, between the ceramic and critical contacts with the attachment or support structure. Thus, for example, windshields in aircraft and the space shuttle are held and sealed by rubber seals, placed where the clamps and rubber seals are removed from much of the environment the windshield may experience. The latter placement is also a common method of raising the typically limited temperature or other environmental range over which such attachment can be used, e.g., placement on the back side of refractory bricks or insulation bats. Another important mechanical attachment method is shrink fitting of a mating metal part around a ceramic part. This entails carefully sizing both parts so on heating the metal allows it to be slipped over the ceramic part on which it shrinks to a tight fit on cooling. There are clear limits on temperatures, sizes, and shapes, e.g., for cylindrical parts for placing ceramic dies or cylinder liners in metal bodies. An extreme of this is casting metal around a ceramic component, e.g., engine port liners, placing demands on the ceramic thermal shock resistance.

TABLE 8.2 Summary of Ceramic Joining Methods[a]

Joining method	Shape flexibility	Temperature/ env. cap.	Vac./ hermiticity	Strength	Cost
Adhesive	M	L	L	L	L
Cementitious	M–H	M	L	L	L
Mechanical	L–M	L–M	L–H	L–M	L–M
Brazing	M–H	M	H	M–H	M
Diffusion welding	L	H	H	H	H
Fusion welding	M	H	H	M–H	M–H

[a]L= low, M= medium, and H= high.

Using organic adhesives is another basically simple and usually low-cost method of joining or attaching ceramics, where environmental and other factors allow. This may be used to bond ceramic components to themselves or other materials directly via their adhesive capabilities. Organic adhesives may also be combined with mechanical attachment; for example, some radomes have been attached to missile structures by bonding a layer of plastic composite to the inside of the dome base such that this layer could be threaded to allow the dome to be screwed onto the missile structure. Limitations of adhesive methods are environmental and stresses. The latter can be substantial if there is considerable temperature excursion of the ceramic-adhesive joint from the temperature of bonding due to typical high-thermal expansions of organic adhesives versus ceramics, e.g., by two- to fivefold. Thus, for example bonding of PZT sonar transducer rings to one another to form elements of a sonar array using an elevated temperature curing epoxy adhesive introduced stresses of about one-half the PZT strength, which together with operational stresses resulted in some ceramic ring failure [2]. With regard to environmental limitations, these, especially temperature limitations, can be relaxed some by placing the adhesive bond in an area of limited temperature or other exposure.

Consider next brazing of ceramics to ceramics or metals with metal or ceramic materials, which also includes glass-to-metal sealing (as well as soldering at lower temperature with lower melting materials). These represent a broad array of materials, applications and techniques based on bonding two similar or dissimilar materials with a material that normally forms a liquid that wets, then bonds the parts being joined on solidification, i.e., generally as in soldering and brazing of metals to themselves. Lower temperature bonding via solders is particularly needed in electronic applications where semiconducting chips are present requiring temperatures that will not damage the semiconductors. Various solders or brazes are also used to give a hierarchy of bonding temperatures such that earlier formed braze or solder joints will not be affected (undone) by subsequent joining operations. Some solders or brazing materials are ceramic especially silicate-based glasses. The latter, which provide a broad range of properties and adjustment of these, are the basis of extensive industrial production of glass to metal seals used to provide electrical insulation of electrical feed throughs that are hermetically sealed to metal housings that provide environmental protection for electrical systems. There is also substantial use of the same or similar solders and brazes used for soldering or brazing metals to ceramics. Brazes are widely used for a variety of materials and applications that include extensive use for electrical and electronic components. Some can also be used for optical materials—e.g., promising strengths (e.g., of 200 MPa) have been achieved in glass brazing of large sapphire window sections (apparently using the same or similar glass used to reduce or eliminate surface polishing [46]; see Sec. 8.3.1). Highest temperature joints by brazing are achieved with noble, reactive (e.g., Ti), or refractory metal (e.g., Mo-based) brazes, which also generally have closer thermal expansions to many ceramics. Note again limitations of materials and component size and shape due to thermal stresses from expansion differences.

An important aspect of some soldering, and especially brazing, operations is reactions that occur between braze constituents and the ceramic being brazed. Understanding and controlling of these is leading to further advances [54]. There has been increasing investigation of other reaction-based joining via reaction processes, particularly by SHS methods (Sec. 6.5) that generate much or all of the heating needed, and often give transient liquid phase(s). While precautions are generally needed to limit effects of thermal stresses, as for fusion welding discussed below, and important issues of outgassing (e.g., of species adsorbed or reactant powder surfaces) during reaction, but there may be promise for some application of such reaction bonding.

There is one large class of longstanding use of special cements (usually involving some reaction) for joining ceramics and two other newer, more specialized, and less developed methods that generally depend on chemical reactions, but may also in part entail some brazing or welding mechanisms. Many refractory bricks and parts are bonded with various cements, e.g., hydroxide or phosphate containing ones, whose decomposition or chemical change on curing aides in developing a bond. Cement bonding is also extensively used to bond metal attachment fixtures to ceramic electrical insulators. One of the newest developments is joining via polymer pyrolysis, where two parts, most likely both ceramic, are joined with a preceramic polymer that is subsequently converted to a ceramic. This technique, in its early development, entails large shrinkages on conversion to ceramic material which must be addressed, e.g., by limiting it to small parts, use of substantial particulate filler in the preceramic polymer, or subsequent densification, e.g., by sintering.

Diffusion bonding or welding has been extensively demonstrated for ceramics. This typically entails joining by placing a ceramic powder joining layer between two ceramic components and heating the components and layer so they sinter together. This is the easiest but generally less useful approach when the two components are in the green state so they and the joining layer simultaneously sinter, and thus limit differential shrinkage between the components and joining layer and resultant problems from such shrinkage differences. It is commonly more desirable to join densified parts, but this presents serious issues of weld shrinkage versus none in the components. This differential shrinkage problem can be eliminated by pressure sintering of the weld by mechanical loading nominally normal to the weld plane, eliminating lateral shrinkage in the weld. However, such hot-press welding or bonding is even more restricted in configurations that can be reasonably joined. Another issue is joining parts of different materials, where both differences in temperature capability of each material and of their properties, e.g., of thermal expansion, may present limitations of both component materials and sizes and shapes to be joined. Some of these issues can be addressed some by using a weld layer that grades its thermal expansion from approximately that of one material to be joined to approximately that of the other material. Thus, while this method can often give weld capabilities approaching those of the ceramic to be joined, there are serious constraints to this useful method, which may be expanded by clever innovations.

Recent developments show substantial advance of diffusion bonding, specifically for sealing the ends of alumina tubes used for sodium and more recently halogen lamps [55]. The greater chemical reactivity of the hot halogen gas precludes use of braze seals like those used with sodium vapor lamps. The solution was to use alumina diffusion bonding to hermetically seal the dense alumina lamp tubes with dense alumina end plugs sintered together by either of two methods. The first is sintering of a green plug in the end of a green tube having a limited but unfilled (e.g., 50 μm) gap between the tube and plug. The closing of this gap between the tube and the plug is accomplished by having the tube body having greater shrinkage, e.g., by 5–10%, due to finer powder, different MgO addition, or lower green density, than the plug. Diffusion welding of already fired fine-grain tubes and plugs can also be accomplished. The key to filling the gap between the plug and the tube is keeping the thickness of the gap to be closed between the two pieces of dense alumina to less than approximately twice the final grain size (e.g., < 50 μm) of the joined pieces, since such grain growth of the parts being joined is the means of closing and sealing the initial gap between them.

The other welding technique is fusion welding, i.e., where two components are joined by temporarily melting the two faces to be joined (or a filler layer between the two faces) such that on solidification they are joined; this is widely used for metals. This is also well established for glasses, since this is essentially the mechanism of joining glass parts, e.g., in conjunction with glass blowing. Such welding of refractory, polycrystalline (and possibly single-crystal) ceramics (that exhibit normal melting behavior, as most, but not all, do) to themselves or refractory metals has been clearly and fairly extensively demonstrated by using electron or laser beam or arc welding, the latter where both materials to be joined are electrical conductors. The key requirement for fusion welding of ceramics, besides congruent melting, is to heat material surrounding the weld to preclude thermal stress cracking during welding or on cooling of the weld, which is also required for glass welding and has been demonstrated by very practical means for the higher heating temperatures needed for refractory ceramics, e.g., 1200°C. Resultant ceramic welds are generally very similar to those of metals, i.e., generally dense (indicating mainly directional solidification as desired) with larger, columnar grains (Fig. 8.5), often with strengths approaching those of some of the parent ceramics. Further, this method can

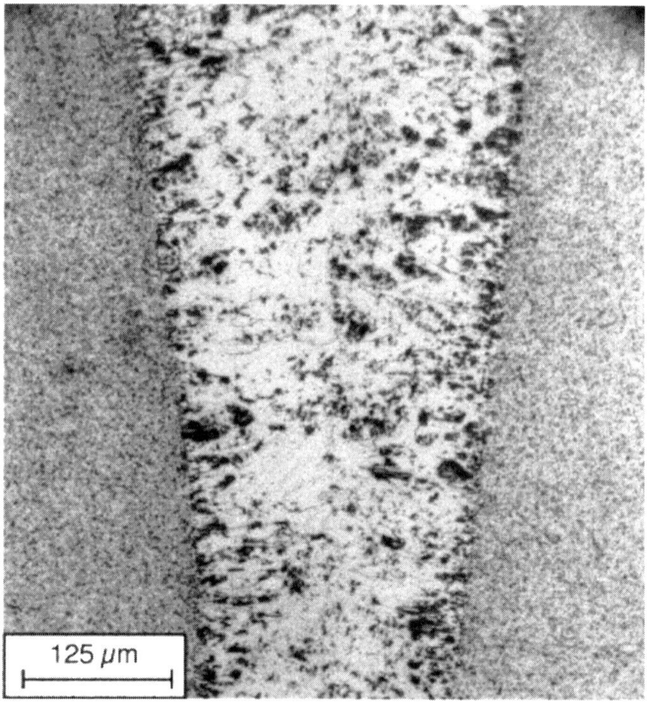

FIGURE 8.5 Micrograph of cross section of an electron beam welding of commercial alumina. (From Ref. 56.)

probably be considerably extended for some composite ceramics or may be of use as weld fillers, such that the weld has a eutectic or composite structure that respectively reduces the effects of grain size on weld strength (Sec. 6.2 and 8.2.3) or limits grain growth in the weld on solidification. There is also some applicability of fusion welding to joining dissimilar materials where the size and shape of the components are within the range allowed by the thermal expansion differences of the materials. However, despite considerable capability having been demonstrated for a number of years, little of no use of this potentially practical method has occurred. This probably reflects such welding using technologies that, while promising, are not very familiar to many in the ceramics field and require some to substantial investment and both considerable development and related uncertainty to be implemented for a specific application. This serves as a reminder of the difficulty of introducing new technology without a clear driving force that requires it or will justify the cost of implementing it.

Finally, an important example of joining is in the fabrication of large telescope mirrors, which not only illustrates the extremes of diffusion bonding, but also illustrates fabrication of some large glass pieces via CVD. Recall that the large, low expansion (borosilicate) glass 5-m-dia. Mt. Palomar telescope mirror was cast in one piece from melt (in 1936, Sec. 6.7.1). However, mirror sizes have further increased, e.g., by > 50%, but glass technology has also changed, e.g., ultralow expansion (ULE) Ti-Si-O glass has been developed. Thus, when an 8.3-m-dia. mirror was required, Corning made it by first fabricating "boules" of ULE glass from gaseous $TiCl_4$ and $SiCl_4$ via a CVD process. Thus, these gasses were oxidized in the vapor state producing oxide particles that formed molten droplets that deposited on a rotating table in the bottom of the furnace to produce boule disks ~ 150 cm in dia. and ~ 14 cm thick. Selected disks were then machined and stacked two high to be diffusion bonded to form disks ~ 150 cm in dia. and 28 cm thick, which were then machined into hexagonal disks (weighing ~ 1 ton). Then 44 such disks were arrayed and diffusion-bonded together (via a thermal cycle over nine days) to form a monolithic mirror blank which was then machined to the mirror profile [57].

8.4 FABRICATION OVERVIEW AND OPPORTUNITES TO IMPROVE MANUFACTURING PROCESSES

Having surveyed the various routes and steps to fabricating ceramics, it is useful to summarize and compare them as well as to address some opportunities to improve manufacturing. Overall powder-based processing via pressureless sintering is, and will remain, the dominant fabrication route for polycrystalline ceramics and some ceramic, mainly particulate, composites. However, there will be continued changes in specific steps in the fabrication, especially in green forming, and possibly some in heating methods. Thus, die pressing remains an important forming process because of its established cost advantages, but is limited both in terms of component shape and size as well as by the relic structure from the spray-drying typically needed to achieve the necessary speed and reliability of die fill. Injection molding is well established, very versatile in shape, but limited in sizes and by binder burnout and residual processing defect structures. Isopressing, both by wet and dry bag methods, has continuously increased in use for some components due to combinations of green body uniformity, large component capability (especially for wet bag pressing), and improving automation (especially for dry bag pressing), but is limited in shape complexity. The various slurry-colloidal processes such as tape and especially slip casting, and its derivatives of pressure or centrifugal casting, have gone through various changes. Previously, slip casting was used for many components but was displaced from some of these due to cost issues such as drying times and mold costs. However, pressure casting has been making gains, e.g., in sanitary ware due to a variety of technical improvements, and centrifugal casting may find use for large tubes. Recent increases in automation (e.g., via robotics [58]) have aided pressure casting, as has earlier automation for sanitary ware, and is promising for other ceramic forming operations where production volumes justify the investment.

Two factors may drive substantial further application of pressure casting. The first is the substantial improvements in mechanical strengths and uniformity due to significantly improved microstructural uniformity, e.g., over injection-molded specimens (Sec. 4.4.1). The other major potential advantage of slip and pressure casting and other slurry processes is that they should be amenable to further improvements in manufacturing quality control. Thus, colloidal mechanisms can limit agglomeration both by breaking agglomerates up or by filtering them out; the latter can also be used to remove larger impurity particles, while magnetic separation can also be used to remove some impurity particles. Further, slurry processing more readily provides opportunities for pH monitoring and particle size measurements, including size distribution, possibly as online monitoring as additional and important quality control tools, complementing NDE of fired and green bodies. Additionally, having the raw materials in slurry form from early stages of processing through that of casting means that they are much better protected from airborne contamination, such as dust and humidity, common problems for ceramics. Certainly some of these quality control measures can be implemented with other processes, but slip processing such as various casting methods, as well as tape, photolithographic, and some ink printer methods for SFF, appears to be particularly advantageous for such quality control. However, cost issues of casting methods, especially of pressure or centrifugal casting equipment, and the times for casting components of any substantial thicknesses, remain important issues to be addressed by further development, especially on engineering factors.

While firing for pressureless sintering is long established, it still presents challenges and opportunities—for example, of firing larger bodies. However, two advances should be noted: (1) Rate-controlled sintering can be an aid and is finding broader application than just laboratory use [59]; (2) newer heating methods for both sintering and hot pressing. Though there are many unknowns, these appear to offer some important opportunities.

Pressure sintering has expanded substantially and is expected to continue to do so, especially for higher value-added components. Hot pressing, though having less shape versatility, is most widely used and will probably grow more since it is simpler and cheaper and has important applications based on both component quality, size, character, and hot pressing incurring shrinkage only parallel with the pressing direction. Thus, hot pressing generally provides near or full density with finer, more homogeneous microstructure than pressureless sintering, and can produce some of the larger ceramic or ceramic composite bodies produced. Hot pressing can also be advantageous to much reaction processing, especially of composites, though homogeneous reaction is significantly preferred to propagating reactions (Sec. 6.5). While the uniaxial densification of hot pressing can yield some anisotropy, which may be a factor of some concern in some cases, this is a major advantage in some, e.g., electronic, applications such as large multilayer ceramic packages, where hot pressing provides much more accurate control of dimensions, especially normal to the hot pressing direction. Hot pressing is also important in fabrication of ceramic composites where better densification is generally obtained in particulate composites, and even more advantages on progressively going to platelet, whisker, and especially uniaxial or biaxial fiber composites, where uniaxial shrinkage only in the hot pressing direction is again an important factor. Further development of the use of binders and their removal in hot pressing, which was important in application to ceramic packages and some large structural ceramics, is expected to be of increasing importance, possibly significantly so. This, for example, might entail injection molding or casting an array of green components connected by small, disposable struts so the array can be efficiently loaded in multicavity dies. The extension of hot pressing to press forging of powder compacts for densification and shaping has progressed to where some specialized use may occur, but without further significant advances to improve its economics (e.g., reduced cycle time), its future remains uncertain. However, press forging of halide crystals to polycrystalline IR windows has been a commercial niche use.

HIPing has also become a process of considerable advanced development and some manufacturing of ceramics and some ceramic composites (mainly particulate, whisker, or platelet) as a result of glass canning and sinter-HIP techniques along with the commercial availability of less costly HIP units. However, costs of glass canning and its removal and issues of interactions of the glass with the component surface, or of the high pressure, especially N_2, in the HIP atmosphere with the component surface in sinter-HIPing (Sec. 6.4) are issues. These issues are in part related to the higher temperatures typically needed for ceramics, especially for nonoxides versus metals, with the higher temperatures for ceramics also meaning higher costs and smaller HIP units, which are a greater constraint of ceramic versus metal HIPing. Such issues and their consequences, such as smaller size HIP units for ceramics and little or no applicability to ceramic fiber composites, limit the scope of ceramic HIPing some.

Powder preparation is a basic step for the above fabrication methods based on powder consolidation. Calcining of salts remains the dominant method for preparation of oxide powders, carbothermic reduction for many nonoxides, and reaction processing for many mixed oxides and particulate composites. There can be serious problems, especially as component size increases and with pressure sintering at lower temperatures, with entrapping residues of the original compounds or of adsorbed species in the dense component. There are also various other reaction processes for making powders, e.g., use of SHS reactions, which have had some unexpected benefits of lower comminution costs. However, there are a variety of other methods of preparation, including newer methods such as sol-gel and preceramic polymer sources, as well as the equivalent of salts for calcining nonoxide powders—though these are constrained by costs, this could be relaxed with further developments. Further, there are trade-offs between prereacting powders for ternary compounds or composites or to carry out the reaction as part of the sintering process, with important differences between doing this with pressureless or pressure sintering—the latter having advantages of better elimination of porosity from the original powder and that generated by the reactions used. There are also opportunities for melt-derived powders as well as for particles for sand and related milling.

Use of additives is common in making some powders, e.g., to break down surface films on particles to be reacted or to catalyze reactions, and are even more prevalent in sintering to aid densification, limit grain size, or both. While improved powder preparation reduces the needs for some additives, they are still in wide use because of cost advantages, e.g., of using cheaper powders or faster/lower temperature densification, or performance advantages via lower porosity, finer grain size, or both, especially at limited use temperatures. Much remains to be understood about additive effects, but enhanced diffusion and, especially, liquid-phase effects on densification are important, with some additives such as LiF in MgO apparently generating a very effective liquid phase for hot pressing, but not enough for similar effects in pressureless sintering. Thus, effective aids for pressure sintering may not be effective for pressureless sintering, but not necessarily vice versa. Grain size limitations arise from both lower temperature or faster densification, or both, as well as the important mechanism of grain growth inhibition of small insoluble second-phase particles, which is often an important factor in many particulate composites. Additives are also used in other processing summarized below, e.g., to aid nucleation and growth of phases in crystallized glasses and some fusion cast refractories, and may be effective for similar purposes in preceramic polymer processing.

Turning to the first of three nonpowder-based fabrication processes, namely chemical vapor deposition (CVD): this is an established commercial process for a number of ceramics and ceramic composites that has significant potential for further applications. The process is well established for coatings on components and more recently for coating of ceramic fibers for composites. It also is established for making bulk components such as earlier reentry ICBM nose tips and rocket engine nozzles as well as for IR-domes and windows, and more recently as one of the methods of producing porous preforms for making optical fibers and telescope mirrors. These applications and the recognition that there are substantial opportunities for CVD processing based on broader materials concepts, such as the porous preforms for fibers, broader use of phase relations for both densification and composite processing, especially for ceramic particulate composites, show excellent opportunities for expanding uses of CVD. These opportunities are expanded by the growing applicability of CVI for fiber composites (Sec. 6.6), e.g., the potential for some important cost savings and making large components with fairly versatile shapes, as well as some further applications of CVI for fabrication of bodies of designed porosity (Sec. 7.3).

The last major fabrication method of melt processing entails very diverse applications ranging from production of glasses, refractories, refractory grain, and glass fibers. Some of these entail very large physical and dollar volumes as well as the largest pieces and product volume of ceramic made, all based on long established processing. It also entails most single-crystal growth, which includes increasing sizes of crystal grown (e.g., to 60-cm dimensions) as well as single-crystal filaments and some other bulk crystal shapes (Sec. 6.7.2). Melt processing also has produced a number of metal-ceramic or all ceramic eutectics as quenched particle, filaments, or bulk bodies, and potential of SFF fabrication from the melt using EFG growth of single crystals or eutectics. These and broader opportunities indicated in layer by layer directional solidification of some zirconia toughened ceramic composite compositions, extension of melt fiber formation by IMS and RIMS indicate further diversification and growth of melt processing. Clearly thermal stresses, entrapment of pores from frequent large liquid to solid volume changes and bubbles from gas entrapment are challenges, clever methods of overcoming such constraints have been demonstrated and more are likely to come, e.g., as suggested by application to SFF. Also, again note the use of melt processing for production of powders and finer pieces such as sand milling media and that melt spraying is a very important coating technique, and has some application for making free standing components, which should have significant further potential.

A few comments are in order on other aspects of fabrication in addition to or reinforcing comments above. Thus, reaction processing of powders has promising applications, but it is generally best to use reactions and actual reactants such that much or all of the densification of the reactants occurs prior to the reaction which generally generates additional porosity. Further, propagation of reactions is generally undesirable, and hot pressing, as well as possibly HIPing make it much easier to achieve full densification, but pressureless sintering may be feasible in some cases. Reaction processing of ceramic fiber composite matrices that can give higher densities/better properties under autoclave conditions deserve more consideration. Reaction processing with preceramic polymers is promising, especially for ceramic, particularly fiber, composites, but costs have been a serious limitation. Increased ceramic yields may aid some, but polymers that go through one or more decomposition stages where some consolidation during these stages may be feasible could be an important an important technical aid.

Solid free-form or rapid prototyping fabrication continues to expand and diversify, but still faces many engineering issues of practicality, scaling, and specific, successful applications. However, the diversity of techniques and possible uses and progress suggests that some substantial applications will occur, but which techniques and applications will survive remains to be seen. Some of the latter may depend on other broader changes; e.g., electronic warehousing of components may depend on substantially reducing the number of materials from which replacement components are to be made.

Single-crystal growth represents significant advances in melt processing of ceramics, as attested to by the sizes and the shapes of crystals grown. Growth of shaped crystal pieces and the recent marriage of one-crystal growth (EFG®) technique with SFF techniques reflect further opportunities, but so do improvements in the efficiency of machining to reduce its costs. In some cases these will compete, but both are likely to add to uses of single-crystal components. There are also important possibilities of using some of these techniques, especially the heat exchanger method (HEM), for directional solidification, e.g., as shown with large Si ingots for photovoltaic devices and other demonstrations of melt casting of CaF_2 and $MgAl_2O_4$ for IR windows which, with their large (centimeter size) grains, had mechanical properties comparable or somewhat advantageous to those of single crystal.

Next consider surface finishing, primarily machining and ceramic coating. The latter reflects a diversity of methods depending in part on whether the coating is made on a metal or ceramic, an area that contains many long-standing applications and more recent ones. The former are illustrated by dip and fire glass-based oxidation resistant coatings for metal heating elements, and the latter by vapor-phase reaction processing of wear and corrosion resistant consumer drill bits and faucet fixtures and doorknobs, as a result of needs, costs, and experience resulting from increased industrial experience.

Some limited chemical and laser methods of machining have seen very limited use, but abrasive machining is the predominant method of surface finishing. It generally adds some to substantial cost (Table 1.2) and is used for dimensional and surface finish requirements, part of which are its ability to give somewhat higher and more reliable strengths. Mechanisms and resultant effects of machining are getting reasonably understood, especially the nature of flaws and less of the residual stresses generated, showing effects of machining direction and of incomplete removal of flaws from the prior stage of machining. Continuing improvements are occurring, driven mainly by production needs; e.g., its impacts on use of single-crystal components and trade-offs in growth methods or parameters.

Joining or attachment of ceramics to metals, or the same or other ceramic is an important need for many uses of ceramics, which is probably increasing. Methods range from mechanical, organic adhesive to various chemical bonding/brazing and actual welding (Table 8.1), which are generally listed in order of increasing temperature capabilities, and to some extent cost. Fusion welding, though demonstrated some with reasonable to good properties (with potential for further improvement) and potentially of moderate cost, has received little or no use. Diffusion welding has produced the best properties, but is generally limited in configurations to which it can be practically applied. However, fabrication of telescope mirrors are an interesting example of use since they have now grown to such sizes that the huge single-piece glass mirror melt castings of the past are no longer big enough. Thus, large mirrors are being made in pieces and being diffusion welded together, for which it is a good application. There is a substantial and growing diversity and development of various metal-based brazing techniques, many based on reactive brazing.

Inspection and quality assurance is a key step, which is generally better done for nonstructural applications, e.g., electrical and, especially, electronic ones. It is much more challenging and less advanced for structural applications since proof testing is very limited in its application, e.g., to rotating components such as grinding wheels. Substantial development of NDE methods has occurred which gives an array of techniques for detecting and characterizing specific flaws to give a quantitative evaluation of the probable strength of the component. However, these methods, while very useful for identifying many defects and thus aiding processing improvement, are still a substantial distance from identifying sufficiently small and diverse flaws with accuracies, reliabilities, and costs needed. Industrial practice is mainly visual inspection, backed by tests such as die penetrants, and other sampling evaluations of components. However, NDE of lower strength (e.g., of some refractory) components and of green bodies as part of process monitoring or control is promising. Such green body monitoring could fit well with monitoring of green body fabrication and of final densification.

Finally, briefly note three important, related topics, two of industrial practice only lightly touched on in this book. The first is use of waste products as ingredients in making ceramic products. Examples are use of pickle liquors as low cost sources of some ferrite powders and of wastes such as oil field sludges to make bricks [60,61]). The second is recycling of used product to be disposed of or materials rejected at various stages in the manufacturing process [62]. High volume ceramic products, such as refractories and building bricks, are examples of this.

The third topic and a good one to close on is the extensive diversification of ceramic fabrication technology, some of which was addressed above. Ceramic fabricatioan technologies include not only more and broader methods of green formation but also densification methods, that is, pressureless sintering and hot pressing or HIPing. They also include broader methods of powder preparation ranging from extending traditional salt calcining for oxides to similar preparation of nonoxides to sol and preceramic polymer preparation of both types of powders to substantial use of various CVD methods of powder preparation to even some and growing use of melt preparation of powders. Other indications of diversification are increasing use of melt spraying for bulk not just coating fabrication, as well as investigation of HIP [63] or CVI [64] densification of plasma-sprayed coatings, which may also have some applicability bulk bodies. Development of improved single-crystal growth, SFF, new heating methods are further examples, while shifts in emphasis on alternate fabrication methods for ceramic composites is particularly significant. The opportunities posed by this diversification are significant, but so are the challenges of selecting among them.

REFERENCES

1. S.-J. L. Kang, K.J. Yoon. Densification of ceramics containing entrapped gases. J. Eur. Cer. Soc. 5:135–139, 1989.
2. R.W. Rice. Failure Analysis of Ceramics.In: J. Varner, G. Quinn, eds. Fractography of Glasses and Ceramics IV. Westerville, OH: Am. Cer. Soc., 2001.
3. D.A. Brosnan. HF emissions in manufacturing traditional ceramics. Am. Cer. Soc. Bul. 76(1):59–61, 1997.
4. D.A. Brosnan. SO_x emissions in firing ceramics. Am. Cer. Soc. Bul. 76(3):53–55, 1997.
5. R.W. Rice. The effect of gaseous impurities on the hot pressing and behavior of MgO, CaO and Al_2O_3. Proc. Brit. Cer. Soc. 12:99–123, 1969.
6. R.W. Rice. Ceramic processing: an overview. AIChE J. 36(4):481–510, 1990.
7. R.W. Rice. Mechanical Properties of Ceramics and Composites: Grain and Particle Effects. New York: Marcel Dekker, 1999.
8. M.H. Leipold, T.H. Nielsen. Fabrication and characterization of isostatically hot-pressed MgO. J. Am. Cer. Soc. 51(2):94–97, 1968.
9. W.H. Rhodes, E.A. Trickett, G.C. Mei. Processing studies for optically transparent La_2O_3-Doped Y_2O_3. Final Report of GTE Labs. Inc., ONR Contract N00014-82-C-0452, 1986.
10. W.H. Rhodes, E.A. Trickett, D.J. Sordelet. Key powder characteristics in sintered optical ceramics.In: G. Messing et al., eds. Ceramic Powder Science III Cer. Trans. V. 12. Westerville, OH: Am. Cer. Soc., 1990, pp. 677–690.
11. I.V. Antipov, L.B. Kuz'min, L.M. Gofman, N.L. Opolchenova, V.S. Kutsev, V.S. Mirochnikov. Removal of chlorine-containing impurities from titanium dioxide obtained by the chlorine method. Neorganicheskie Materialy 14(5):896–899, 1978.
12. A. Briggs. The formation of hydrogen-filled cavities in MgO crystals annealed in reducing atmospheres. J. Mat. Sci. 10:737–746, 1975.
13. Yu.A. Borodenko, B.L. Timan, A.M. Mel'nikova. Gas inclusions in sapphire upon horizontal, directional crystallization. Neorganicheskie Materialy 22(7):1122–1126, 1986.
14. S. Kitaoka, H. Matsubara, H. Kawamoto, H. Yanagida, M. Matsumoto, M. Kanno. Gas evolution from advanced ceramics during fracture under high vacuum. Cer. Eng. Sci. Proc. 18(4):1997.

15. W.A. Bryant. High-temperature strength stability of three forms of chemically vapor deposited tungsten. J. Vac. Sci. Tech. 11(4):695, 1974.
16. D.L. Johnson. Microwave and plasma sintering of ceramics. Cer. Intl. 17:295–300, 1991.
17. H. Su, D.L. Johnson. Sintering of alumina in microwave-induced oxygen plasma. J. Am. Cer. Soc. 79(12):3211–3217, 1996.
18. W.H. Sutton. Microwave processing of ceramic materials. Am. Cer. Soc. Bul. 68(12):376–386, 1989.
19. S. Das, T.R. Curlee. Microwave sintering of ceramics: can we save energy? Am. Cer. Soc. Bul. 66(7):1093–1094, 1987.
20. L.M. Shepard. Manufacturing ceramics with microwaves: the potential for economical production. Am. Cer. Soc. Bul. 67(10):1656–1661, 1988.
21. A.C. Metaxes, J.G.P. Binner. Microwave processing of ceramics. In: J.G.P. Binner, ed. Advanced Ceramic Processing and Technology. Vol. 1. Park Ridge, NJ: Noyes Publications, 1990, 285–367.
22. J.D. Katz, R.D. Blake. Microwave sintering of multiple alumina and composite components. Am. Cer. Soc. Bul. 67(10):1656–1061, 1988.
23. M. Henrichsen, J. Hwang, V.P. Dravid, D.L. Johnson. Ultrarapid phase in β-alumina tubes. J. Am. Cer. Soc. 83(11):2861–2862, 2000.
24. A. Birnboim, Y. Carmel. Simulation of microwave sintering of ceramic bodies with complex geometry. J. Am. Cer. Soc. 70(8):1304–1308, 1991.
25. Do-H. Kim, C.H. Kim. Entrapped gas effect in the fast firing of yttria-doped zirconia. J. Am. Cer. Soc. 75(3):716–718, 1992.
26. R. Ruginets, R. Fischer. Microwave sintering of boron carbide composites. Am. Cer. Soc. Bul. 74(1):16–58, 1995.
27. R.C. Dalton, I. Ahmad, D.E. Clark. Combustion synthesis using microwave energy. Cer. Eng. Sci. Proc. 11(9–10):1729–1742, 1990.
28. S.-T. Oh, K.-i., M. Ando, T. Ohji. Strengthening of porous alumina by pulse electric current sintering and nanocomposite processing. J. Am. Cer. Soc. 83(5):1314–1316, 2000.
29. J.R. Groza, J.D. Curtis, M. Krämer. Field-assisted sintering of nanocrystalline titanium nitride. J. Am. Cer. Soc. 83(5):1281–1283, 2000.
30. R.W. Rice, D. Lewis III. Ceramic fiber composites based upon refractory polycrystalline ceramic matrices.In: S.M. Lee, ed. Reference Book for Composites Technology. Vol. 1. Lancaster, PA: Technomic Pub. Co., Inc., 1989, pp. 117–142.
31. K.M. Prewo, J.J. Brennan. Fiber reinforced glasses and glass ceramics for high performance applications", In: S.M. Lee, ed. Reference Book for Composites Technoloty. Vol. 1. Lancaster, PA:Technomic Pub. Co., Inc., 1989, pp. 97–116.
32. R. Rice. Processing of ceramic composites. In: J.G.P. Binner, ed. Advanced ceramic processing and technology. Vol. 1. Park Ridge, NJ: Noyes Pubs, 1990, pp. 123–213.
33. R.W. Rice. Effects of amount, location, and character of porosity on stiffness and strength of ceramic fiber composites via different processing. J. Mat. Sci. 34:2769–2772, 1999.
34. R.W. Rice. Porosity of Ceramics. New York: Marcel Dekker, 1998.

35. R. W. Rice. Machining flaw model evaluation: monolithic and composite ceramic property-microstructural effects. J. Eur. Cer. Soc. 22. 2202, 1411–1424.
36. R.W. Rice. Machining of Ceramics. In: J.J. Burke, A.E. Gorum, R.N. Katz eds. Ceramics for High Performance Applications, Proc. of 2nd Army Mat. Tech., Metals and Ceramic Info. Center, Columbus, OH, 1974, pp. 287–343.
37. A.G. Evans, D.B. Marshall. Wear mechanisms in ceramics. In: D.A. Ringney, ed. Fundamentals of Friction and Wear of Materials, 1980 ASM Materials Science Seminar, American Soc. for Metals, Metals Park, OH, 1981, pp. 439–452
38. B. Marshall. Failure from surface flaws. In: A.G. Evans ed. Fracture in Ceramic Materials—Toughening Mechanisms, Machining Damage, Shock. New York: Noyes Publications, 1984, pp. 190–220.
39. R.W. Rice. The effect of grinding direction on the strength of ceramics. In: S.J. Schneider, R.W. Rice, eds. The Science of Ceramic Machining and Surface Finishing, NBS Special Pub. 348. U.S. Govt. Printing Office, Wash. D.C. 1972, pp. 365–376.
40. J.J. Mecholsky, Jr., S.W. Freiman, R.W. Rice. Effect of flaw geometry and fracture of glass. J. Am. Cer. Soc. 60(3-4):114–117, 1977.
41. R.W. Rice, J.J. Mecholsky, Jr. The nature of strength controlling machining flaws in ceramics. In: B.J. Hockey, R.W. Rice eds. The Science of Ceramic Machining and Surface Finishing II. U.S. Govt. Printing Office, Wash. D.C., 1979, pp. 351–378.
42. R.W. Rice. Machining Flaws and the Strength-Grain Size Behavior of Ceramics. In: B.J. Hockey, R.W. Rice, eds. The Science of Ceramic Machining and Surface Finishing II. NBS Special Pub. 562. 1979, pp. 429–454.
43. R.W. Rice, J.J. Mecholsky, Jr., P.F. Becher. The effect of grinding direction on flaw character and strength of single crystal and polycrystalline ceramics. J. Mat. Sci. 16:853–862, 1981.
44. R.W. Rice. Effects of ceramic microstructural character on machining direction—strength anisotropy. In: S. Jahanmir ed. Machining of Advanced Materials. NIST Special Pub. 847, U.S. Govt. Printing Office, Wash. D.C. 1993, pp. 185–204.
45a. R.W. Rice. Correlation of machining-grain-size effects on tensile strength with tensile strength-grain-size behavior. J. Am. Cer. Soc. 76(4):1068–1070, 1993.
45b. RW. Rice. Porosity effects on machining direction-strength anisotropy and failure mechanisms. J. Am. Cer. Soc. 77(8):2232–2236,1994.
46a. P. McGuire, B. Pazol, R. Gentilman, J. Askinazi, J. Locher. Large area edge-bonded flat and curved sapphire windows. In: R.W. Tustison, ed. Window and Dome Technologies and Materials VII. Bellingham WA: Proc. SPIE V 4375, 2001, pp. 12–19.
46b. R. Gentilman, personal communication, 2001.
47. L.M. Shepard. Evolution of NDE continues for ceramics. Am. Cer. Soc. Bul. 70(8):1265–1279, 1991.
48. C.H. Schilling, J.N. Gray. NDE: needs, opportunities in ceramic processing. Am. Cer. Soc. Bul. 78(12):74–80, 1999.
49. R.W. Rice. Joining of Ceramics. In: J.J. Burke, A.E. Gorum, A. Tarpinian, eds. Advances in Joining Technology. Chestnut Hill, Mass.: Brook Hill Publishing Co., 1976, pp. 69–111.

50. M.U. Goodyear, A. Ezis. Joining of Turbine Engine Ceramics. In: J.J. Burke, A.E. Gorum, A. Tarpinian, eds. Advances in Joining Technology. Chestnut Hill, Mass.: Brook Hill Publishing Co., 1976, pp. 113–153.
51. J.E. Kelley, D.H. Sumner, H.J. Kelly. Systems for Uniting Refractory Materials. In: J.J. Burke, A.E. Gorum, A. Tarpinian, eds. Advances in Joining Technology. Chestnut Hill, Mass.: Brook Hill Publishing Co., 1976, pp. 155–183.
52. J.F. Burgess, C.A. Neugebauer, G. Flanagan, R.E. More. Direct bonding of metals to ceramics for electronic applications. In: J.J. Burke, A.E. Gorum, A. Tarpinian, eds. Advances in Joining Technology. Chestnut Hill, Mass.: Brook Hill Publishing Co., 1976, pp. 185–197.
53a. M.G. Nicholas, ed. Joining of Ceramics. London: Chapman and Hall, 1990.
53b. I.E. Reimanis. Joining of ceramics. In: M.N. Rahaman, ed. Handbook of Ceramic Engineering. New York: Marcel Dekker, Inc., 2001.
54a. C. Suryanarayana, J.J. Moore, R.P. Radtke. Novel methods of brazing dissimilar materials. Adv. Matls. & Procs. 2001, pp. 29–31.
54b. V. Greenhut. TEPP a new technology for ceramic-metal bonding and metallization: part I. Processing and Structure. Part II. Properties and performance. (in press) Cer. Eng. Sci. Proc., 2002.
55. C. Scott, General Electric, Cincinnati, private communication, 2000.
56. R.W. Rice. Advanced Ceramic Materials and Processes. In: D.L. Cocke, A. Clearfield, eds. Design of New Materials. New York: Plenum Publishing Corp., 1987, pp. 169–194.
57. CI Staff. Corning produces monolithic mirror blank. Cer. Ind. 1994, pp. 36–40.
58. J. Ronger. Pressure casting with robots. Am. Cer. Soc. Bul. 79(12):62–64, 2000.
59. M.L. Huckabee, M.J. Paisley, R.L. Russell, H. Palmour III. RCS–Taking the mystery out of densification profiles. Am. Cer. Soc. Bul. 73(9):82–86, 1994.
60. N.J. Saikia, P. Sengupta, D.K. Dutta, P.C. Saikia, P.C. Borthakur. Oil field sludge used to make brick. Am. Cer. Soc. Bul. 79(7):71–74, 2000.
61. A.R. Boccaccini, M. Petitmermet, E. Wintermantel. Glass-ceramics from municipal incinerator fly ash. Am Cer. Soc. Bul. 76(11):75–78, 1997.
62. J.P. Bennett, K.-S. Kwong, S.W. Sikich. Recycling and disposal of refractories. Am Cer. Soc. Bul. 74(12):71–77, 1995.
63. Y. Ishiwata, Y. Ito, H. Kashiwaya. Densification of plasma-sprayed ceramic coatings by HIP treatment and their cracking behavior. J. Cer. Soc. Jpn. Intl. Ed. 99:665, 1991.
64. T. Mäntylä, P. Vuoristo, P. Kettunen. Chemical vapour deposition densification of plasma-sprayed oxide coatings. Thin Solid Films 118:437–444, 1984.

Index